A First Course
in Stochastic
Calculus

Pure and Applied
UNDERGRADUATE TEXTS · 53

A First Course in Stochastic Calculus

Louis-Pierre Arguin

AMERICAN
MATHEMATICAL
SOCIETY
Providence, Rhode Island USA

2020 *Mathematics Subject Classification.* Primary 60G07, 60H05, 91G20, 91G60.

For additional information and updates on this book, visit
www.ams.org/bookpages/amstext-53

Library of Congress Cataloging-in-Publication Data

Names: Arguin, Louis-Pierre, 1979- author.
Title: A first course in stochastic calculus / Louis-Pierre Arguin.
Description: Providence, Rhode Island : American Mathematical Society, [2022] | Series: Pure
  and applied undergraduate texts, 1943-9334 ; volume 53 | Includes bibliographical references
  and index.
Identifiers: LCCN 2021032211 | ISBN 9781470464882 (paperback) | ISBN 9781470467784 (ebook)
Subjects: LCSH: Stochastic analysis. | Calculus. | Stochastic processes. | AMS: Probability theory
  and stochastic processes – Stochastic processes – General theory of processes. | Probability
  theory and stochastic processes – Stochastic analysis – Stochastic integrals. | Game theory,
  economics, social and behavioral sciences – Mathematical finance – Derivative securities. |
  Game theory, economics, social and behavioral sciences – Mathematical finance – Numerical
  methods (including Monte Carlo methods).
Classification: LCC QA274.2 .A74 2022 | DDC 519.2/2–dc23
LC record available at https://lccn.loc.gov/2021032211

Contents

Foreword

As a former trader and quant, and now a teacher of quants, I can say that the book I have been waiting for has finally arrived: an introduction to stochastic calculus with financial applications in mind, shorn of unnecessary details. The respect that Louis-Pierre Arguin (L-P to his friends) clearly holds for the reader of this book is evident throughout; while the writing is conversational, it is also concise. This book is "action-packed" — every page has something of real value.

It is self-evident that stochastic calculus is a prerequisite to understand the current mathematical finance literature, in particular the literature on option pricing. And these days, not only quants but also traders, programmers, and even some salespeople are familiar with this literature. This book provides the best possible introduction to stochastic calculus, as suitable for those encountering the subject for the first time as it is for people like me who may wish to refresh their understanding of a particular topic. Also, it is worth emphasizing that while this book is introductory, everything is treated with rigor and written in the language of the contemporary literature, preparing readers well for the challenges of further reading.

Mathematicians and physicists may argue over the value of intuition, but in financial practice, intuition is absolutely key. For example, finance practitioners often need to be able to quickly judge how to alter the terms of a transaction, typically to reduce costs while still meeting the needs of the client. So how does one acquire good intuition? Experience obviously is one part of the answer. But having a good teacher is more efficient and less expensive: L-P is renowned as a great teacher, and this book fully lives up to his reputation. His intuitive explanations of stochastic calculus are backed up and reinforced by the many numerical projects and exercises he provides.

I clearly recall my first introduction to the central limit theorem from an undergraduate experiment with a radioactive source in a physics lab at the University of Glasgow. Still today, in my mind's eye, I can see the histogram of counts changing from Poisson to Gaussian as I bring the source closer to the detector. There is no substitute for this kind of training; I have the feeling that L-P betrays his background as a physicist with his emphasis on simulation in the numerical projects. L-P builds the kind of intuition that can be felt in your bones.

Another prerequisite in the practice of quantitative finance is programming skill, and the language of choice is Python. All of the numerical projects use Python and many Python hints are provided. I suspect that many readers of this book may be more comfortable writing Python than reading or writing mathematical proofs — for them, this book will be a godsend.

L-P is to be congratulated for writing this wonderful book. I can see myself referring to it for many years to come.

Jim Gatheral
Presidential Professor of Mathematics
Baruch College, City University of New York

Preface

There is a tendency in each profession to veil the mysteries in a language inaccessible to outsiders. Let us oppose this. Probability theory is used by many who are not mathematicians, and it is essential to develop it in the plainest and most understandable way possible. [1]
– Edward Nelson (1932–2014)

The development of stochastic calculus, and in particular of Itô calculus based on Brownian motion, is one of the great achievements of twentieth-century probability. This book was developed primarily to open the field to undergraduate students. The idea for such a book came when I was assigned to teach the undergraduate course *Stochastic Calculus for Finance* in 2016 at Baruch College, City University of New York. Some of the material was also developed while teaching the course *Probability and Stochastic Processes in Finance* for the Master's in Financial Engineering program at Baruch College in 2016 and in 2019. The end product is a mathematical textbook for readers who are interested in taking a first step towards advanced probability. Specifically, it is targeted at:

(A) **Advanced undergraduate students in mathematics.** As such, the book assumes a good background in multivariable calculus (Calculus 3), in linear algebra (a first course in undergraduate linear algebra in $\mathbb{R}^d$ with some exposure to vector spaces is sufficient), and probability (e.g., a first introductory course in probability covering joint distributions). Prior knowledge of basic stochastic processes or advanced calculus (i.e., mathematical analysis) can be helpful but is not assumed.

(B) **Master's students in financial engineering.** The book may serve as a first introduction to stochastic calculus with applications to option pricing in mind. Applications are given in the last chapter. However, as detailed below, they can be introduced in parallel with the other chapters. It can also be used as a self-study guide for the basics of stochastic calculus in finance.

---

[1] From the preface to the Russian edition of *Radically Elementary Probability* [**Nel95, Nel87, KK15**].

Nelson's quote above sums it all: The challenge in presenting stochastic calculus at the undergraduate level is to avoid measure theory concepts (whenever possible) without reducing the subject to a litany of recipes. With this in mind, the book approaches the subject from two directions. First and foremost, there is a strong emphasis on

NUMERICS!

Today's undergraduate math students might not be versed in analysis, but they are usually acquainted with basic coding. This is a powerful tool to introduce the students to stochastic calculus. Every chapter contains numerical projects that bring the theoretical concepts of the chapter to life. It is strongly suggested that the students write the codes in Python in a Jupyter notebook (https://jupyter.org). The coding is elementary and the necessary Python knowledge can be picked up on the way. All figures in the book were generated this way. The minimal time invested to learn the necessary basic coding will open the gates to the marvels of stochastic calculus. It might not be obvious to prove Itô's formula, the arcsine law for Brownian motion, Girsanov's theorem, and the invariant distribution of the CIR model, but they are easily verified with a couple of lines of codes, as the reader will find out.

Second, from a theoretical point of view, the reader is invited to consider random variables as elements of a linear space. This allows us to rigorously introduce many fundamental results of stochastic calculus and to prove most of them. This approach is also useful to build an intuition for the subject as it relies on a geometric point of view very close to the one in $\mathbb{R}^d$. Of course, some measure theory concepts (such as convergence theorems) need to be introduced to seriously study this subject. These are presented as needed.

The general objective of the book is to present the theory of Itô calculus based on Brownian motion and Gaussian processes. In particular, the goal is to construct stochastic processes using the Itô integral and to study their properties. The last chapter presents the applications of the theory to option pricing in finance. Each chapter contains many exercises, mixing both theory and concrete examples. Exercises marked with a ⋆ are more advanced. The remarks in the text also relate the material to more sophisticated concepts.

Here are the two suggested approaches to reading this book:

(A) How to use this book for an undergraduate course. This book offers *two* possible pathways for a course at the undergraduate level depending on whether the student's interest is in *mathematical finance* or in *advanced probability*. The common path of these two options consists of Chapters 1 to 5, with a road fork after. The blueprint is based on a 14-week, 4-credit class with no exercise sessions.

- Chapter 1 (1.1–1.4): It reviews the elementary notions of probability theory needed along the way. This usually covers one week of material and offers sufficient opportunity to get acquainted with basic commands in Python.

- Chapter 2 (2.1–2.4): The chapter introduces the notion of Gaussian processes, based on the notion of Gaussian vectors. This is where we learn the Cholesky decomposition to express jointly Gaussian random variables in terms of IID standard Gaussians. This concept is very useful to numerically sample a plethora of Gaussian processes, such as Brownian motion, the Ornstein-Uhlenbeck process, and fractional Brownian motion. It also sets the table for the notion of *orthogonal projection* and its relation to conditional expectation. This geometric point of view is introduced in Section 2.4. Around two weeks of classes are dedicated to this chapter.

- Chapter 3 (3.1–3.2, 3.4): This chapter studies the properties of Brownian motion in more detail. In particular it introduces the notion of quadratic variation that is central to Itô calculus. The Poisson process is also presented as a point of comparison with Brownian motion. This chapter takes up around one to two weeks.

- Chapter 4 (4.1–4.5): The class of stochastic processes known as martingales is introduced here. It is built on the conditional expectation, which is defined as a projection in the space of random variables with finite variance. Elementary martingales, such as geometric Brownian motion, are given as examples. One of the powers of Itô calculus is to give a systematic way to construct martingales using Brownian motion. Martingales are useful, as some probabilistic computations are simplified in this framework. For example, solving the gambler's ruin problem using martingales is illustrated in Section 4.4. This chapter is longer and takes about two weeks to cover.

- Chapter 5 (5.1–5.5): The Itô integral is constructed as a limit of a martingale transform of Brownian motion. The martingale transform is analogous to Riemann sums in standard calculus. Itô's formula, which can be seen as the *fundamental theorem of Itô calculus*, is also proved and numerically verified. This is where we start to explore the beautiful interplay between partial differential equations (PDE) and stochastic processes. Around two weeks are needed to study Chapter 5.

At this point, the student should have a pretty good grasp of the basic rules of Itô calculus. There are then two possible subsequent paths:

- *Mathematical finance road* (Chapters 7 and 10). The ultimate objective here is to study the basics of option pricing through the lens of stochastic calculus. The reader may jump to Chapter 7 where Itô processes, and in particular diffusions, are studied. It is then fairly straightforward to study them using the rules of Itô calculus. An important part of this chapter is the introduction of stochastic differential equations (SDE). The emphasis here should be on numerically sampling diffusions using SDEs. Suggested sections are 7.1, 7.2, and 7.4. This takes only around one week, as the reader should be fluent with the rules of Itô calculus by that point.

  The course ends with Chapter 10 on the applications in finance. The suggested sections are 10.1 to 10.6 (with minimal inputs from Section 9.2). At this level, the focus should be on the Black-Scholes model. The pricing of options can be given using the Black-Scholes PDE and the risk-neutral pricing. This is an excuse to present the Cameron-Martin theorem of Chapter 9 which ensures the existence of a risk-neutral probability for the model. A good numerical project there is the biased sampling of paths based on the Cameron-Martin theorem. If time permits, Section 10.7 on exotic options is a good ending.

- *Probability and PDE road* (Chapters 6 and 7). The ultimate objectives here are the study of the Dirichlet problem and SDEs. It is particularly suited for students that love multivariate calculus. Suggested sections are 6.1 to 6.4 and 7.1 to 7.5. The generalization of Itô's formula for multidimensional Brownian motion is done in Chapter 6. It is used to prove the transience of Brownian motion in higher dimensions. The course can then move on to Itô processes and SDEs and include multivariate examples of such processes. If time permits, the Markov property of diffusions and its relation to PDEs can be presented in Chapter 8.

(B) How to use this book for a master's course in mathematical finance. Besides some more advanced subjects, the main difference in a master's course is that the finance concepts could be presented concurrently with the probability material. With this in mind, here are some suggestions for how to incorporate the material of Chapter 10 within the text.

- Sections 10.1, 10.2, and 10.3 on market models, derivative products, and the no-arbitrage pricing can be presented right at the beginning of the class. It may serve as a motivation for the need of stochastic calculus. This is also a possibility at the undergraduate level.

- Chapters 2 and 3 should be covered right after, with an emphasis on numerics to instill some intuition on Brownian motion and other stochastic processes. This is also a good time to understand the notion of projection for random variables, which is necessary in Chapter 4.

- Sections 10.4, 10.5, and 10.6 on the Black-Scholes model, the Greeks, and risk-neutral pricing for Black-Scholes model (with a brief mention of the Cameron-Martin theorem) are a good fit concurrently with Chapters 4 and 5 and Sections 7.1, 7.2, and 7.4. By then, the student has all the tools to derive the Black-Scholes PDE. Exotic options from Section 10.7 can be done there too.

- Sections 10.6, 10.8, and 10.9 on general risk-neutral pricing, interest rate models, and stochastic volatility models are good companions to Chapters 8 and 9.

Acknowledgments. The development of this book was made possible by the financial support of the National Science Foundation CAREER grant 1653602. I am grateful for the support of my colleagues in the Department of Mathematics at Baruch College. More specifically, I would like to thank Jim Gatheral for having graciously accepted to write a foreword to the book. I am also grateful to Warren Gordon and Dan Stefanica for giving me the opportunity to develop this course at the undergraduate level and at the master's level. The students of MTH5500 and MTH9831 made important contributions to this book by their inputs, questions, and corrections. Special thanks to Jaime Abbario for some numerical simulations and to Joyce Jia and Yosef Cohen for the careful reading of early versions of the manuscript. I would like to thank Ina Mette from the AMS for believing in this project and Andrew Granville for good advice on writing a math textbook. I am indebted to Eli Amzallag for his careful reading and his inputs on the first draft and to Alexey Kuptsov for great tips on Chapter 10. On a more personal note, I would like to thank Alessandra for her love and dedication, Mariette and Pierre-Alexandre for making every day a joy, Louise for having shown me how to be a good parent and a good academic, Jean-François for his constant support, and Ronald and Patricia for their generous help in the last few years.

Chapter 1

Basic Notions of Probability

In this chapter, we review the notions of probability that will be needed to study stochastic calculus. The concept of probability space, that gives a mathematical framework to random experiments, is introduced in Section 1.1. Random variables and their distributions are studied in Section 1.2. This is where we review the concepts of cumulative distribution function (CDF) and probability density function (PDF). Section 1.3 explains the basic properties of the expectation of a random variable. In particular, we introduce the moment generating function. Finally, we review some basic inequalities relating probabilities and expectations in Section 1.4.

1.1. Probability Space

Probability theory provides a framework to study random experiments, i.e., experiments where the exact outcome cannot be predicted accurately. Such random experiments involve a *source of randomness*. Elementary random experiments include the roll of a die and consecutive coin tosses. In applications, the source of randomness could be, for example, all the transactions that take place at the New York Stock Exchange in a given time, the times of decay of a radioactive material, and pseudorandom numbers generated by computers. The latter will be the most important example of random experiments in this book.

The theoretical framework of a random experiment is as follows. The set of all possible outcomes of the random experiment is called the *sample space* and is denoted by Ω. An outcome of the random experiment is then represented by an element ω of Ω. An *event* is a subset of Ω. An event can be described by either enumerating all the elements contained in the subset or by giving the properties that characterize the elements of the subset. Note that the empty set $\emptyset$ and Ω are subsets of Ω and are thus events.

Example 1.1. For the random experiment consisting of generating an ordered pair of 0's and 1's, the sample space is

$$\Omega = \{(0,0), (0,1), (1,0), (1,1)\}.$$

An example of an event is $A = \{(0,0), (0,1)\}$, that is, the set of outcomes whose first number is 0. Note that Ω can also be written as $\{0,1\} \times \{0,1\}$ where $\times$ stands for the Cartesian product between two sets (like $\mathbb{R} \times \mathbb{R}$).

To quantify the randomness in a random experiment, we need the notion of probability on the sample space. In Example 1.1, we can define a probability on Ω by

$$(1.1) \qquad\qquad \mathbf{P}(\{\omega\}) = 1/4 \ \text{ for all } \omega.$$

Since there are four outcomes in the sample space, this is the equiprobability on the sample space.

More generally, a probability on a general sample space Ω is defined as follows.

Definition 1.2. A probability $\mathbf{P}$ is a function on events of Ω with the following properties:

(1) For any event $A \subseteq \Omega$, the probability of the event denoted $\mathbf{P}(A)$ is a number in $[0,1]$:

$$\mathbf{P}(A) \in [0,1].$$

(2) $\mathbf{P}(\emptyset) = 0$ and $\mathbf{P}(\Omega) = 1$.

(3) *Additivity*: If $A_1, A_2, A_3, \ldots$ is an infinite sequence of events in Ω that are *mutually exclusive* or *disjoint*, i.e., $A_i \cap A_j = \emptyset$ if $i \neq j$, then

$$\mathbf{P}(A_1 \cup A_2 \cup A_3 \cup \cdots) = \mathbf{P}(A_1) + \mathbf{P}(A_2) + \mathbf{P}(A_3) + \cdots.$$

It is not hard to check, see Exercise 1.2, that if Ω has a finite number of outcomes, then the equiprobability defined by

$$\mathbf{P}(A) = \frac{\#A}{\#\Omega},$$

where $\#A$ stands for the cardinality of the set A, is a probability as defined above. It is very important to keep in mind that there might be many probabilities that can be defined on a given sample space Ω! [1]

The defining properties of a probability have some simple consequences.

Proposition 1.3. *The defining properties of a probability $\mathbf{P}$ on Ω imply the following:*

(1) *Finite additivity: If two events A, B are disjoint, then $\mathbf{P}(A \cup B) = \mathbf{P}(A) + \mathbf{P}(B)$.*

(2) *For any event A, $\mathbf{P}(A^c) = 1 - \mathbf{P}(A)$.*

(3) *For any events A, B, $\mathbf{P}(A \cup B) = \mathbf{P}(A) + \mathbf{P}(B) - \mathbf{P}(A \cap B)$.*

(4) *Monotonicity: If $A \subseteq B$, $\mathbf{P}(A) \leq \mathbf{P}(B)$.*

Proof. See Exercise 1.1. □

A less elementary albeit very useful property is the continuity of a probability.

---

[1] This fact plays an important role in Chapter 9.

Lemma 1.4 (Continuity of a probability). *Consider* $\mathbf{P}$ *a probability on* Ω. *If* A_1, A_2, $A_3, \ldots$ *is an infinite sequence of increasing events, i.e.,*

$$A_1 \subseteq A_2 \subseteq A_3 \subseteq \cdots,$$

then

$$\mathbf{P}(A_1 \cup A_2 \cup A_3 \cup \cdots) = \lim_{n \to \infty} \mathbf{P}(A_n).$$

Similarly, if A_1, A_2, $A_3, \ldots$ *is an infinite sequence of decreasing events, i.e.,*

$$A_1 \supseteq A_2 \supseteq A_3 \supseteq \cdots,$$

then

$$\mathbf{P}(A_1 \cap A_2 \cap A_3 \cap \cdots) = \lim_{n \to \infty} \mathbf{P}(A_n).$$

Remark 1.5. Note that the limit of $\mathbf{P}(A_n)$ exists a priori as $n \to \infty$ since the sequence of numbers $\mathbf{P}(A_n)$ is increasing, when the events are increasing by Proposition 1.3 (or decreasing, when the events are decreasing). It is also bounded below by 0 and above by 1.

Proof. The continuity for decreasing events follows from the claim for increasing events by taking the complement. For increasing events, this is a consequence of the additivity of a probability. We construct a sequence of mutually exclusive events from the A_n's. Let's set $A_0 = \emptyset$. Take $B_1 = A_1 = A_1 \setminus A_0$. Then consider $B_2 = A_2 \setminus A_1$ so that B_1 and B_2 do not intersect. More generally, it suffices to take $B_n = A_n \setminus A_{n-1}$ for $n \geq 1$. By construction, the B_n's are disjoint. Moreover, we have $\bigcup_{n \geq 1} B_n = \bigcup_{n \geq 1} A_n$. Therefore, we have

$$\mathbf{P}(A_1 \cup A_2 \cup A_3 \cup \cdots) = \mathbf{P}(B_1 \cup B_2 \cup B_3 \cup \cdots) = \mathbf{P}(B_1) + \mathbf{P}(B_2) + \mathbf{P}(B_3) + \cdots.$$

For each $n \geq 1$, we have $\mathbf{P}(B_n) = \mathbf{P}(A_n \setminus A_{n-1}) = \mathbf{P}(A_n) - \mathbf{P}(A_{n-1})$. Therefore, the infinite sum can be written as a limit of a telescopic sum:

$$\lim_{N \to \infty} \sum_{n=1}^{N} \mathbf{P}(B_n) = \lim_{N \to \infty} \sum_{n=1}^{N} \{\mathbf{P}(A_n) - \mathbf{P}(A_{n-1})\} = \lim_{N \to \infty} \mathbf{P}(A_N).$$

This proves the claim. $\square$

Remark 1.6. If Ω is finite or countably infinite (i.e., its elements can be enumerated such as $\mathbb{N}$, $\mathbb{Z}$, or $\mathbb{Q}$), then a probability can always be defined on every subset of Ω. If Ω is uncountable however (such as $\mathbb{R}$, $[0, 1]$, $2^{\mathbb{N}}$), there might be subsets on which the probability cannot be defined. For example, let $\Omega = [0, 1]$ and consider the *uniform probability* $\mathbf{P}((a, b]) = b - a$ for $(a, b] \subset [0, 1]$. The probability is then the "length" of the interval. It turns out that there exist subsets of $[0, 1]$ for which the probability does not make sense! In other words, there are subsets of $[0, 1]$ for which the concept of length does not have meaning.

For the reasons laid out in the remark above, it is necessary to develop a consistent probability theory to restrict the probability to "good subsets" of the sample space on which the probability is well-defined. In probability terminology, these subsets are said to be *measurable*. In this book, measurable subsets will simply be called *events*. Let's denote by $\mathcal{F}$ the collection of events of Ω on which $\mathbf{P}$ is defined. It is good to think of

$\mathcal{F}$ as the domain of the probability **P**. Of course, we want the probability to be defined when we take basic operations of events such as unions and complement. Because of this, it it reasonable to demand that the collection of events $\mathcal{F}$ has the following properties:

- The sample space Ω is in $\mathcal{F}$.
- If A is in $\mathcal{F}$, then A^c is also in $\mathcal{F}$.
- If $A_1, A_2, \ldots, A_n, \ldots$ is a sequence of events in $\mathcal{F}$, then the union $\bigcup_{n \geq 1} A_n$ is also in $\mathcal{F}$.

A collection of subsets of Ω with the above properties is called a *sigma-field* of Ω. We will go back to this notion in Chapter 4 where we will study martingales; see Definition 4.9.

Example 1.7 (The power set of Ω). What is an appropriate sigma-field for the sample space $\Omega = \{0, 1\} \times \{0, 1\} = \{(0, 0), (0, 1), (1, 0), (1, 1)\}$ in Example 1.1? In this simple case, there are a finite number of subsets of Ω. In fact, there are $2^4 = 16$. The set of all subsets of a sample space Ω is called the *power set* and is usually denoted by $\mathcal{P}(\Omega)$. There is no problem of defining a probability on each of the subsets of Ω here by simply adding up the probabilities of each outcome in a particular event, since there can only be a finite number of outcomes in an event. More generally, if Ω is countable, finite or infinite, a probability can always be defined on all subsets of Ω by adding the probability of each outcome. Note that the power set $\mathcal{P}(\Omega)$ is a sigma-field as it satisfies all the properties above.

For now, let's just think of a sigma-field $\mathcal{F}$ as the subsets of Ω for which **P** is defined. With these notions, we have completed the theoretical framework of a random experiment in terms of a *probability space*.

Definition 1.8. A probability space $(\Omega, \mathcal{F}, \mathbf{P})$ is a triplet consisting of a sample space Ω, a sigma-field $\mathcal{F}$ of events of Ω, and a probability **P** that is well-defined on events in $\mathcal{F}$.

1.2. Random Variables and Their Distributions

A *random variable X* is a function from a sample space Ω taking values in $\mathbb{R}$. (To be precise, this definition needs to be slightly refined; cf. Remark 1.13.) This is written in mathematical notation as

$$X : \Omega \to \mathbb{R}$$

$$\omega \mapsto X(\omega).$$

Example 1.9. Consider Example 1.1. An example of random variables is $X(\omega)$ that gives the number of 0's in the outcome ω. For example, if $\omega = (0, 1)$, then $X(\omega)$ is equal to 1, and if $\omega = (0, 0)$, then $X(\omega) = 2$.

Remark 1.10. Why do we call X a random variable when it is actually a function on the sample space? It goes back to the terminology of *dependent variable* for a function whose input is an *independent* variable. (Think of $y = f(x)$ in calculus.) Here, since the input of X is random, we call the function X a *random variable*.

How can we construct a random variable on Ω? If we have a good description of the sample space, then we can build the function for each ω as we did in Example 1.9. An important example of random variables that we can easily construct is an *indicator function*.

Example 1.11 (Indicator functions as random variables). Let $(\Omega, \mathcal{F}, \mathbf{P})$ be a probability space. Let A be some event in $\mathcal{F}$. We define the random variable called the *indicator function* of the event A as follows:

$$\mathbf{1}_A(\omega) = \begin{cases} 1 & \text{if } \omega \in A, \\ 0 & \text{if } \omega \notin A. \end{cases}$$

In words, $\mathbf{1}_A(\omega) = 1$ if the event A occurs, i.e., the outcome $\omega \in A$, and $\mathbf{1}_A(\omega) = 0$ if the event A does not occur, i.e., the outcome $\omega \notin A$.

More generally, if X is a random variable on the probability space $(\Omega, \mathcal{F}, \mathbf{P})$, then any "reasonable" function of X is also a random variable. For example if $g : \mathbb{R} \to \mathbb{R}$ is a continuous function (like $g(x) = x^2$ for example), then the composition $g(X)$ is also a random variable. Clearly, there are many different random variables that can be defined on a given sample space.

Consider a probability space $(\Omega, \mathcal{F}, \mathbf{P})$ modelling some random experiment. It is often difficult to have a precise knowledge of all the outcomes of Ω and of the specific function X. If the experiment is elementary, such as a die roll or coin tosses, then there is no problem in enumerating the elements of Ω and to construct random variables explicitly as we did in Example 1.9. However, in more complicated modelling situations, such as models of mathematical finance, the sample space might be very large and hard to describe. Furthermore, the detailed relations between the source of randomness and the observed output (e.g., the Dow Jones index at closing on a given day) might be too complex to write down. This is one of the reasons why it is often more convenient to study the *distribution of a random variable* rather than to study the random variable as a function on the source of randomness.

To illustrate the notion of distribution, consider the probability space $(\Omega, \mathcal{F}, \mathbf{P})$ in Example 1.1 with the equiprobability (1.1) and $\mathcal{F}$ being all the subsets of Ω. Take X to be the random variable in Example 1.9. This random variable takes three possible values: 0, 1, or 2. The exact value of $X(\omega)$ is random since it depends on the input ω. In fact, it is not hard to check that

$$\mathbf{P}(\{\omega : X(\omega) = 0\}) = 1/4, \quad \mathbf{P}(\{\omega : X(\omega) = 1\}) = 1/2, \quad \mathbf{P}(\{\omega : X(\omega) = 2\}) = 1/4.$$

Now, if one is only interested in the values of X and not on the particular outcome, then X can serve as a source of randomness itself. This source of randomness is in fact a *probability* on the possible outcomes $\{0, 1, 2\}$. However, this is not the equiprobability on $\{0, 1, 2\}$! Therefore, the right way to think of the distribution of X is as a probability on $\mathbb{R}$.

Definition 1.12. Consider a probability space $(\Omega, \mathcal{F}, \mathbf{P})$ and a random variable X on it. The distribution of X is a probability on $\mathbb{R}$ denoted by ρ_X such that for any interval $(a, b]$ in $\mathbb{R}$, $\rho_X((a, b])$ is given by the probability that X takes value in $(a, b]$. In other words, we have

$$\rho_X((a, b]) = \mathbf{P}(\{\omega \in \Omega : X(\omega) \in (a, b]\}).$$

To lighten the notation, we write the events involving random variables by dropping the ω's. For example, the probability above is written

$$\rho_X((a, b]) = \mathbf{P}(\{\omega \in \Omega : X(\omega) \in (a, b]\}) = \mathbf{P}(X \in (a, b]).$$

But always keep in mind that a probability is evaluated on a subset of the outcomes!

We stress that the distribution ρ_X is a probability on subsets of $\mathbb{R}$. In particular, it satisfies the properties in Definition 1.2; see Exercise 1.3.

Remark 1.13 (A refined definition of a random variable). In the above definition of distribution, how can we be sure the event $\{\omega \in \Omega : X(\omega) \in (a, b]\}$ is in $\mathcal{F}$ so that the probability is well-defined? In general, we are not! This is why it is necessary to be more precise when building rigorous probability theory. With this in mind, the correct definition of a random variable X is a function $X : \Omega \to \mathbb{R}$ such that for any interval $(a, b] \subset \mathbb{R}$, the pre-image $\{\omega \in \Omega : X(\omega) \in (a, b]\}$ is an event in $\mathcal{F}$. If we consider a function of X, $g(X)$, then we must ensure that events of the form $\{\omega \in \Omega : g(X(\omega)) \in (a, b]\} = \{\omega \in \Omega : X(\omega) \in g^{-1}((a, b])\}$ are in $\mathcal{F}$. This is the case in particular if g is continuous. More generally, a function whose pre-image of intervals $(a, b]$ are reasonable subsets of $\mathbb{R}$ is called *Borel measurable*. In this book, when we write $g(X)$, we will always assume that g is Borel measurable.

Note that by the properties of a probability, we have for any interval $(a, b]$

$$\mathbf{P}(X \in (a, b]) = \mathbf{P}(X \le b) - \mathbf{P}(X \le a).$$

This motivates the following definition.

Definition 1.14. The cumulative distribution function (CDF) of a random variable X on a probability space $(\Omega, \mathcal{F}, P)$ is a function $F_X : \mathbb{R} \to [0, 1]$ defined by

$$F_X(x) = \mathbf{P}(X \le x).$$

Note that we also have, by definition, that $F_X(x) = \rho_X((-\infty, x])$. Clearly, if we know F_X, we know the distribution of X for any interval $(a, b]$. It turns out that the CDF determines the distribution of X for any (Borel measurable) subset of $\mathbb{R}$. These distinctions will not be important at this stage; only the fact that the CDF *characterizes* the distribution of a random variable will be important to us.

Proposition 1.15. *Let $(\Omega, \mathcal{F}, \mathbf{P})$ be a probability space. If two random variables X and Y have the same CDF, then they have the same distribution. In other words,*

$$\rho_X(B) = \rho_Y(B),$$

for any (Borel measurable) subset of $\mathbb{R}$.

We will not prove this result here, but we will use it often. Here are some important examples of distributions and their CDF.

Example 1.16.

(i) *Bernoulli distribution.* A random variable X has Bernoulli distribution with parameter $0 \le p \le 1$ if the CDF is

$$F_X(x) = \begin{cases} 0 & \text{if } x < 0, \\ 1 - p & \text{if } 0 \le x < 1, \\ 1 & \text{if } x \ge 1. \end{cases}$$

A Bernoulli random variable takes the value 0 with probability $1 - p$, and it takes value 1 with probability p.

(ii) *Binomial distribution.* A random variable X is said to have binomial distribution with parameters $0 \le p \le 1$ and $n \in \mathbb{N}$ if

$$\mathbf{P}(X = k) = \binom{n}{k} p^k (1 - p)^{n-k}, \qquad k = 0, 1, \dots, n.$$

In this case, the CDF is

$$F_X(x) = \begin{cases} 0 & \text{if } x < 0, \\ \sum_{j=0}^{k} \binom{n}{j} p^j (1 - p)^{n-j} & \text{if } k \le x < k + 1, 0 \le x < n, \\ 1 & \text{if } x \ge n. \end{cases}$$

(iii) *Poisson distribution.* A random variable X has Poisson distribution with parameter $\lambda > 0$ if

$$\mathbf{P}(X = k) = \frac{\lambda^k}{k!} e^{-\lambda}, \quad k = 0, 1, 2, 3, \dots.$$

In this case, the CDF is

$$F_X(x) = \begin{cases} 0 & \text{if } x < 0, \\ \sum_{j=0}^{k} \frac{\lambda^j}{j!} e^{-\lambda} & \text{if } k \le x < k + 1, k \in \mathbb{N} \cup \{0\}. \end{cases}$$

(iv) *Uniform distribution.* A random variable X has uniform distribution on $[0, 1]$ if

$$F_X(x) = \begin{cases} 0 & \text{if } x < 0, \\ x & \text{if } 0 \le x < 1, \\ 1 & \text{if } x \ge 1. \end{cases}$$

(v) *Exponential distribution.* A random variable X has exponential distribution with parameter $\lambda > 0$ if

$$F_X(x) = \begin{cases} 0 & \text{if } x < 0, \\ \int_0^x \lambda e^{-\lambda y} \, dy & \text{if } x \ge 0. \end{cases}$$

(vi) *Gaussian or normal distribution.* A random variable Z has Gaussian distribution with mean 0 and variance 1 if

$$F_Z(x) = \int_{-\infty}^{x} \frac{1}{\sqrt{2\pi}} e^{-z^2/2} \, dz.$$

A Gaussian distribution with mean 0 and variance 1 is said to be *standard*. If Z has a standard Gaussian distribution, then the random variable $X = \sigma Z + m$ for some $\sigma > 0$ and $m \in \mathbb{R}$ has Gaussian distribution with mean m and variance σ^2. Conversely, if X has Gaussian distribution with mean m and variance σ^2, then $Z = (X - m)/\sigma$ has standard Gaussian distribution.

As mentioned earlier, if X is a random variable on Ω, then any reasonable function of X (for example a continuous function on $\mathbb{R}$) is also a random variable. For example, X^2 is also a random variable. Then X^2 also defines a distribution on $\mathbb{R}$. This new distribution can be obtained by writing down the CDF of X^2 and expressing it in terms of the CDF of X; see Exercise 1.6.

Example 1.17. Let Z be a standard Gaussian variable. For $m \in \mathbb{R}$ and $\sigma > 0$, define the random variable $X = \sigma Z + m$ as in Example 1.16(vi). Then the CDF of X can be obtained from the one of Z:

$$F_X(x) = \mathbf{P}(\sigma Z + m \leq x) = \int_{-\infty}^{(x-m)/\sigma} \frac{1}{\sqrt{2\pi}} e^{-z^2/2}\, \mathrm{d}z = \int_{-\infty}^{x} \frac{1}{\sqrt{2\pi}\sigma} e^{-(y-m)^2/(2\sigma^2)}\, \mathrm{d}y,$$

where the last equality follows by the change of variable $y = \sigma z + m$.

Example 1.18. Let X be a random variable on the probability space $(\Omega, \mathcal{F}, \mathbf{P})$ with uniform distribution on $[0, 1]$. We consider the random variable $Y = e^X$. The distribution of Y is no longer uniform. Clearly, the possible values of Y are $[1, e]$. Therefore, $F_Y(y) = 0$ if $y < 1$ and $F_Y(y) = 1$ for $y \geq e$. For $y \in [1, e]$ we have

$$F_Y(y) = \mathbf{P}(Y \leq y) = \mathbf{P}(e^X \leq y) = \mathbf{P}(X \leq \log y) = \log y.$$

The properties of a CDF are a consequence of the properties of a probability.

Proposition 1.19 (Properties of a CDF). *Let $F_X : \mathbb{R} \to [0, 1]$ be a CDF of some random variable X on $(\Omega, \mathcal{F}, \mathbf{P})$. Then:*

- *F_X is an increasing function (but not necessarily strictly increasing).*
- *F_X is right-continuous: for all $x_0 \in \mathbb{R}$, $\lim_{x \to x_0^+} F_X(x) = F_X(x_0)$.*
- *$\lim_{x \to -\infty} F_X(x) = 0$ and $\lim_{x \to +\infty} F_X(x) = 1$.*

Proof. If $x \leq y$, then clearly $(-\infty, x] \subseteq (-\infty, y]$. Since the distribution ρ_X is a probability on $\mathbb{R}$, we get

$$F_X(x) = \rho_X((-\infty, x]) \leq \rho_X((-\infty, y]) = F_X(y).$$

This proves that F_X is increasing. For the right-continuity, we have by definition

$$\lim_{n \to \infty} F_X(x + 1/n) = \lim_{n \to \infty} \rho_X((-\infty, x + 1/n]).$$

Note that the sets $(-\infty, x + 1/n]$ are decreasing. Therefore, we have by the continuity of the probability ρ_X in Lemma 1.4

$$\lim_{n \to \infty} \rho_X((-\infty, x + 1/n]) = \rho_X\left(\bigcap_{n \geq 1}(-\infty, x + 1/n]\right) = \rho_X((-\infty, x]) = F_X(x).$$

This proves the right-continuity of F_X. Let $a_n \to +\infty$ be an increasing sequence. Consider the sets $(-\infty, a_n]$. These sets are increasing. Therefore, we have again by the continuity of the probability

$$\lim_{n \to \infty} F_X(a_n) = \lim_{n \to \infty} \rho_X\big((-\infty, a_n]\big) = \rho_X\Big(\bigcup_{n \geq 1}(-\infty, a_n]\Big) = \rho_X((-\infty, \infty)) = 1.$$

The proof for $x \to -\infty$ is similar with $a_n \to -\infty$ a decreasing sequence. $\qquad\square$

A CDF is not necessarily left-continuous at a point. If it is left-continuous at $a \in \mathbb{R}$, then we must have $\mathbf{P}(X = a) = 0$. In general, we always have that

$$(1.2) \qquad \mathbf{P}(X = a) = F_X(a) - \lim_{x \to a^-} F_X(x) = F_X(a) - F_X(a^-);$$

see Exercise 1.8. In other words, the jumps of a CDF occurs exactly at the values of X of positive probability. Moreover, the size of the jump gives the probability of that value. This leads to the following definition.

Definition 1.20. A random variable X with CDF F is said to be:

- *Continuous* if F is continuous.

- *Continuous with a probability density function* (PDF) if $F(x) = \int_{-\infty}^{x} f(y)\, dy$ for some function $f \geq 0$ with

$$\int_{-\infty}^{\infty} f(y)\, dy = 1.$$

  If a random variable is continuous with PDF f, then the fundamental theorem of calculus implies the following relation between the CDF and the PDF:

$$(1.3) \qquad \frac{d}{dx}F(x) = f(x).$$

- *Discrete* if F is piecewise constant.

It is good to classify the distributions of Example 1.16 according to the above terminology. The Bernoulli and Poisson distributions are discrete. The exponential, uniform, and Gaussian distributions are continuous with a PDF. (What are their respective PDFs?) The random variable Y constructed in Example 1.18 is continuous with PDF $f(y) = 1/y$ if $y \in [1, e]$ and $f(y) = 0$ otherwise.

Remark 1.21. It is important to note that if a random variable has a PDF, then the CDF is automatically continuous. However the converse is not true! There exist CDFs that are continuous that do not have a PDF. A classic example is *the Cantor function*.

Example 1.22 (Apppproximating a PDF using a histogram). The distribution of a random variable can be approximated numerically by repeating the experiment represented by the probability space $(\Omega, \mathcal{F}, \mathbf{P})$ and by drawing the histogram of the values of $X(\omega)$ for the different sampled outcomes ω; see Figure 1.1 and Numerical Project 1.1. This can be done with any random variable. Let's work with a uniform random variable X on $[0, 1]$ defined on $(\Omega, \mathcal{F}, \mathbf{P})$. The distribution ρ_X can be approximated as follows. Divide the interval $[0, 1]$ into m bins $B_1, \ldots, B_m$ of equal length $1/m$: $B_1 = [0, 1/m]$ and

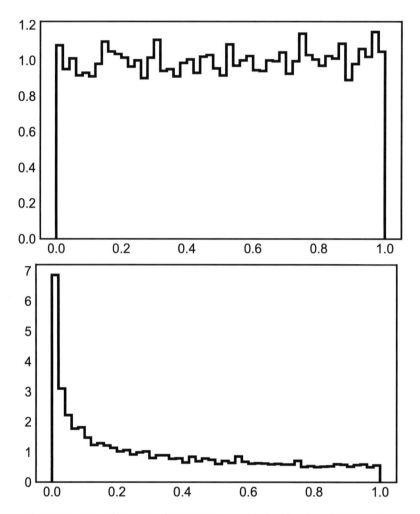

Figure 1.1. Top: A histogram of 10,000 points sampled uniformly with 50 bins. Bottom: A histogram for the same sampled points squared.

more generally, $B_j = ((j-1)/m, j/m]$ for $1 \leq j \leq m$. Then, generate numerically N random numbers $X_1, \ldots, X_N$ with uniform distribution on $[0, 1]$. Then the probability that X takes value in the bin B_j, $\rho_X(B_j)$, $j \leq m$, is approximated by the number of points of the simulations that fell in B_j divided by N:

(1.4) $$\rho_X(B_j) \approx \frac{\#\{i \leq N : X_i \in B_j\}}{N}.$$

This is a consequence of the law of large numbers (more on this in Chapters 2 and 3). If the bins are made arbitrarily small by taking the limit $m \to \infty$, then one recovers the PDF after normalizing by the length. This is because of equation (1.3):

$$\lim_{m \to \infty} \frac{\rho_X(B_j)}{1/m} = \lim_{m \to \infty} \frac{1}{1/m}(F_X(j/m) - F_X((j-1)/m)) = \frac{d}{dx}F_X(x).$$

This fact motivates the term *probability density*.

Example 1.23 (Generating a random variable from a uniform one). Let U be a uniform random variable on $[0, 1]$. Let X be a random variable with CDF $F_X : \mathbb{R} \to (0, 1)$ so that F_X tends to 0 and 1 when $x \to -\infty$ and $+\infty$, respectively, but never reaches it. Suppose also that F_X is continuous and strictly increasing. In that case, it is easy to see that the inverse function $F_X^{-1} : (0, 1) \to \mathbb{R}$ exists. Indeed, since the function is strictly increasing, it is one-to-one, hence invertible on its range. Here the range of F_X is $(0, 1)$ since $\lim_{x \to +\infty} F(x) = 1$ and $\lim_{x \to -\infty} F(x) = 0$.

Now, consider the random variable

$$Y = F_X^{-1}(U).$$

We claim that this random variable has the same distribution as X. To see this, let's compute the CDF:

$$\mathbf{P}(Y \leq y) = \mathbf{P}(F_X^{-1}(U) \leq y).$$

Now, we can apply F_X to both sides of the inequality without changing the probability, since F_X is strictly increasing. We get

$$\mathbf{P}(Y \leq y) = \mathbf{P}(U \leq F_X(y)) = F_X(y),$$

since $F_U(x) = x$ for $x \in [0, 1]$. This proves the claim.

This shows how to simulate a continuous random variable with strictly increasing CDF using a uniform random variable. The same idea can be applied to any random variable, even discrete. For example, how would you generate a Bernoulli random variable with $p = 1/2$ from a uniform random variable?

1.3. Expectation

The *expectation* or *mean* of a random variable is simply its average value weighted by the probability of that value. Concretely, if X is a random variable on $(\Omega, \mathcal{F}, \mathbf{P})$ and Ω is a finite sample space, then the expectation is

$$\mathbf{E}[X] = \sum_{\omega \in \Omega} X(\omega) \mathbf{P}(\{\omega\}).$$

The expectation is therefore a *sum over the outcomes ω*.

Example 1.24. If we generate a random integer in $\{1, 2, 3, 4, 5, 6\}$ with equal probability, then the probability space is $\Omega = \{1, 2, 3, 4, 5, 6\}$ with $\mathbf{P}(\{\omega\}) = 1/6$ for any outcome. Therefore, if X is the random variable defined by $X(\omega) = +1$ if ω is even and $X(\omega) = -1$ if ω is odd, then

$$\mathbf{E}[X] = (-1) \cdot \frac{1}{6} + 1 \cdot \frac{1}{6} + (-1) \cdot \frac{1}{6} + 1 \cdot \frac{1}{6} + (-1) \cdot \frac{1}{6} + 1 \cdot \frac{1}{6} = 0.$$

Another way to compute the expectation is as a *sum over the values of X*. This uses the distribution of X. In the above example, the random variable X takes only two values. By regrouping the ω's with a common value for X we get

$$\mathbf{E}[X] = 1 \cdot \frac{1}{2} + (-1) \cdot \frac{1}{2} = 0.$$

More generally, if X takes a finite number of values, we can express the expectation as a sum over values of X as

$$\mathbf{E}[X] = \sum_{a:\, \text{values of } X} a\, \mathbf{P}(X = a).$$

The above definition also makes sense as an infinite sum if Ω is countably infinite or if X takes a countably infinite number of values (such as Poisson random variables) granted that the infinite sum converges.

If X is a continuous random variable with PDF f, then we take as a definition that

$$\mathbf{E}[X] = \int_{-\infty}^{\infty} y f(y)\, dy.$$

Note that the expectation here is an integral over the values taken by X. Since an integral is a limit of Riemann sums, this definition keeps the spirit of the finite case. It is still morally a sum over the values of X weighted by their probability as the term $f(y)\, dy$, being the probability density times dy, can be interpreted as the probability that X takes value in the infinitesimal interval $(y, y + dy]$.

Remark 1.25. If X does not have a PDF, the expectation can be made sense of as the following integral over the values of X:

$$\mathbf{E}[X] = \int_{-\infty}^{\infty} y\, dF(y).$$

The differential term $dF(y)$ must be given a new meaning when F is not differentiable. This can be done for any CDF using *Lebesgue integration*. See Exercise 1.13 where the above formula is written differently using integration by parts.

Similarly, we also have the expectation as a limit of sums over the outcomes in the general case where Ω is uncountable. However, the abstract framework of *measure theory* is needed to make sense of this. The good news is that even in the general setting, the expectation keeps the basic properties of a sum (see Theorem 1.28).

The sums and integrals entering in the definition of the expectation could be problematic. This could happen if the sum is infinite or if the integral is improper. The convergence of the sums and integrals defining the expectation is an important question. In other words:

Not every random variable has a defined expectation!

If X is always positive, then either X has finite expectation or X has infinite expectation, $\mathbf{E}[X] = +\infty$. However, if X takes positive and negative values, the expectation might not even be defined.

Example 1.26. Consider the random variable with distribution given by the PDF

$$f(x) = \frac{1}{\pi} \frac{1}{1 + x^2}, \quad x \in \mathbb{R}.$$

This distribution is called a *Cauchy distribution*. The histogram is plotted in Figure 1.2 and compared to the Gaussian one. The CDF can be obtained by integrating:

$$F_X(x) = \frac{1}{\pi} \arctan x + \frac{1}{2}.$$

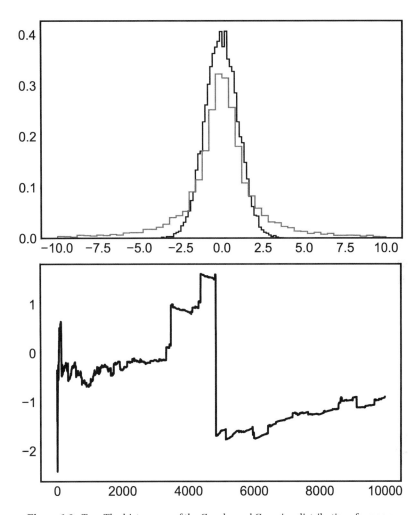

Figure 1.2. Top: The histograms of the Cauchy and Gaussian distributions for a sample of 10,000 points. Bottom: The values of the empirical mean S_N for $N \leq 10,000$ sampled from Cauchy random variables.

Note that $\lim_{x \to -\infty} F_X(x) = 0$ and $\lim_{x \to +\infty} F_X(x) = 1$. The expectation of X is not well-defined since the improper integral

$$\int_{-\infty}^{\infty} \frac{1}{\pi} \frac{x}{1 + x^2} \, dx$$

does not converge. (Why?) The Cauchy distribution is an example of a *heavy-tailed distribution*. More on this in Example 1.34 and Numerical Project 1.4.

A sufficient condition (not proved here) for the expectation of a random variable X to exist is $\mathbf{E}[|X|] < \infty$. This is reminiscent of absolute convergence of infinite sums and integrals.

Definition 1.27. Let X be a random variable on $(\Omega, \mathcal{F}, P)$. It is said to be *integrable* if $\mathbf{E}[|X|] < \infty$.

Whenever the random variable X is integrable, we have the following simple but powerful properties.

Theorem 1.28 (Properties of expectation). *Let X, Y be two integrable random variables on $(\Omega, \mathcal{F}, \mathbf{P})$. Then:*

- *Positivity: If $X \geq 0$, then $\mathbf{E}[X] \geq 0$.*
- *Monotonicity: If $X \leq Y$, then $\mathbf{E}[X] \leq \mathbf{E}[Y]$.*
- *Linearity: For any $a, b \in \mathbb{R}$, then $\mathbf{E}[aX + bY] = a\mathbf{E}[X] + b\mathbf{E}[Y]$.*

We will not prove this theorem since the general proof necessitates measure theory. Note, however, that they all are basic and intuitive properties of a sum. The properties also apply when $X \geq 0$ and $Y \geq 0$ under the extra conditions that $a, b \geq 0$. In this case, the expectation is always defined if we accept $+\infty$ as a value.

Example 1.29 (Approximating the expectation with the empirical mean). The expectation of a random variable X, if it exists, can be approximated numerically by computing the *empirical mean*. Specifically, we can generate N values of X, say $X_1, \ldots, X_N$ and compute

$$(1.5) \qquad \mathbf{E}[X] \approx \frac{1}{N}(X_1 + \cdots + X_N) = \frac{1}{N}\sum_{j=1}^{N} X_j.$$

It turns out that the empirical mean on the right converges to $\mathbf{E}[X]$ in a suitable sense. This is again a consequence of the law of large numbers; see Numerical Project 1.2. Note that equation (1.4) is a particular case of (1.5) where the random variable X is the indicator function $\mathbf{1}_{\{X \in B\}}$ of the event $\{X \in B\}$. What if the expectation does not exist as in the case of Cauchy random variables in Example 1.26? In that case, the empirical mean never converges to a given value and wanders around as N increases; see Figure 1.2.

Let $g : \mathbb{R} \to \mathbb{R}$ be a nice function, say continuous. Then $Y = g(X)$ is also a random variable. The expectation of Y, if it exists, can be computed from its distribution as usual. The expectation of Y can also be computed from the distribution of X. For example, if X has PDF f, then

$$\mathbf{E}[Y] = \mathbf{E}[g(X)] = \int_{-\infty}^{\infty} g(x)f(x)\,\mathrm{d}x.$$

Here are some important quantities computed as an expectation of a given random variable:

- *Variance*: $\mathrm{Var}(X) = \mathbf{E}[(X - \mathbf{E}[X])^2]$. Here the function is $g(x) = (x - \mathbf{E}[X])^2$. Recall that $\mathbf{E}[X]$ is a number! The variance is the average square-distance between X and its mean $\mathbf{E}[X]$. By linearity of expectation, we get

$$\mathrm{Var}(X) = \mathbf{E}[(X - \mathbf{E}[X])^2] = \mathbf{E}[X^2] - (\mathbf{E}[X])^2.$$

  The *standard deviation* is the square root of the variance.

- *Moments*: The n-th moment of X is $\mathbf{E}[X^n]$ for $n \in \mathbb{N}$. Here, $g(x) = x^n$.

- *Probability as an expectation*: Let B be some subset of $\mathbb{R}$. We consider the indicator function of the event $\{X \in B\}$ as defined in Example 1.11. Then

(1.6) $$\mathbf{E}[\mathbf{1}_{\{X \in B\}}] = \mathbf{E}[\mathbf{1}_B(X)] = \mathbf{P}(X \in B).$$

Here the function is $g(x) = \mathbf{1}_B(x)$ that equals 1 for $x \in B$ and 0 elsewhere.

- *Moment Generating Function (MGF)*: Let $\lambda \in \mathbb{R}$. The moment generating function of the random variable X is the function of λ defined by

$$\phi(\lambda) = \mathbf{E}[e^{\lambda X}].$$

Here we have $g(x) = e^{\lambda x}$.

Keep in mind that the variance and the MGF of a random variable might be infinite. If the MGF is finite for every λ, then it characterizes the distribution like the CDF.

Proposition 1.30. *Let $(\Omega, \mathcal{F}, \mathbf{P})$ be a probability space. If two random variables X and Y have the same well-defined MGF, then they have the same CDF, and thus the same distribution.*

We will not prove this result here, but we will use it often.

Example 1.31 (The MGF of a Gaussian). The moment generating function of a Gaussian random variable Z with mean 0 and variance 1 is

$$\mathbf{E}[e^{\lambda Z}] = e^{\lambda^2/2}.$$

Note the resemblance with the PDF! To see this, we compute the expectation using the PDF of a standard Gaussian random variable

$$\mathbf{E}[e^{\lambda Z}] = \int_{-\infty}^{\infty} e^{\lambda z} \frac{e^{-z^2/2}}{\sqrt{2\pi}} \, dz = \int_{-\infty}^{\infty} \frac{e^{-\frac{1}{2}(z^2 - 2\lambda z + \lambda^2 - \lambda^2)}}{\sqrt{2\pi}} \, dz,$$

by completing the square. This is equal to

$$e^{\lambda^2/2} \int_{-\infty}^{\infty} \frac{e^{-\frac{1}{2}(z-\lambda)^2}}{\sqrt{2\pi}} \, dz.$$

The integral is 1, since it is the integral over $\mathbb{R}$ of the PDF of a Gaussian random variable of mean λ and variance 1.

This also implies that the MGF of a Gaussian random variable X with mean m and variance σ^2 is

$$\mathbf{E}[e^{\lambda X}] = e^{\lambda m + \lambda^2 \sigma^2/2}.$$

This is simply by noting X has the same distribution as $\sigma Z + m$ where Z is the standard Gaussian; see Example 1.17.

Example 1.32 (Moments from the MGF). As the name suggests, the moments of a random variable X can be read from the MGF, assuming it exists. To see this, recall the Taylor expansion of the exponential function around 0:

$$e^x = 1 + x + \frac{x^2}{2!} + \cdots + \frac{x^n}{n!} + \cdots = \sum_{n=0}^{\infty} \frac{x^n}{n!}.$$

If we replace x by λX for some random variable X in the above and take expectation on both sides, we get

$$(1.7) \qquad \mathbf{E}[e^{\lambda X}] = \sum_{n=0}^{\infty} \frac{\lambda^n}{n!} \mathbf{E}[X^n].$$

Note that we have interchanged the infinite sum with the expectation here, a manipulation that necessitates some care in general. We will say more on this in Chapter 3. From equation (1.7) we see that the n-th moment can be obtained by taking the n-th derivative of the MGF evaluated at $\lambda = 0$:

$$\frac{\mathrm{d}}{\mathrm{d}\lambda} \mathbf{E}[e^{\lambda X}]\Big|_{\lambda=0} = \mathbf{E}[X^n].$$

If we have the Taylor expansion of $\mathbf{E}[e^{\lambda X}]$ already in hand, then the moments can be read off the coefficients readily.

For example, the MGF of a standard Gaussian in Example 1.31 is

$$\mathbf{E}[e^{\lambda Z}] = e^{\lambda^2/2} = \sum_{j=0}^{\infty} \frac{\lambda^{2j}}{2^j j!}.$$

Therefore, comparing this with equation (1.7), we get that the standard Gaussian moments are $\mathbf{E}[Z^n] = 0$ if n is odd, and if $n = 2j$, it is even:

$$(1.8) \qquad \mathbf{E}[Z^{2j}] = \frac{(2j)!}{2^j j!} = (2j-1) \cdot (2j-3) \cdot (\dots) \cdot 5 \cdot 3 \cdot 1.$$

For example, we have $\mathbf{E}[Z^4] = 3$ and $\mathbf{E}[Z^6] = 15$. If X is Gaussian with mean 0 and variance σ^2, then it suffices to write $X = \sigma Z$ as in Example 1.16(vi) to get

$$(1.9) \qquad \mathbf{E}[X^{2j}] = \frac{(2j)!}{2^j j!} \sigma^{2j} = (2j-1) \cdot (2j-3) \cdot (\dots) \cdot 5 \cdot 3 \cdot 1 \cdot \sigma^{2j}.$$

Exercise 1.10 offers another way to compute the Gaussian moments.

1.4. Inequalities

We end this chapter by deriving important inequalities relating probabilities of large values of a random variable to its expectation. These relations highlight the fact that the finiteness of the expectation of a random variable (and of functions of a random variable) is intimately connected with the probabilities that the random variable takes large values.

Corollary 1.33 (Markov's inequality). *Let X be a random variable on $(\Omega, \mathcal{F}, \mathbf{P})$ that is positive, $X \geq 0$. Then for any $a > 0$, we have*

$$(1.10) \qquad \mathbf{P}(X > a) \leq \frac{1}{a} \mathbf{E}[X].$$

Proof. The inequality is clear if $\mathbf{E}[X] = +\infty$. Therefore, we can assume that X is integrable: $\mathbf{E}[X] < \infty$. This is a nice consequence of the properties of expectations. Indeed, consider the indicator functions $\mathbf{1}_{\{X>a\}}$ and $\mathbf{1}_{\{X\leq a\}}$. Clearly, we must have $\mathbf{1}_{\{X>a\}} + \mathbf{1}_{\{X\leq a\}} = 1$. The linearity of expectation then implies

$$\mathbf{E}[X] = \mathbf{E}[X\mathbf{1}_{\{X>a\}}] + \mathbf{E}[X\mathbf{1}_{\{X\leq a\}}].$$

The second term is greater than or equal to 0. For the first term, note that $X\mathbf{1}_{\{X>a\}} \geq a\mathbf{1}_{\{X>a\}}$. Therefore, by monotonicity of expectation, we have

$$\mathbf{E}[X] \geq \mathbf{E}[a\mathbf{1}_{\{X>a\}}] = a\mathbf{P}(X > a),$$

as claimed. □

Two other important inequalities can be derived directly from this: one is the Chebyshev inequality,

(1.11) $$\mathbf{P}(|X| > a) \leq \frac{1}{a^2}\mathbf{E}[X^2].$$

This follows from Markov's inequality with the random variable X^2. We also have the Chernoff bound,

(1.12) $$\mathbf{P}(X > a) \leq e^{-\lambda a}\mathbf{E}[e^{a\lambda X}],$$

obtained from Markov's inequality with the random variable $e^{\lambda X}$, $\lambda > 0$.

Example 1.34 (Tail probability and moments). Equations (1.10), (1.11), and (1.12) highlight the intimate connection between the finiteness of moments and the tail probability of a random variable. By tail probability of a random variable X, we mean the probability that X takes large values, say $\mathbf{P}(|X| > x)$ for x large, or simply $\mathbf{P}(X > x)$ if we are interested in positive values only. For example, if $\mathbf{E}[X^2] < \infty$, then Chebyshev's inequality ensures that the probability that $|X|$ is larger than x decays faster than $1/x^2$. Chernoff's bound guarantees exponential decay of the tail probabilities whenever the MGF is finite. See Exercise 1.12 for an instance where the exponential decay can be lifted to a Gaussian decay whenever the MGF is bounded by the Gaussian MGF. Heavy-tailed random variables are such that the tail probability decays so slowly that it prohibits finiteness of moments, as in Example 1.26.

1.5. Numerical Projects and Exercises

1.1. **Distributions as histograms**. The goal is to reproduce Figure 1.1.
  (a) Sample $N = 10{,}000$ random numbers uniformly on $[0, 1]$.
    This can be done using the command `numpy.random` *in Python.*
  (b) Plotting the PDF: Divide the interval $[0, 1]$ into m bins $B_1, \ldots, B_{50}$ of equal length $1/50$. Plot the histogram of the values for 50 bins where the value at each bin j is
    $$\frac{\#\{i \leq N : X_i \in B_j\}}{N}.$$
    The command `hist` *in* `matplotlib.pyplot` *is useful here. The option* `density` *gives the PDF.*
  (c) Plotting the CDF: Plot the cumulative histogram of the values for 50 bins where the value at each bin j is
    $$\frac{\#\{i \leq N : X_i \leq j/50\}}{N}.$$
    The option `cumulative` *does the trick here.*
  (d) Re-do items (b) and (c) for the square of each number. This would approximate the PDF of X^2 where X is uniformly distributed on $[0, 1]$.

1.2. **The law of large numbers.** This project is about getting acquainted with the law of large numbers seen in equation (1.5). We experiment with exponential random variables, but any distributions would do as long as the expectation is defined.

    (a) **The strong law.** Sample $N = 10{,}000$ random numbers $X_1, \ldots, X_N$ exponentially distributed with parameter 1. Use the command `numpy.cumsum` to get the empirical mean $\frac{S_N}{N} = \frac{1}{N}(X_1 + \cdots + X_N)$. Plot the values for $N = 1, \ldots, 10{,}000$. What do you notice?

    (b) **The weak law.** Define a function in Python using the command `def` that returns the empirical mean of a sample of size N as above. This will allow you to sample the empirical mean $\frac{S_N}{N}$ as many times as needed for a given N. Plot the histograms (PDF and CDF) of a sample of size 10,000 of the empirical mean for $N = 100$ and $N = 10{,}000$. What do you notice?

1.3. **The central limit theorem.** The approximation (1.5) of the expectation in terms of the empirical mean is not exact for finite N. The error is controlled by the central limit theorem. For a sample $X_1, \ldots, X_N$ of the random variable X of mean $\mathbf{E}[X]$ and variance σ^2, this theorem says that the sum $S_N = X_1 + \cdots + X_N$ behaves like

$$S_N \approx N\mathbf{E}[X] + \sqrt{N}\sigma Z,$$

where Z is a standard Gaussian random variable. More precisely, this means that

(1.13) $$\lim_{N \to \infty} \frac{S_N - N\mathbf{E}[X]}{\sigma\sqrt{N}} = Z.$$

The limit should be understood here as the convergence in distribution. Practically speaking, this means that the histogram of the random variable $\frac{S_N - N\mathbf{E}[X]}{\sigma\sqrt{N}}$ should resemble the one of a standard Gaussian variable when N is large. We check this numerically.

    (a) Let $S_N = X_1 + \cdots + X_N$ where the X_i are exponentially distributed random variables of parameter 1. Define a function in Python using the command `def` that returns for a given N the value $Y_N = \frac{S_N - N\mathbf{E}[X]}{\sigma\sqrt{N}}$.

    (b) Plot the histograms (PDF) of a sample of size 10,000 of Y_N for $N = 100$. What do you notice?

    (c) Compare the above to the histogram of a sample of size 10,000 of points generated using the standard Gaussian distribution.

1.4. **Sampling Cauchy random variables.** Here, we reproduce Figure 1.2. In particular, it shows that there is no law of large numbers (weak or strong) for Cauchy random variables.

    (a) Let F_X^{-1} be the inverse of the CDF of the Cauchy distribution in Example 1.26. Plot the histogram of $F_X^{-1}(U)$ where U is a uniform random variable for sample of 10,000 points. Use 100 bins in the interval [-10,10].
       The command `random` *in* `numpy` *also has a Cauchy option, if you want to skip this step.*

    (b) Compare the above to the histogram of a sample of size 10,000 of points generated using the standard Gaussian distribution.

(c) Let $(C_n, n \leq 10{,}000)$ be the values obtained in (a). Plot the empirical mean $\frac{S_N}{N} = \frac{1}{N} \sum_{n \leq N} C_n$ for $N = 1, \ldots, 10{,}000$. What do you notice?

(d) Define a function in Python using the command def that returns the empirical mean $\frac{S_N}{N}$ of a sample of size N as above. Plot the histograms (PDF) of a sample of size 10,000 of the empirical mean for $N = 10$ and $N = 100$. What do you notice compared to the histograms in Project 1.2?

Exercises

1.1. Basic properties of a probability. Prove Proposition 1.3.
Hint: For the finite additivity, use the additivity for an infinite sequence and the fact that $\mathbf{P}(\emptyset) = 0$.

1.2. Equiprobability. Let Ω be a sample space with a finite number of outcomes. Show that the equiprobability defined by $\mathbf{P}(A) = \#A/\#\Omega$, $A \subseteq \Omega$, has the three defining properties of a probability.

1.3. Distribution as a probability on $\mathbb{R}$. Let ρ_X be the distribution of a random variable X on some probability space $(\Omega, \mathcal{F}, \mathbf{P})$. Show that ρ_X has the properties of a probability on $\mathbb{R}$.

1.4. Distribution of an indicator function. Let $(\Omega, \mathcal{F}, \mathbf{P})$ be a probability space and A an event in $\mathcal{F}$ with $0 < \mathbf{P}(A) < 1$. We consider the random variable $\mathbf{1}_A$. What is the distribution of $\mathbf{1}_A$?

1.5. Events of probability one. Consider a probability space $(\Omega, \mathcal{F}, \mathbf{P})$. Let A_1, A_2, A_3, ... be an infinite sequence of events in $\mathcal{F}$ such that $\mathbf{P}(A_n) = 1$ for all $n \geq 1$. Show that their intersection also has probability one.

1.6. Constructing a random variable from another one. Let X be a random variable on $(\Omega, \mathcal{F}, \mathbf{P})$ that is uniformly distributed on $[-1, 1]$. Consider $Y = X^2$.
(a) Find the CDF of Y. Plot the graph.
(b) Find the PDF of Y. Plot the graph.

1.7. Sum of integrable variables. Let X and Y be two integrable random variables on the same probability space. Argue that we also have that the random variable $aX + bY$ is integrable for any $a, b \in \mathbb{R}$.

1.8. Jumps and probabilities. Let X be a random variable and let F_X be its CDF. Use the properties of a probability to show equation (1.2)

1.9. Memory loss property. Let Y be an exponential random variable with parameter λ. Show that for any $s, t > 0$

$$\mathbf{P}(Y > t + s | Y > s) = \mathbf{P}(Y > t).$$

Recall that the conditional probability of A given the event B is

$$\mathbf{P}(A|B) = \mathbf{P}(A \cap B)/\mathbf{P}(B).$$

1.10. Gaussian integration by parts.

(a) Let Z be a standard Gaussian random variable. Show using integration by parts that for a differentiable function g,

(1.14) $$\mathbf{E}[Zg(Z)] = \mathbf{E}[g'(Z)],$$

where we assume that both expectations are well-defined.

(b) Use this to recover the expression for the moments given in equation (1.8).

1.11. MGF of exponential random variables. Show that the MGF of an exponential random variable of parameter $\lambda > 0$ is

$$\mathbf{E}[e^{tX}] = \frac{\lambda}{\lambda - t}, \quad t < \lambda.$$

Deduce from this $\mathbf{E}[X]$ and $\mathrm{Var}(X)$. What happens if $t \geq \lambda$?

1.12. Gaussian tail. Consider a random variable X with finite MGF such that

$$\mathbf{E}[e^{\lambda X}] \leq e^{\lambda^2/2} \text{ for all } \lambda \in \mathbb{R}.$$

Prove that for $a > 0$

$$\mathbf{P}(X > a) \leq e^{-a^2/2}.$$

Hint: Optimize λ.

1.13. ★ Expectation from CDF. Consider X a random variable that is nonnegative, $X \geq 0$, and has a PDF. Prove that the expectation of X can be written as

$$\mathbf{E}[X] = \int_0^\infty \mathbf{P}(X > x)\,\mathrm{d}x.$$

Now, take a random variable X, possibly negative, with PDF such that $\mathbf{E}[|X|] < \infty$ so that the expectation is well-defined. Prove that

$$\mathbf{E}[X] = \int_0^\infty \mathbf{P}(X > x)\,\mathrm{d}x - \int_{-\infty}^0 \mathbf{P}(X \leq x)\,\mathrm{d}x.$$

Hint: Write $X = X\mathbf{1}_{\{X \geq 0\}} + X\mathbf{1}_{\{X < 0\}}$.

1.14. ★ Characteristic function. The characteristic function of a random variable X is defined by $\mathbf{E}[e^{itX}]$ where $i = \sqrt{-1}$ so $i^2 = -1$. Like the MGF, it determines the distribution of X. The advantage of the characteristic function over the MGF is that it is always defined.

(a) Argue that $\mathbf{E}[e^{itX}] = \mathbf{E}[\cos tX] + i\mathbf{E}[\sin tX]$ by a Taylor expansion.

(b) Argue that the characteristic function of Gaussian of mean m and variance σ^2 is $\mathbf{E}[e^{itX}] = e^{itm - \sigma^2 t^2/2}$.

1.15. ★ When $\mathbf{E}[X] < \infty$. Let X be a random variable on $(\Omega, \mathcal{F}, \mathbf{P})$ with $X \geq 0$. Show that if $\mathbf{E}[X] < \infty$, then we must have $\mathbf{P}(X < \infty) = 1$.
Hint: Write the event $\{X = \infty\}$ as $\{X = \infty\} = \bigcap_{n \geq 1}\{X > n\}$. Use an inequality and Lemma 1.4.

1.16. ★ When $\mathbf{E}[X] = 0$. Let X be a random variable on $(\Omega, \mathcal{F}, \mathbf{P})$ with $X \geq 0$. Show that if $\mathbf{E}[X] = 0$, then we must have $\mathbf{P}(X = 0) = 1$.
Hint: Follow a similar strategy as the above.

1.6. Historical and Bibliographical Notes

For the reader interested in learning more about probability from the point of view of measure theory, a good introductory book is [**Wal12**]. Kolmogorov is usually credited for laying the rigorous foundations of probability theory using measure theory [**Kol50**]. The full power of measure theory is needed when one tries to prove Theorem 1.28 in the general case. The reason is that the expectation is an integral over the abstract space Ω. The most famous example of a nonmeasurable set in $\mathbb{R}$ is a *Vitali set*, whose construction invokes the axiom of choice. The proofs that the CDF and the MGF determine the distribution of a random variables are done in most graduate textbooks in probability, e.g., in [**Bil95**].

Heavy-tailed random variables are not singular objects. They play a fundamental role in the theory of probability and in applications to the theory in the real world. The rare events that have a small probability of occurring, but that have a large influence on the empirical mean (as the jumps in Figure 1.2), are sometimes called *black swans*. The term was coined by Nassim Taleb in the popular book [**Tal08**].

Chapter 2

Gaussian Processes

In this chapter, we explore the properties of an important class of stochastic processes: the Gaussian processes. Brownian motion is one example of Gaussian processes. In Section 2.1, we start by looking at random vectors, i.e., a finite collection of random variables. Section 2.2 puts an emphasis on Gaussian vectors also known as *jointly Gaussian random variables*. This is where we see the *Cholesky decomposition*, which allows us to decompose the random variables in terms of independent identically distributed standard Gaussians. We then define Gaussian processes using Gaussian vectors in Section 2.3. We introduce important examples: Brownian motion, Ornstein-Uhlenbeck process, Brownian bridge, and fractional Brownian motion. We finish with Section 2.4 that presents an important geometric point of view for the space of random variables with finite variance, the so-called *square-integrable random variables*. This point of view will be fundamental in the next chapter on conditional expectation and martingales.

2.1. Random Vectors

Consider a probability space $(\Omega, \mathcal{F}, \mathbf{P})$. We can define several random variables on Ω. A n-tuple of random variables on this space is called a *random vector*. For example, if $X_1, \ldots, X_n$ are random variables on $(\Omega, \mathcal{F}, \mathbf{P})$, then the n-tuple $(X_1, \ldots, X_n)$ is a random vector on $(\Omega, \mathcal{F}, \mathbf{P})$. The vector is also said to be n-dimensional because it contains n variables. We will sometimes denote a random vector by X.

A good point of view is to think of a random vector $X = (X_1, \ldots, X_n)$ as a random variable in $\mathbb{R}^n$. In other words, for an outcome $\omega \in \Omega$, $X(\omega)$ is a point sampled in $\mathbb{R}^n$, where $X_j(\omega)$ represents the j-th coordinate of the point. The *distribution* of X, denoted by ρ_X, is a probability on $\mathbb{R}^n$ defined by the events related to the values of X:

$$\mathbf{P}(X \in A) = \rho_X(A) \text{ for a subset } A \text{ in } \mathbb{R}^n.$$

In other words, $\mathbf{P}(X \in A) = \rho_X(A)$ is the probability that the random point X falls in A. The distribution of the vector X is also called the *joint distribution* of $(X_1, \ldots, X_n)$.

We will mostly be interested in the examples where the distribution is given by a *joint probability density function* or *joint PDF* for short. The joint PDF $f(x_1, \ldots, x_n)$ of a random vector X is a function $f : \mathbb{R}^n \to \mathbb{R}$ such that the probability that X falls in a subset A of $\mathbb{R}^n$ is expressed as the integral of $f(x_1, \ldots, x_n)$ over A:

$$\mathbf{P}(X \in A) = \int_A f(x_1, \ldots, x_n) \, dx_1 \ldots dx_n.$$

Note that we must have that the integral of f over the whole of $\mathbb{R}^n$ is 1.

Example 2.1 (Sampling uniformly in the unit disc). Consider the random vector $X = (X, Y)$ corresponding to a random point chosen uniformly in the unit disc $\{(x, y) : x^2 + y^2 \leq 1\}$. In this case, the joint PDF is 0 outside the disc and $\frac{1}{\pi}$ inside the disc:

$$f(x, y) = \frac{1}{\pi} \quad \text{if } x^2 + y^2 \leq 1.$$

The random point (X, Y) has x-coordinate X and y-coordinate Y. Each of these are random variables and their PDFs can be computed. For X, it is not hard to check that its CDF is given by

$$\mathbf{P}(X \leq a) = \int_{\{(x,y) \in \mathbb{R}^2 : x \leq a\}} f(x, y) \, dx \, dy = \int_{-1}^{a} \int_{-\sqrt{1-x^2}}^{\sqrt{1-x^2}} \frac{1}{\pi} \, dy \, dx = \int_{-1}^{a} \frac{2}{\pi} \sqrt{1 - x^2} \, dx.$$

The PDF of X is obtained by differentiating the CDF as usual:

$$(2.1) \qquad\qquad f_X(x) = \frac{2}{\pi} \sqrt{1 - x^2}, \quad \text{for } -1 \leq x \leq 1.$$

Not surprisingly the distribution of the x-coordinate is no longer uniform! See Figure 2.1. Why is this to be expected?

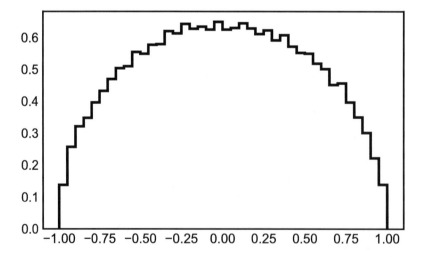

Figure 2.1. A histogram of the x-coordinate of $N = 100{,}000$ points sampled uniformly on the unit disk.

If $(X_1, \ldots, X_n)$ is a random vector, the distribution of a single coordinate, say X_1, is called a *marginal distribution*. In Example 2.1, the marginal distribution of X is determined by the PDF (2.1).

Random variables $X_1, \ldots, X_n$ defined on the same probability space are said to be *independent* if for any intervals $A_1, \ldots, A_n$ in $\mathbb{R}$, the probability factors:

$$\mathbf{P}(X_1 \in A_1, \ldots, X_n \in A_n) = \mathbf{P}(X_1 \in A_1) \times \cdots \times \mathbf{P}(X_n \in A_n).$$

We say that the random variables $X_1, \ldots, X_n$ are *independent and identically distributed* (IID) if they are independent and their marginal distributions are the same.

When the random vector $(X_1, \ldots, X_n)$ has a joint PDF $f(x_1, \ldots, x_n)$, the independence of the variables $X_1, \ldots, X_n$ is equivalent to saying that the joint PDF is given by the product of the marginal PDFs:

$$f(x_1, \ldots, x_n) = f_1(x_1) \times \cdots \times f_n(x_n), \quad \text{where } \mathbf{P}(X_i \in A_i) = \int_{A_i} f_i(x)\,dx.$$

In particular, if the expectation is over a product of functions, each depending on a single distinct variable, the expectation factors. Note that the variables X and Y in Example 2.1 are not independent.

Here are some important expectations of a function of a random vector. We will constantly use these two examples.

Example 2.2 (Covariance). If (X, Y) is a random vector, then the covariance is given by

$$\mathrm{Cov}(X, Y) = \mathbf{E}[(X - \mathbf{E}[X])(Y - \mathbf{E}[Y])] = \mathbf{E}[XY] - \mathbf{E}[X]\mathbf{E}[Y],$$

where the second equality is by linearity of expectation. Note that $\mathrm{Cov}(X, X) = \mathrm{Var}(X)$. If we have a vector $X = (X_1, \ldots, X_n)$, then we can encode all the covariances into an $n \times n$ matrix $\mathcal{C}$ called the *covariance matrix*, where the ij-entry is the covariance of the variables X_i and X_j:

$$\mathcal{C}_{ij} = \mathrm{Cov}(X_i, X_j).$$

It is easy to check (see Exercise 2.7) that a covariance matrix $\mathcal{C}$ is always symmetric and *positive semidefinite*; i.e., using matrix notation:

$$\text{for any } a \in \mathbb{R}^n, \quad a^T \mathcal{C} a = \sum_{i,j=1}^{n} a_i a_j \mathcal{C}_{ij} \geq 0.$$

The notation $a^T b = a_1 b_1 + \cdots + a_n b_n$ stands for the scalar or dot product between two elements a and b in $\mathbb{R}^n$ [1]. Thus, in the above, $a^T \mathcal{C} a$ is the dot product between the vector a and the vector $\mathcal{C} a$.

We say that the matrix is *positive definite* if $a^T \mathcal{C} a > 0$ whenever a is not the zero vector. We say that two variables are *uncorrelated* if their covariance is 0. We will also say sometimes that they are *orthogonal*. This is because the covariance has an important geometric meaning for random variables that is close to the dot product in $\mathbb{R}^n$; see Section 2.4. Clearly, if two variables are independent, then they are uncorrelated.

---

[1] We think here of b as a column vector and a^T as a row vector. This notation for the dot product is useful for matrix computations later on.

(Why?) However, two variables that are uncorrelated might not be independent; see Exercise 2.1.

Example 2.3 (Joint moment generating function (MGF)). The joint MGF of a random vector $X = (X_1, \ldots, X_n)$ on $(\Omega, \mathcal{F}, \mathbf{P})$ is the function on $\mathbb{R}^n$ defined by

$$\phi(a) = \mathbf{E}[\exp(a_1 X_1 + \cdots + a_n X_n)], \quad \text{for } a = (a_1, \ldots, a_n) \in \mathbb{R}^n.$$

If the random variables are independent and the MGF of each marginal distribution exists, then the joint MGF is simply the product of the MGFs of each single variable. (Why?)

The following result will be stated without proof. It will be useful when studying Gaussian vectors.

Proposition 2.4. *Let $(\Omega, \mathcal{F}, \mathbf{P})$ be a probability space. Two random vectors X and Y that have the same moment generating function have the same distribution.*

An important example of random vectors for our purpose is the case where the coordinates are IID Gaussian random variables.

Example 2.5. Consider (X, Y) a random vector with value in $\mathbb{R}^2$ such that X and Y are IID with standard Gaussian distribution. Then the joint PDF is

$$f(x, y) = \frac{1}{\sqrt{2\pi}} e^{-x^2/2} \times \frac{1}{\sqrt{2\pi}} e^{-y^2/2} = \frac{1}{2\pi} e^{-(x^2+y^2)/2}.$$

The moment generating function is obtained by independence and Example 1.31:

$$(2.2) \qquad \mathbf{E}[\exp(a_1 X + a_2 Y)] = \mathbf{E}[\exp(a_1 X)] \times \mathbf{E}[\exp(a_2 Y)] = e^{(a_1^2 + a_2^2)/2}.$$

More generally, we can consider n IID random variables with standard Gaussian distribution. We then have the joint PDF

$$f(x_1, \ldots, x_n) = \frac{e^{-(x_1^2 + \cdots + x_n^2)/2}}{(2\pi)^{n/2}},$$

with MGF

$$\mathbf{E}[\exp(a_1 X_1 + \cdots + a_n X_n)] = e^{(a_1^2 + \cdots + a_n^2)/2}.$$

The next example is good practice to get acquainted with the formalism of random vectors.

Example 2.6 (Computations with random vectors). Let (X, Y) be two IID standard Gaussian random variables. We can think of (X, Y) as a random point in $\mathbb{R}^2$ with x-coordinate X and y-coordinate Y.

First, let's compute the probability that the point (X, Y) is in the disc $\{x^2 + y^2 \leq 1\}$. The probability is given by the double integral

$$\iint_{\{x^2+y^2 \leq 1\}} \frac{1}{2\pi} e^{-(x^2+y^2)/2} \, \mathrm{d}x \, \mathrm{d}y.$$

We change the integral to polar coordinates (r, θ), remembering that $dx\, dy = r\, dr\, d\theta$. We get

$$\int_0^{2\pi} \int_0^1 \frac{1}{2\pi} e^{-r^2/2} r\, dr\, d\theta = 1 - e^{-1/2} = 0.393 \ldots.$$

Consider now the random variable $R = (X^2 + Y^2)^{1/2}$ giving the random distance of the point to the origin. Let's compute $\mathbf{E}[R]$. Again by polar coordinates, this reduces to the integral

$$\int_0^{2\pi} \int_0^\infty \frac{1}{2\pi} e^{-r^2/2} r^2\, dr\, d\theta = \int_0^\infty r^2 e^{-r^2/2}\, dr.$$

This integral is similar to the integral of the variance of a standard normal

$$\int_{-\infty}^\infty x^2 \frac{e^{-x^2/2}}{\sqrt{2\pi}}\, dx = 1.$$

From this, we get

$$\int_0^\infty r^2 e^{-r^2/2}\, dr = \frac{1}{2}\sqrt{2\pi} = \sqrt{\frac{\pi}{2}}.$$

More generally, the CDF of R is by a change of coordinates

$$\mathbf{P}(R \le r) = \iint_{\{\sqrt{x^2+y^2} \le r\}} \frac{1}{2\pi} e^{-(x^2+y^2)/2}\, dx\, dy = \int_0^{2\pi} \int_0^r \frac{1}{2\pi} e^{-s^2/2} s\, ds\, d\theta$$

$$= \int_0^r s e^{-s^2/2}\, ds.$$

So by taking the derivative with respect to r, we get that the PDF is $r e^{-r^2/2}$, $r \ge 0$. Consider now the random angle that the random point makes with the x-axis, i.e., the random variable $\Theta = \arctan\frac{Y}{X}$. It is not hard to compute the joint PDF of (R, Θ). Indeed, we have

$$\mathbf{P}(R \le r, \Theta \le \theta) = \iint_{\{\sqrt{x^2+y^2} \le r,\ \tan\theta \le \frac{y}{x}\}} \frac{1}{2\pi} e^{-(x^2+y^2)/2}\, dx\, dy$$

$$= \int_0^\theta \int_0^r \frac{1}{2\pi} e^{-s^2/2} s\, ds\, d\phi.$$

By taking the derivative in r and θ, we get that the joint PDF is $f(r, \theta) = \frac{1}{2\pi} r e^{-r^2/2}$, $r \ge 0, 0 \le \theta < 2\pi$. In particular, the variables (R, Θ) are independent since the joint PDF is the product of the marginals! Note that the marginal distribution of the random angle Θ is uniform on $[0, 2\pi)$.

The above example gives an interesting method to generate a pair of IID standard Gaussian random variables. This is called the *Box-Mueller method*. Let U_1, U_2 be two independent uniform random variables on $[0, 1]$. Define the random variables (Z_1, Z_2) as follows:

(2.3) $\qquad Z_1 = \sqrt{-2\log U_1}\cos(2\pi U_2), \qquad Z_2 = \sqrt{-2\log U_1}\sin(2\pi U_2).$

The variables (Z_1, Z_2) are IID Gaussians. See Numerical Project 2.1 and Exercise 2.4.

2.2. Gaussian Vectors

Example 2.5 is the simplest example of Gaussian vectors. We generalize the definition.

Definition 2.7. A n-dimensional random vector $X = (X_1, \ldots, X_n)$ is said to be *Gaussian* if and only if any linear combination of $X_1, \ldots, X_n$ is a Gaussian random variable [2]. In other words, the random variable

$$a_1 X_1 + \cdots + a_n X_n \text{ is Gaussian for any } a_1, \ldots, a_n \in \mathbb{R}.$$

We will also say that the random variables $X_1, \ldots, X_n$ are *jointly Gaussian*.

Note that Example 2.5 with (X, Y) IID standard Gaussians is a Gaussian vector with this definition. This is because equation (2.2) implies that $a_1 X + a_2 Y$ is a Gaussian random variable of mean 0 and variance $a_1^2 + a_2^2$.

As a simple consequence of the definition, we also get that each coordinate of a Gaussian vector is a Gaussian random variable, by setting all a's but a single one to 0. More generally, we have the following useful property:

Lemma 2.8. *If X is an n-dimensional Gaussian vector and M is some $m \times n$ matrix, then the vector MX is an m-dimensional Gaussian vector.*

Proof. See Exercise 2.8. □

An equivalent definition can be stated in terms of the joint MGF since it determines the distribution by Proposition 2.4. Before introducing the second definition, we first make two important observations about the mean and the variance of a linear combination of random variables. First, the mean of a linear combination is

$$\mathbf{E}[a_1 X_1 + \cdots + a_n X_n] = a_1 m_1 + \cdots + a_n m_n = a^T m,$$

where $m = (m_i, i \le n)$ is the *mean vector* with $m_i = \mathbf{E}[X_i]$. (It is useful to think of the mean vector as a column vector again to use the notation $a^T m$ for the dot product.) The variance is obtained with a short calculation using the linearity of expectation:

$$\mathrm{Var}(a_1 X_1 + \cdots + a_n X_n) = \mathbf{E}[(a_1(X_1 - m_1) + \cdots + a_n(X_n - m_n))^2]$$

$$= \sum_{i,j=1}^{n} a_i a_j \, \mathrm{Cov}(X_i, X_j) = a^T \mathcal{C} a.$$

The above two expressions, Proposition 2.4, and Definition 2.7 directly give the following equivalent definition:

Proposition 2.9. *A random vector $X = (X_1, \ldots, X_n)$ is Gaussian if and only if the moment generating function of X is*

$$(2.4) \qquad \mathbf{E}[\exp(a^T X)] = \exp\left(a^T m + \frac{1}{2} a^T \mathcal{C} \, a\right),$$

where m is the mean vector with $m_i = \mathbf{E}[X_i]$ and $\mathcal{C}$ is the covariance matrix of X.

---

[2] It is convenient to see the degenerate random variable that is 0 with probability one as a Gaussian random variable of mean 0 and variance 0.

Note the similarity to the moment generating function with one variable in Example 1.31. It was mentioned in Example 2.2 that independent random variables are uncorrelated, but the converse is not true in general. One of the most useful properties of Gaussian vectors is that uncorrelated implies independent.

Proposition 2.10. *Let $X = (X_1, \ldots, X_n)$ be a Gaussian vector. Then the covariance matrix is diagonal if and only if the variables are independent.*

Proof. The *if* direction is clear. In the *only if* direction, suppose that the covariance matrix $\mathcal{C}$ is diagonal. We use the moment generating function. Note that for any a in $\mathbb{R}^n$ we have

$$a^T \mathcal{C} \, a = \sum_{i,j=1}^{n} a_i a_j \mathcal{C}_{ij} = \sum_{i=1}^{n} a_i^2 \mathcal{C}_{ii}, \quad \text{since } \mathcal{C}_{ij} = 0 \text{ for } i \neq j.$$

Therefore, the MGF in (2.4) is

$$\mathbf{E}[\exp(a^T X)] = \exp\left(\sum_{i=1}^{n} a_i m_i + \frac{1}{2} \sum_{i=1}^{n} a_i^2 \mathcal{C}_{ii} \right) = \prod_{i=1}^{n} \exp\left(a_i m_i + \frac{1}{2} a_i^2 \mathcal{C}_{ii} \right).$$

This is the same as the MGF for independent Gaussian variables with mean m_i and variance $\mathcal{C}_{ii}$, $i = 1, \ldots n$. Since the moment generating function determines the distribution by Proposition 2.4, we conclude that the variables are independent Gaussians. □

Before writing the joint PDF of a Gaussian vector in terms of the mean vector and the covariance matrix, we need to introduce the important notion of degenerate vector. We say a Gaussian vector is *degenerate* if the covariance matrix $\mathcal{C}$ is such that its determinant is zero: $\det \mathcal{C} = 0$. We say it is nondegenerate if $\det \mathcal{C} \neq 0$.

Example 2.11. Consider (Z_1, Z_2, Z_3) IID standard Gaussian random variables. We define $X = Z_1 + Z_2 + Z_3$, $Y = Z_1 + Z_2$, and $W = Z_3$. Clearly, (X, Y, W) is a Gaussian vector. It has mean 0 and covariance

$$\begin{pmatrix} 3 & 2 & 1 \\ 2 & 2 & 0 \\ 1 & 0 & 1 \end{pmatrix}.$$

It is easy to check that $\det \mathcal{C} = 0$; thus (X, Y, W) is a degenerate Gaussian vector.

The above example is helpful to illustrate the notion. Note that we have the linear relation $X - Y - W = 0$ between the random variables. Therefore, the random variables are *linearly dependent*. In other words, one vector is redundant, say X, in the sense that its value can be recovered from the others for any outcome. The relation between degeneracy and linear dependence is general.

Lemma 2.12. *Let $X = (X_1, \ldots, X_n)$ be a Gaussian vector. Then X is degenerate if and only if the coordinates are linearly dependent; that is, there exist $c_1, \ldots, c_n$ (not all zeros) such that $c_1 X_1 + \cdots + c_n X_n = 0$ (with probability one).*

Proof. This is Exercise 2.12. □

We are now ready to state the form of the PDF for Gaussian vectors.

Proposition 2.13 (Joint PDF of Gaussian vectors). *Let* $X = (X_1, \ldots, X_n)$ *be a non-degenerate Gaussian vector with mean vector m and covariance matrix* $\mathcal{C}$. *Then the joint distribution of X is given by the joint PDF*

$$f(x_1, \ldots, x_n) = \frac{\exp\left(-\frac{1}{2}(x-m)^T \mathcal{C}^{-1}(x-m)\right)}{(2\pi)^{n/2}(\det \mathcal{C})^{1/2}}, \quad x = (x_1, \ldots, x_n).$$

Note that the inverse of $\mathcal{C}$ exists from the assumption of nondegeneracy. Observe also that if $\mathcal{C}$ is diagonal, then the joint density is the product of the marginals, as expected from Proposition 2.10, since the variables are then independent.

Example 2.14. Consider a Gaussian vector (X_1, X_2) of mean 0 and covariance matrix $\mathcal{C} = \begin{pmatrix} 2 & 1 \\ 1 & 2 \end{pmatrix}$. The inverse of $\mathcal{C}$ is $\mathcal{C}^{-1} = \begin{pmatrix} 2/3 & -1/3 \\ -1/3 & 2/3 \end{pmatrix}$ and $\det \mathcal{C} = 3$. By doing the matrix operations, the joint PDF is

$$f(x, y) = \frac{\exp(\frac{-1}{2}x^T \mathcal{C}^{-1} x)}{2\pi\sqrt{3}} = \frac{\exp(\frac{-1}{3}x^2 + \frac{1}{3}xy - \frac{1}{3}y^2)}{2\pi\sqrt{3}}.$$

Probabilities and expectations can be computed using the PDF. For example, the probabilities that $X > 2$ and $Y < 3$ are given by the double integral

$$\mathbf{P}(X_1 > 2, X_2 < 3) = \int_2^\infty \int_{-\infty}^3 \frac{\exp(\frac{-1}{3}x^2 + \frac{1}{3}xy - \frac{1}{3}y^2)}{2\pi\sqrt{3}} \, dy \, dx.$$

The double integral can be evaluated using a software and is equal to 0.0715 ....

We will not prove Proposition 2.13 yet. Instead, we will take a short detour and derive it from a powerful decomposition of Gaussian vectors as a linear combination of IID Gaussians. The decomposition is the generalization of making a random variable *standard*. Suppose X is Gaussian with mean 0 and variance σ^2. Then we can write as in Example 1.17

$$X = \sigma Z,$$

where Z is a standard Gaussian variable. (This makes sense even when X is *degenerate*; i.e., $\sigma^2 = 0$.) If $\sigma^2 \neq 0$, then we can reverse the relation to get

$$Z = X/\sigma.$$

We generalize this procedure to Gaussian vectors.

Proposition 2.15 (Decomposition into IID). *Let* $X = (X_j, j \leq n)$ *be a Gaussian vector of mean 0. If X is nondegenerate, there exist n IID standard Gaussian random variables* $Z = (Z_j, j \leq n)$ *and an invertible* $n \times n$ *matrix A such that*

$$X = AZ, \quad Z = A^{-1}X.$$

Here again, we think of X and Z as column vectors when performing the matrix multiplication. The choice of Z's, and thus the matrix A, is generally not unique as the following simple example shows.

Example 2.16. Consider the Gaussian vector (X_1, X_2) given by

$$X_1 = Z_1 + Z_2,$$
$$X_2 = Z_1 - Z_2,$$

where (Z_1, Z_2) are IID standard Gaussians. The matrix A is $\begin{pmatrix} 1 & 1 \\ 1 & -1 \end{pmatrix}$. The covariance matrix of X is $\begin{pmatrix} 2 & 0 \\ 0 & 2 \end{pmatrix}$. In particular, the random variables are independent by Proposition 2.10. So another choice of decomposition is simply $W_1 = X_1/\sqrt{2}$ and $W_2 = X_2/\sqrt{2}$. These random variables are also IID standard Gaussians.

Proof of Proposition 2.15. This is done using the same Gram-Schmidt procedure as for $\mathbb{R}^n$. The idea is to take the variables one by one and to subtract the components in the directions of the previous ones using the covariance. Lemma 2.12 ensures that no variables are linear combinations of the others. To start, we take $Z_1 = X_1/\sqrt{\mathcal{C}_{11}}$. Clearly, Z_1 is a standard Gaussian. Then, we define Z_2' as

(2.5) $$Z_2' = X_2 - \mathbf{E}[X_2 Z_1 | Z_1].$$

Observe that Z_2' is uncorrelated with Z_1:

$$\mathbf{E}[Z_2' Z_1] = \mathbf{E}[X_2 Z_1] - \mathbf{E}[X_2 Z_1]\mathbf{E}[Z_1^2] = 0.$$

Moreover, the vector (Z_1, Z_2') is Gaussian, being a linear transformation of the Gaussian vector (X_1, X_2). Therefore, by Proposition 2.10, they are independent! Of course, Z_2' might not have variance 1. So let's define Z_2 to be $Z_2'/\sqrt{\mathrm{Var}(Z_2')}$. In the same way, we take Z_3' to be

(2.6) $$Z_3' = X_3 - \mathbf{E}[X_3 Z_2 | Z_2] - \mathbf{E}[X_3 Z_1 | Z_1].$$

Again, it is easy to check that Z_3' is independent of Z_2 and Z_1. As above, we define Z_3 to be Z_3' divided by the square root of its variance. This procedure is carried on until we run out of variables. Note that since $\mathcal{C}$ is nondegenerate, none of the variances of the Z_i' will be 0, and therefore, they can be standardized. □

The proof can also be applied to the case of degenerate vectors, if we apply the procedure only to the maximal linearly independent subset of the vectors. This yields a similar statement where the matrix A is not invertible; see Exercise 2.13.

The covariance matrix can be written in terms of A. Write $a_{ij} = (A)_{ij}$ for the ij-th entry of the matrix A. By the relation $X = AZ$, we have

$$\mathbf{E}[X_i X_j] = \mathbf{E}\left[\sum_{k,l} a_{ik} a_{jl} Z_k Z_l\right] = \sum_k a_{ik} a_{jk} = (AA^T)_{ij}.$$

Thus we have

$$\boxed{\mathcal{C} = AA^T.}$$

This is called the *Cholesky decomposition* of the positive definite matrix $\mathcal{C}$. Note that this implies that a vector is nondegenerate if and only if A is invertible; see Exercise 2.12. For applications and numerical simulations, it is important to get the matrix A from the covariance matrix. This is done in Examples 2.17, 2.18, and 2.19. The decomposition is the exact analogue of the decomposition of a vector in $\mathbb{R}^3$ (say) written as a sum of orthonormal basis vectors. In particular, the condition of being nondegenerate is equivalent to linear independence. This geometric point of view will be generalized further in Section 2.4

We have now all the tools to prove Proposition 2.13.

Proof of Proposition 2.13. Without loss of generality, we suppose that $m = (0, \ldots, 0)$; otherwise we just need to subtract it from X. We use the decomposition in Proposition 2.15. First, note that since $\mathcal{C} = AA^T$, the determinant of $\mathcal{C}$ is

$$\det \mathcal{C} = (\det A)^2.$$

In particular, since X is nondegenerate, we have $\det A \neq 0$, so A is invertible. We also have by the decomposition that there exist IID standard Gaussians Z such that $X = AZ$; therefore

$$\mathbf{P}(X \in B) = \mathbf{P}(Z \in A^{-1}B).$$

But we know the PDF of Z by Example 2.5. This gives

$$\mathbf{P}(X \in B) = \int \cdots \int_{A^{-1}B} \frac{\exp\left(-\frac{1}{2}z^T z\right)}{(2\pi)^{n/2}} \, dz_1 \ldots dz_n,$$

because $z = (z_1, \ldots, z_n)$ and $z^T z = z_1^2 + \cdots + z_n^2$. It remains to do the change of variable $x = Az$ where $x = (x_1, \ldots, x_n)$. This means that $z \in A^{-1}B$ is equivalent to $x \in B$. Moreover, we have $z^T z = (A^{-1}x)^T A^{-1}x = x^T \mathcal{C}^{-1}x$ since $\mathcal{C} = AA^T$. Finally, $dz_1 \ldots dz_n = |\det(A^{-1})| \, dx_1 \ldots dx_n$ because the Jacobian of the transformation $z = A^{-1}x$ is A^{-1}. The final expression is obtained by putting this altogether and noticing that $\det(A^{-1}) = (\det A)^{-1} = (\det \mathcal{C})^{-1/2}$. $\qquad\qquad\square$

We now explore three ways to find the matrix A in the decomposition of Gaussian vectors of Proposition 2.15. We proceed by example.

Example 2.17 (Cholesky by Gram-Schmidt). This is the method suggested by the proof of Proposition 2.15. It suffices to successively go through the X's by subtracting the projection of a given X_i onto the previous random variables.

Consider the random vector $X = (X_1, X_2)$ with mean 0 and covariance matrix

$$\mathcal{C} = \begin{pmatrix} 2 & 1 \\ 1 & 2 \end{pmatrix}.$$

It is easy to check that X is nondegenerate. Take

$$Z_1 = X_1/\sqrt{2}.$$

This is obviously a standard Gaussian variable. For Z_2, consider first

$$Z_2' = X_2 - \mathbf{E}[X_2 Z_1]Z_1.$$

It is straightforward to check that Z_1 and Z_2' are jointly Gaussian, since they arise from a linear transformation of X, and also independent because $\mathbf{E}[Z_1 Z_2'] = 0$. Note that

$$Z_2' = X_2 - \mathbf{E}[X_2 Z_1]Z_1 = X_2 - \mathbf{E}\left[X_2 \frac{X_1}{\sqrt{2}}\right] \cdot \frac{X_1}{\sqrt{2}} = X_2 - \frac{1}{2}\mathbf{E}[X_1 X_2]X_1 = X_2 - \frac{1}{2}X_1.$$

In particular, we have by linearity of expectation

$$\mathbf{E}[(Z_2')^2] = \mathbf{E}[X_2^2] - \mathbf{E}[X_1X_2] + \frac{1}{4}\mathbf{E}[X_1^2] = 2 - 1 + \frac{1}{2} = \frac{3}{2},$$

by reading off the covariances from the covariance matrix. To get a variable of variance 1 that is a multiple of Z_2', we take $Z_2 = \sqrt{\frac{2}{3}}Z_2'$. Altogether we get

$$Z_1 = \frac{X_1}{\sqrt{2}}, \quad Z_2 = -\frac{1}{2}\sqrt{\frac{2}{3}}X_1 + \sqrt{\frac{2}{3}}X_2.$$

We thus constructed two standard IID Gaussians from the X's. In particular, we have

$$A^{-1} = \begin{pmatrix} \frac{1}{\sqrt{2}} & 0 \\ -\frac{1}{\sqrt{6}} & \sqrt{\frac{2}{3}} \end{pmatrix}, \quad A = \begin{pmatrix} \sqrt{2} & 0 \\ \frac{1}{\sqrt{2}} & \sqrt{\frac{3}{2}} \end{pmatrix}.$$

You can check that $AA^T = \mathcal{C}$ as expected. Note that the probability in Example 2.14 can be evaluated differently using the PDF of (Z_1, Z_2) and A. Indeed, we have

(2.7)
$$\begin{aligned}
\mathbf{P}(X > 2, Y < 3) &= \mathbf{P}\left(\sqrt{2}Z_1 > 2, \frac{1}{\sqrt{2}}Z_1 + \sqrt{\frac{3}{2}}Z_2 < 3\right) \\
&= \int_{\sqrt{2}}^{\infty} \int_{\sqrt{\frac{2}{3}} - \frac{1}{\sqrt{3}}z_1}^{\infty} \frac{e^{\frac{-1}{2}(z_1^2 + z_2^2)}}{2\pi} \, \mathrm{d}z_2 \, \mathrm{d}z_1.
\end{aligned}$$

Example 2.18 (Cholesky by solving a system of equations). Consider the same example as above. Write $A = \begin{pmatrix} a & b \\ c & d \end{pmatrix}$. Then the relation $\mathcal{C} = AA^T$ yields the three equations

$$a^2 + b^2 = 2, \quad c^2 + d^2 = 2, \quad ac + bd = 1.$$

There are several solutions. One of them is $a = b = 1$ and $c = \frac{1+\sqrt{3}}{2}, d = \frac{1-\sqrt{3}}{2}$. Another one, as in Example 2.17, is $a = \sqrt{2}, b = 0, c = \frac{1}{\sqrt{2}}$, and $d = \sqrt{\frac{3}{2}}$.

Example 2.19 (Cholesky by diagonalization). This method takes advantage of the symmetry of the covariance matrix. It is a fact that if a matrix $\mathcal{C}$ is symmetric, then it is diagonalizable; i.e., it can be expressed as a diagonal matrix in a suitable basis. In other words, there exists a change-of-basis matrix M such that

$$M^{-1}\mathcal{C}M = D.$$

The columns of the matrix M are given by the eigenvectors of $\mathcal{C}$ (normalized to have norm 1). The entries of the diagonal matrix D are given by the eigenvalues of $\mathcal{C}$. (Recall that an eigenvector of $\mathcal{C}$ with eigenvalue λ is a nonzero vector v such that $\mathcal{C}v = \lambda v$.) Furthermore, since $\mathcal{C}$ is symmetric, its eigenvectors are orthogonal. This implies that the matrix M is orthogonal; that is, its inverse is simply its transpose. We then have

$$M^T\mathcal{C}M = D,$$

and since $\mathcal{C} = AA^T$, we get

$$M^TAA^TM = D \iff AA^T = MDM^T.$$

It suffices to take

$$A = MD^{1/2},$$

where M is the matrix with the columns given by the eigenvectors of $\mathcal{C}$ and $D^{1/2}$ is the diagonal matrix with the square root of the eigenvalues on the diagonal. Why are we sure that the eigenvalues are nonnegative?

For the above example, where the covariance matrix is $\begin{pmatrix} 2 & 1 \\ 1 & 2 \end{pmatrix}$, it can be checked directly that $(1/\sqrt{2}, 1/\sqrt{2})$ and $(1/\sqrt{2}, -1/\sqrt{2})$ are eigenvectors. The matrix M is therefore

$$M = \begin{pmatrix} 1/\sqrt{2} & 1/\sqrt{2} \\ 1/\sqrt{2} & -1/\sqrt{2} \end{pmatrix}.$$

The corresponding eigenvalues are 3 and 1, respectively, so that

$$D^{1/2} = \begin{pmatrix} \sqrt{3} & 0 \\ 0 & 1 \end{pmatrix}.$$

We conclude that

$$A = MD^{1/2} = \begin{pmatrix} \sqrt{\frac{3}{2}} & \frac{1}{\sqrt{2}} \\ \sqrt{\frac{3}{2}} & \frac{-1}{\sqrt{2}} \end{pmatrix}.$$

It is not hard to verify that $AA^T = \mathcal{C}$ as expected. Note that the entries of A satisfy the equation of Example 2.18, as they should.

Example 2.20 (IID decomposition). Let $X = (X_1, X_2, X_3)$ be a Gaussian vector with mean 0 and covariance matrix

$$\mathcal{C} = \begin{pmatrix} 1 & 1 & 1 \\ 1 & 2 & 2 \\ 1 & 2 & 3 \end{pmatrix}.$$

Let's find a matrix A such that $X = AZ$ for $Z = (Z_1, Z_2, Z_3)$ IID standard Gaussians. The vector is not degenerate since $\det \mathcal{C} \neq 0$. If we do a Gram-Schmidt procedure, we get $Z_1 = X_1$ and $Z_2 = X_2 - \mathbf{E}[Z_1 X_2] X_1 = X_2 - X_1$. Note that Z_2 has norm $\mathbf{E}[Z_2^2] = \mathbf{E}[X_2^2] - 2\mathbf{E}[X_2 X_1] + \mathbf{E}[X_1^2] = 1$. Also $Z_3 = X_3 - \mathbf{E}[X_3 Z_2]Z_2 - \mathbf{E}[X_3 Z_1]Z_1$. Thus $Z_3 = X_3 - X_2$ is again of norm 1. An IID decomposition of this vector is therefore $X_1 = Z_1, X_2 = Z_1 + Z_2$, and $X_3 = Z_1 + Z_2 + Z_3$ with

$$A = \begin{pmatrix} 1 & 0 & 0 \\ 1 & 1 & 0 \\ 1 & 1 & 1 \end{pmatrix}.$$

As we will see in Definition 2.25 in the next section, this random vector corresponds to the position of a Brownian motion at time 1, 2, and 3.

2.3. Gaussian Processes

We are now in a position to introduce the family of Gaussian processes. In general, a *stochastic process* is an infinite collection of random variables on a probability space

$(\Omega, \mathcal{F}, \mathbf{P})$. The collection can be countable or uncountable. We are interested mostly in the case where the variables are indexed by time; for example

$$X = (X_t, t \in \mathcal{T}),$$

where $\mathcal{T}$ can be discrete or continuous. We will often take $\mathcal{T} = \{0, 1, 2, 3, \dots\}$, $\mathcal{T} = [0, T]$ for some fixed $T > 0$ or $\mathcal{T} = [0, \infty)$.

In the case where $\mathcal{T} = [0, \infty)$ or $[0, T]$, the realization of the process $X(\omega)$ can be thought of as a function of time for each outcome ω:

$$X(\omega) : [0, \infty) \to \mathbb{R}$$
$$t \mapsto X_t(\omega).$$

This function is sometimes called a *path* or a *trajectory* of the process. With this in mind, we can think of the process X as a random function, as each outcome produces a function.

How can we compute probabilities for a stochastic process? In other words, what object captures its distribution? The most common way (there are others) is to use the *finite-dimensional distributions*. The idea here is to describe the probabilities related to any finite set of time. More precisely, the finite-dimensional distributions are given by

$$\mathbf{P}(X_{t_1} \in B_1, \dots, X_{t_n} \in B_n),$$

for any $n \in \mathbb{N}$, any choice of $t_1, \dots, t_n \in \mathcal{T}$, and any events $B_1, \dots, B_n$ in $\mathbb{R}$. Of course for any fixed choice of the t's, $(X_{t_1}, \dots, X_{t_n})$ is a random vector as in the previous section. The fact that we can control the probabilities for the whole random function comes from the fact that we have the distributions of these vectors for any n and any choice of t's.

Remark 2.21. Here we are avoiding technicalities related to events for random functions. It turns out that finite-dimensional distributions (by taking countable union and intersection) only describe probabilities involving a countable number of times

$$\mathbf{P}(X_{t_j} \in B_j, j \in \mathbb{N}).$$

This is a problem if $\mathcal{T}$ is uncountable, since to fully describe a function say on $[0, \infty)$ we need in general more than a countable number of points! Luckily, such technicalities can be avoided if we add the extra assumption that the function $t \mapsto X_t(\omega)$ is continuous for every ω. Indeed, a continuous function can be described at every point if we know its values on a countable dense subset of $[0, \infty)$, for example at the rational points.

Some important types of stochastic processes include Markov processes, martingales, and Gaussian processes. We will encounter them along the way. Let's start with Gaussian processes.

Definition 2.22. A *Gaussian process* $X = (X_t, t \in \mathcal{T})$ is a stochastic process whose finite-dimensional distributions are jointly Gaussian. In other words, for any $n \in \mathbb{N}$ and any choice of $t_1 < \cdots < t_n$ we have that $(X_{t_1}, \dots, X_{t_n})$ is a Gaussian vector. In particular, its distribution is determined by the *mean function* $m(t) = \mathbf{E}[X_t]$ and the *covariance function* $\mathcal{C}(s, t) = \mathrm{Cov}(X_t, X_s)$.

In the same vein as Lemma 2.8, linear combinations of Gaussian processes remain Gaussian.

Lemma 2.23. *Let $X^{(1)}, X^{(2)}, \dots, X^{(m)}$ be m Gaussian processes on $[0, \infty)$ defined on the same probability space. Then any process constructed by taking linear combinations is also a Gaussian process:*

$$a_1 X^{(1)} + \cdots + a_n X^{(m)} = (a_1 X_t^{(1)} + \cdots + a_m X_t^{(m)}, t \geq 0),$$

where $a_1, \dots, a_m$ can depend on t.

Proof. It suffices to take the Gaussian vector for given times $t_1, \dots, t_n$ and to apply Lemma 2.8. □

The most important example of a Gaussian process is Brownian motion.

Example 2.24 (Standard Brownian motion or Wiener process). The *standard Brownian motion* $(B_t, t \geq 0)$ is the Gaussian process with mean $m(t) = \mathbf{E}[B_t] = 0$ for all $t \geq 0$ and covariance

$$\mathrm{Cov}(B_t, B_s) = \mathbf{E}[B_t B_s] = s \wedge t,$$

where $s \wedge t$ stands for the minimum between s and t. It turns out that the process can be constructed so that the *Brownian paths* given by the functions $t \mapsto B_t(\omega)$ are continuous for a set of ω of probability 1. See Figure 2.2 for a simulation of 10 paths of Brownian motion on $[0, 1]$. Brownian motion has a myriad of properties that we will explore in more depth in Chapter 3.

This is such a central process to the study of Itô calculus that its definition deserves to be stated on its own.

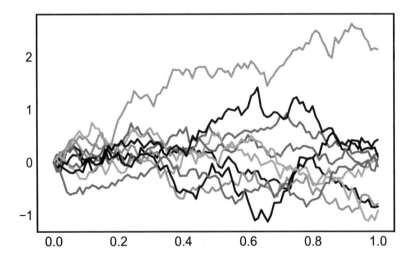

Figure 2.2. A simulation of 10 Brownian paths on $[0, 1]$.

Definition 2.25. A process $(B_t, t \geq 0)$ has the distribution of a standard Brownian motion if and only if the following hold:

(1) It is a Gaussian process.

(2) It has mean 0 and covariance $\mathbf{E}[B_t B_s] = s \wedge t$.

(3) Its paths $t \mapsto B_t(\omega)$ are continuous for ω in a set of probability one.

We state four more examples of Gaussian processes that will accompany us in our journey on the road of stochastic calculus.

Example 2.26 (Brownian motion with drift). For $\sigma > 0$ (called the *volatility* or the *diffusion coefficient*) and $\mu \in \mathbb{R}$ (called the *drift*), we define the process

$$X_t = \sigma B_t + \mu t,$$

where $(B_t, t \geq 0)$ is a standard Brownian motion. This is a Gaussian process because it is a linear transformation of Brownian motion, which is itself a Gaussian process, by Lemma 2.23. A straightforward computation shows that the mean is $\mathbf{E}[X_t] = \mu t$ and the covariance is $\mathrm{Cov}(X_t, X_s) = \sigma^2 \, s \wedge t$.

Example 2.27 (Brownian bridge). The Brownian bridge is a Gaussian process $(Z_t, t \in [0,1])$ defined by the mean $\mathbf{E}[Z_t] = 0$ and the covariance $\mathrm{Cov}(Z_t, Z_s) = s(1-t)$ if $s \leq t$. Note that by construction $Z_0 = Z_1 = 0$. It turns out that if $(B_t, t \in [0,1])$ is a standard Brownian motion on $[0,1]$, then the process

$$(2.8) \qquad Z_t = B_t - tB_1, \qquad t \in [0,1],$$

has the distribution of a Brownian bridge; see Exercise 2.14. Figure 2.3 gives a sample of 10 paths of Brownian bridge.

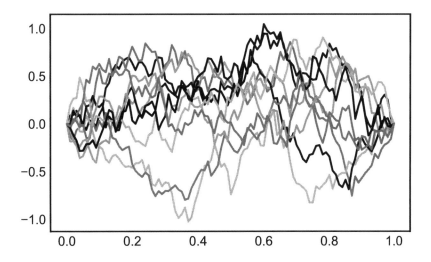

Figure 2.3. A simulation of 10 paths of Brownian bridge on $[0,1]$.

Example 2.28 (Fractional Brownian motion). The fractional Brownian motion $(B_t^{(H)},$ $t \geq 0)$ with index H, $0 < H < 1$ (called the *Hurst index*), is the Gaussian process with mean 0 and covariance

$$\text{Cov}(Y_s, Y_t) = \mathbf{E}[B_t^{(H)} B_s^{(H)}] = \frac{1}{2}(t^{2H} + s^{2H} - |t - s|^{2H}).$$

The case $H = 1/2$ corresponds to Brownian motion. (Why?) Figure 2.4 gives a sample of its paths for two choices of H.

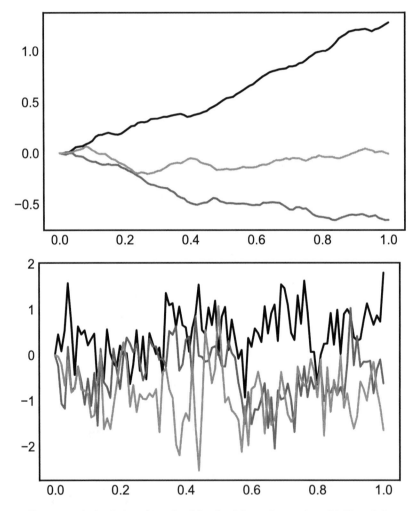

Figure 2.4. A simulation of 3 paths of fractional Brownian motion with Hurst index 0.9 (top) and 0.1 (bottom).

Example 2.29 (Ornstein-Uhlenbeck process). The Ornstein-Uhlenbeck process $(Y_t, t \geq 0)$ starting at $Y_0 = 0$ is the Gaussian process with mean $\mathbf{E}[Y_t] = 0$ and covariance

$$\text{Cov}(Y_s, Y_t) = \frac{e^{-2(t-s)}}{2}(1 - e^{-2s}), \quad \text{for } s \leq t.$$

If the starting point Y_0 is random, specifically Gaussian with mean 0 and variance 1/2, then we have that $\mathbf{E}[Y_t] = 0$ and

$$\text{Cov}(Y_s, Y_t) = \frac{e^{-2(t-s)}}{2}, \quad \text{for } s \leq t.$$

The covariance only depends on the difference of time! This means that the process $(Y_t, t \geq 0)$ has the same distribution if we shift time by an amount a for any $a \geq 0$: $(Y_{t+a}, t \geq 0)$. Processes with this property are called *stationary*. As can be observed from Figure 2.5, the statistics of stationary processes do not change over time.

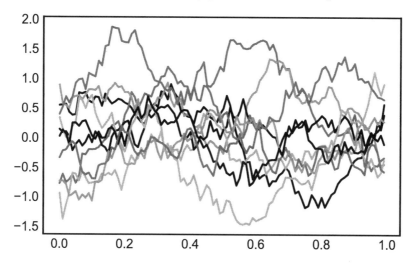

Figure 2.5. A simulation of 10 Ornstein-Uhlenbeck paths on $[0, 1]$ with stationary distribution.

Example 2.30 (Sampling a Gaussian process using the Cholesky decomposition). The IID decomposition in Proposition 2.15 is useful for generating a sample of a Gaussian process X. More precisely, suppose we want to generate a sample of the Gaussian process $(X_t, t \in [0, T])$. First, we need to fix the *discretization* or *step size*. Take for example a step size of 0.01, meaning that we approximate the process by evaluating the position at every 0.01. This is given by the Gaussian vector

$$(X_{\frac{j}{100}}, j = 1, \ldots, 100T).$$

This Gaussian vector has a covariance matrix $\mathcal{C}$ and a matrix A from the IID decomposition. Note that we start the vector at 0.01 and not 0. This is because in some cases (like standard Brownian motion) the value at time 0 is 0. Including it in the covariance matrix would result in a degenerate covariance matrix. You can always add the position 0 at time 0 after performing the Cholesky decomposition. It then suffices to sample $100T$ IID standard Gaussian random variable $Z = (Z_1, \ldots, Z_{100T})$ and to apply the deterministic matrix A to the sample vector

$$(2.9) \qquad\qquad (X_{\frac{j}{100}}, j = 1, \ldots, 100T) = AZ.$$

Numerical Projects 2.2, 2.3, 2.4, and 2.5 offer some practice on generating processes this way. We will see other methods in Chapters 5 and 7.

We will later revisit the examples of Gaussian processes, possibly adding some more parameters. At this point, we stress that they are processes that are defined by taking an explicit function of an underlying Brownian motion. This is not so clear at this point for the Ornstein-Uhlenbeck process, but we will see in Chapter 5 that we can write it as a stochastic integral over Brownian motion. This is the spirit of stochastic calculus in general:

> From a well-understood underlying source of randomness (usually a Brownian motion), what can we say about the distribution of a process that is constructed from it?

Itô calculus provides extremely powerful tools to deal with this question.

2.4. A Geometric Point of View

Before turning to studying Gaussian processes in more detail, it is worthwhile to spend some time to further explore the analogy between random vectors and vectors in $\mathbb{R}^d$. It turns out that this geometric structure applies not only to Gaussian random variables, but to all random variables with finite second moment. The big difference is that, unlike for Gaussians, uncorrelated random variables are not in general independent.

Definition 2.31. For a given probability space $(\Omega, \mathcal{F}, \mathbf{P})$, the space $L^2(\Omega, \mathcal{F}, \mathbf{P})$ (or L^2 for short) consists of all random variables defined on $(\Omega, \mathcal{F}, \mathbf{P})$ such that

$$\mathbf{E}[X^2] < \infty.$$

Such random variables are called *square-integrable*.

In the same spirit, the space of integrable random variables defined in Definition 1.27 is denoted by $L^1(\Omega, \mathcal{F}, \mathbf{P})$; see Exercise 2.16. We will see in Example 4.23 that any square-integrable random variable must be integrable. In other words, $L^2(\Omega, \mathcal{F}, \mathbf{P})$ is a subset of $L^1(\Omega, \mathcal{F}, \mathbf{P})$. In particular, square-integrable random variables have a well-defined expectation. This means that we can think of L^2 as the set of random variables on a given probability space with *finite variance*. Clearly, random variables on $(\Omega, \mathcal{F}, \mathbf{P})$ with Gaussian distribution are in L^2.

The space L^2 has an important geometric structure. First, it is a *linear space* or *vector space*. In general, a linear space is a space with two operations: addition of elements of the space and multiplication of an element of the space by a real number. An important example of linear space to keep in mind is the space $\mathbb{R}^d$ of d-tuple real numbers: $u = (u_1, \ldots, u_d)$ with $u_i \in \mathbb{R}$ for $i = 1, \ldots, d$. The space $\mathbb{R}^d$ has the following property as a linear space:

- If $u, v \in \mathbb{R}^d$ and $a, b \in \mathbb{R}$, then the linear combination $au + bv = (au_1 + bv_1, \ldots, au_d + bv_d)$ is also an element of $\mathbb{R}^d$.

- The vector $0 = (0, \ldots, 0)$ is the zero vector. This means that for any $a \in \mathbb{R}$, $a0 = 0$ and $0 + 0 = 0$.

It turns out that L^2 is also a linear space. Indeed:

- If X, Y are two random variables in L^2, then the linear combination $aX + bY$ is also a random variable in L^2 for any $a, b \in \mathbb{R}$. This is by linearity of expectation and the elementary inequality $(aX + bY)^2 \leq 2a^2 X^2 + 2b^2 Y^2$.

- The zero element of the linear space L^2 is the random variable $X = 0$ (with probability one). [3]

Example 2.32. Consider $(\Omega, \mathcal{P}(\Omega), \mathbf{P})$ of Example 1.1 where $\Omega = \{0, 1\} \times \{0, 1\}$, $\mathbf{P}$ is the equiprobability, and $\mathcal{P}(\Omega)$ is the power set of Ω, i.e., all the subsets of Ω. An example of a random variable is $X = 2\mathbf{1}_{\{(0,0)\}}$ where $\mathbf{1}_{\{(0,0)\}}$ is the indicator function of the event $\{(0,0)\}$. In other words, X takes the value 2 on the outcome $(0,0)$ and 0 for the other outcomes. Of course, we can generalize this construction by taking a linear combination of multiples of indicator functions. Namely, consider the random variable

(2.10) $$X = a\mathbf{1}_{\{(0,0)\}} + b\mathbf{1}_{\{(1,0)\}} + c\mathbf{1}_{\{(0,1)\}} + d\mathbf{1}_{\{(1,1)\}},$$

for some fixed $a, b, c, d \in \mathbb{R}$. Clearly, any random variable on this probability space can be written in this form. Moreover, any random variable of this form will have a finite variance. Therefore, the space $L^2(\Omega, \mathcal{P}(\Omega), \mathbf{P})$ in this example consists of random variables of the form (2.10). This linear space has dimension 4, since we can write any random variables as a linear combination of the four indicator functions. In general, if Ω is finite, the space L^2 is finite-dimensional as a linear space. However, if Ω is infinite, the space $L^2(\Omega, \mathcal{F}, \mathbf{P})$ might be infinite-dimensional.

Similar to $\mathbb{R}^d$, the space $L^2(\Omega, \mathcal{F}, \mathbf{P})$ has a *norm* or *length*: for a random variable X in L^2, its norm $\|X\|$ is given by [4]

$$\|X\| = \mathbf{E}[X^2]^{1/2}.$$

Note that this is very close in spirit to the length for a vector v in $\mathbb{R}^d$ given by $\|v\| = \sqrt{x_1^2 + \cdots + x_d^2}$, since the expectation is heuristically a sum over outcomes. This definition is good since $\|aX\| = a\|X\|$ for any $a \geq 0$ and $\|X\| = 0$ if and only if $X = 0$ with probability one (see Exercise 1.16). Moreover, it satisfies the *triangle inequality*, see Exercise 2.15,

$$\|X + Y\| \leq \|X\| + \|Y\|.$$

These are the expected properties of a norm.

Even better, like $\mathbb{R}^d$, the space L^2 has a *dot product* or *scalar product* between two elements X, Y of the space. It is given by

$$\mathbf{E}[XY].$$

More generally, this operation is called an *inner product*. It has the same properties as the dot product in $\mathbb{R}^d$:

- *Symmetric*: $\mathbf{E}[XY] = \mathbf{E}[YX]$.

- *Linear*: $\mathbf{E}[(aX + bY)Z] = a\mathbf{E}[XZ] + b\mathbf{E}[YZ]$ for any $a, b \in \mathbb{R}$.

---

[3] When working in L^2, we identify random variables that differ only on a set of outcomes of probability 0. In other words, we say that $X = Y$ whenever $X(\omega) = Y(\omega)$ on a set of ω of probability one.

[4] The norm in L^2 is usually denoted by $\| \cdot \|_2$ with an index 2 to emphasize the exponent. Since this will be the norm we work with most of the time for random variables, we drop the index for conciseness.

- *Positive definite*: If $X = Y$, then $\mathbf{E}[XY] \geq 0$. Moreover, we have $\mathbf{E}[X^2] = 0$ if and only if $X = 0$ with probability one (again by Exercise 1.16).

As for the dot product in $\mathbb{R}^d$, the inner product $\mathbf{E}[XY]$ satisfies the Cauchy-Schwarz inequality.

Proposition 2.33 (Cauchy-Schwarz inequality). *If $X, Y \in L^2(\Omega, \mathcal{F}, \mathbf{P})$, then*

$$|\mathbf{E}[XY]| \leq \mathbf{E}[X^2]^{1/2} \times \mathbf{E}[Y^2]^{1/2}.$$

Note that for $\mathbb{R}^d$, the analogous inequality for vectors $u, v \in \mathbb{R}^d$ is as expected

$$u^T v \leq \|u\| \, \|v\|.$$

The Cauchy-Schwarz inequality readily implies, by replacing X with $X - \mathbf{E}[X]$ and Y with $Y - \mathbf{E}[Y]$, that

$$|\operatorname{Cov}(X, Y)| \leq \sqrt{\operatorname{Var} X} \times \sqrt{\operatorname{Var} Y}.$$

This is sometimes written in terms of the *correlation coefficient*

$$|\rho(X, Y)| \leq 1, \quad \text{where } \rho(X, Y) = \frac{\operatorname{Cov}(X, Y)}{\sqrt{\operatorname{Var} X} \sqrt{\operatorname{Var} Y}} = \operatorname{Cov}(X/\|X\|, Y/\|Y\|).$$

Proof of Proposition 2.33. Note that we always have the inequality $|xy| \leq x^2/2 + y^2/2$. Therefore, since X, Y are in L^2 we must have by the properties of expectations that

$$\mathbf{E}[|XY|] \leq \frac{1}{2}\mathbf{E}[X^2] + \frac{1}{2}\mathbf{E}[Y^2] < \infty.$$

Thus we have that XY is integrable.

Now consider the random variable $X - tY$ for some $t \in \mathbb{R}$. Then we have by linearity of expectation (here we use the fact that XY is integrable)

$$0 \leq \mathbf{E}[(X - tY)^2] = \mathbf{E}[X^2] - 2t\mathbf{E}[XY] + t^2\mathbf{E}[Y^2].$$

The right side is a quadratic function in t. It is minimized at $t^* = \frac{\mathbf{E}[XY]}{\mathbf{E}[Y^2]}$. The result is obtained by putting this back and taking the square root. $\square$

The above proof is close in spirit to the proof of Proposition 2.15. We see that the random variable

$$(2.11) \qquad\qquad X^\perp = X - \frac{\mathbf{E}[XY]}{\mathbf{E}[Y^2]} Y$$

appears in both. This random variable is uncorrelated with Y or orthogonal to Y, in the sense that the inner product with Y is 0: indeed, by linearity of expectation,

$$\mathbf{E}\left[YX^\perp\right] = \mathbf{E}[XY] - \frac{\mathbf{E}[XY]\mathbf{E}[Y^2]}{\mathbf{E}[Y^2]} = 0.$$

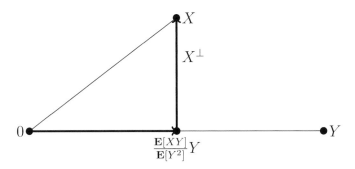

Figure 2.6. A representation of the decomposition of the random variable X in terms of its projection on Y and the component $X^{\perp}$ orthogonal to Y.

Effectively what is shown in the proof of the inequality is that the random variable $\frac{\mathbf{E}[XY]}{\mathbf{E}[Y^2]}Y$ is the random variable of the form tY, $t \in \mathbb{R}$, that is the *closest to X* in the L^2-sense. We will make this more precise when we define the conditional expectation of a random variable in Chapter 4. For now, we simply note that these considerations imply the important decomposition

$$(2.12) \qquad\qquad X = X^{\perp} + \frac{\mathbf{E}[XY]}{\mathbf{E}[Y^2]}Y.$$

See Figure 2.6. The random variable $X^{\perp} = X - \frac{\mathbf{E}[XY]}{\mathbf{E}[Y^2]}Y$ is the *component of X orthogonal to Y*. The random variable

$$\mathrm{Proj}_Y(X) = \frac{\mathbf{E}[XY]}{\mathbf{E}[Y^2]}Y$$

is called the *orthogonal projection* of the random variable X onto Y. Put another way, this is the *component of X in the direction of Y*. This is the equivalent of the orthogonal projection in $\mathbb{R}^d$ of a vector u in the direction of a vector v given by $\frac{u^T v}{\|v\|^2}\,v$.

Example 2.34. Going back to Example 2.32, let's define the random variables $Y = 2\mathbf{1}_{\{(0,0)\}} + \mathbf{1}_{\{(1,0)\}}$ and $W = \mathbf{1}_{\{(0,0)\}}$. Then, the orthogonal projection of Y onto W is

$$\frac{\mathbf{E}[YW]}{\mathbf{E}[W^2]}W = \frac{2\mathbf{P}(\{(0,0)\})}{\mathbf{P}(\{(0,0)\})}W = 2W.$$

The orthogonal decomposition of Y is simply

$$Y = 2W + (Y - 2W).$$

The notion of norm induces a notion of distance between random variables in L^2 given by $\|X - Y\| = \mathbf{E}[(X - Y)^2]^{1/2}$. In particular, we see that the orthogonal projection of Y onto X is the closest point from X among all the multiples of Y. This is what the proof of the Cauchy-Schwarz inequality does! This is also consistent with Figure 2.6. The L^2-distance also gives rise to a notion of convergence.

Definition 2.35 (L^2-convergence of random variables). We say that a sequence of random variables $(X_n, n \geq 1)$ in $(\Omega, \mathcal{F}, \mathbf{P})$ converges in L^2 to a random variable X if the distance between them goes to zero; that is,

$$\|X_n - X\| = \mathbf{E}[(X_n - X)^2]^{1/2} \to 0 \quad \text{as } n \to \infty.$$

This is also called the *convergence in mean square*. Note that this is equivalent to $\mathbf{E}[(X_n - X)^2] \to 0$ as $n \to \infty$.

It turns out that the limit random variable X of convergent sequence $(X_n, n \geq 1)$ in L^2 is guaranteed to be in L^2. This is because L^2 is *complete*; see Exercise 5.20. This property is crucial for the construction of the Itô stochastic integral in Chapter 5.

Example 2.36 (A version of the weak law of large numbers). Consider a sequence of random variables $X_1, X_2, X_3, \ldots$ in $L^2(\Omega, \mathcal{F}, \mathbf{P})$ such that $\mathbf{E}[X_i] = 0$, $\mathbf{E}[X_i^2] = \sigma^2 < \infty$ for all $i \geq 1$, and that they are orthogonal to each other; i.e., $\mathbf{E}[X_i X_j] = 0$ for all $i \neq j$. We show that the empirical mean

$$\frac{1}{n} S_n = \frac{1}{n}(X_1 + \cdots + X_n) \text{ converges to } 0 \text{ in the } L^2\text{-sense.}$$

It suffices to apply the definition of L^2-convergence and show $\mathbf{E}[\frac{1}{n^2} S_n^2] \to 0$. We have by linearity of expectation

$$\frac{1}{n^2} \mathbf{E}[S_n^2] = \frac{1}{n^2} \sum_{i,j \leq n} \mathbf{E}[X_i X_j] = \frac{1}{n^2} \sum_{i \leq n} \mathbf{E}[X_i^2] = \frac{\sigma^2}{n},$$

since $\mathbf{E}[X_i^2] = \sigma^2 < \infty$ and $\mathbf{E}[X_i X_j] = 0$ for all $i \neq j$. Therefore, the above goes to 0 as claimed.

The properties of the space L^2 are summarized and compared to the ones of $\mathbb{R}^3$ in Table 2.1.

Table 2.1. A summary of the analogous concepts between $\mathbb{R}^3$ and $L^2(\Omega, \mathcal{F}, \mathbf{P})$.

Notions	$\mathbb{R}^3$	$L^2(\Omega, \mathcal{F}, \mathbf{P})$
Elements	$x = (x_1, x_2, x_3)$	$X : \Omega \to \mathbb{R}$ with $\mathbf{E}[X^2] < \infty$
Zero vector	$(0, 0, 0)$	$X = 0$
Linear space	$ax + by \in \mathbb{R}^3$	$aX + bY \in L^2$
Norm	$\|x\| = (x_1^2 + x_2^2 + x_3^2)^{1/2}$	$\|X\| = (\mathbf{E}[X^2])^{1/2}$
Metric (Distance)	$\|x - y\|$	$\|X - Y\| = (\mathbf{E}[(X - Y)^2])^{1/2}$
Inner product	$x^T y = x_1 y_1 + x_2 y_2 + x_3 y_3$	$\mathbf{E}[XY]$
Orthogonal projection	$\text{Proj}_y(x) = \frac{x^T y}{\|y\|^2} y$	$\text{Proj}_Y(X) = \frac{\mathbf{E}[XY]}{\mathbf{E}[Y^2]} Y$
Convergence	$x_n \to x \Leftrightarrow \|x_n - x\| \to 0$	$X_n \to X \Leftrightarrow \|X_n - X\| \to 0$

2.5. Numerical Projects and Exercises

2.1. **The Box-Mueller method.**

(a) Generate a sample of 10,000 pairs (Z_1, Z_2) following the Box-Mueller method given in equation (2.3).

(b) Plot the histogram for Z_1 and Z_2 and compare to the standard Gaussian distribution.

(c) Estimate the covariance $\mathbf{E}[Z_1 Z_2]$ by computing the empirical mean

$$\mathbf{E}[Z_1 Z_2] \approx \frac{1}{10,000} \sum_{j=1}^{10,000} Z_1^{(j)} Z_2^{(j)},$$

where j indexes the sample.

2.2. **Simulating Brownian motion.** The goal of the project is to simulate 100 paths of Brownian motion on $[0, 1]$ using a step size of 0.01 using the Cholesky decomposition as outlined in Example 2.30.

(a) Construct the covariance matrix of $(B_{j/100}, 1 \leq j \leq 100)$ using a `for` loop. Recall that for Brownian motion $\mathcal{C}(s, t) = s \wedge t$ with mean 0.

(b) The command `numpy.linalg.cholesky` in Python gives the Cholesky decomposition of a covariance matrix $\mathcal{C}$. Use this to find the matrix A.

(c) Define a function using `def` whose output is a sample of N standard Gaussian random variables and whose input is N.

(d) Use the above to plot 100 paths of Brownian motion on $[0, 1]$ with step size of 0.01. Do not forget B_0!

2.3. **Simulating the Ornstein-Uhlenbeck process.** Follow the same steps as Project 2.2 to generate 100 paths with step size of 0.01 of the following processes on $[0, 1]$:

(a) Ornstein-Uhlenbeck process: $\mathcal{C}(s, t) = \frac{e^{-2(t-s)}}{2}(1 - e^{-2s})$, for $s \leq t$, with mean 0 (so that $Y_0 = 0$).

(b) Stationary Ornstein-Uhlenbeck process: $\mathcal{C}(s, t) = \frac{e^{-2(t-s)}}{2}$, for $s \leq t$, with mean 0 (so that Y_0 is a Gaussian random variable of mean 0 and variance 1/2).

2.4. **Simulating fractional Brownian motion.** Follow the same steps as Project 2.2 to generate 100 paths with step size of 0.01 of fractional Brownian on $[0, 1]$ with Hurst index $H = 0.1$, $H = 0.5$, and $H = 0.9$.

2.5. **Simulating the Brownian bridge.**

(a) Follow the steps of Project 2.2 to generate 100 paths with step size of 0.01 of Brownian bridge on $[0, 1]$: $\mathcal{C}(s, t) = s(1 - t)$ for $s \leq t$ with mean 0.
Hint: Be careful because the variance is 0 at time 0 and 1 here.

(b) Another way to construct a Brownian bridge is to use equation (2.8). Generate 100 paths this way using Brownian paths.

(c) Compare a single path of Brownian bridge constructed in (a) and in (b) using the same source of randomness, that is, the same sample of IID standard Gaussians. What do you notice?
To work with the same random sample, the command `np.random.seed` *could be useful.*

Exercises

2.1. **An example of uncorrelated random variables that are not independent.** Let X be a standard Gaussian. Show that $\mathrm{Cov}(X, X^2) = 0$, yet X and X^2 are not independent.

2.2. **Sum of Exponentials is Gamma**. Use Exercise 1.11 to show that the sum of n IID exponential random variables of parameter λ has the Gamma distribution with PDF

$$f(x) = \frac{\lambda^n}{(n-1)!} x^{n-1} e^{-\lambda x}, \quad x \geq 0.$$

2.3. **Why $\sqrt{2\pi}$?**
Use polar coordinates to prove that

$$\int_{-\infty}^{\infty} \int_{-\infty}^{\infty} e^{-x^2} e^{-y^2} \, dx \, dy = 2\pi.$$

Conclude that this implies that

$$\int_{-\infty}^{\infty} \frac{e^{-x^2}}{\sqrt{2\pi}} \, dx = 1.$$

2.4. **Box-Muller method.** This method was stated in equation (2.3). Let U_1, U_2 be two independent uniform random variables on $[0, 1]$. Show that the random variables (Z_1, Z_2) defined by

$$Z_1 = \sqrt{-2 \log U_1} \cos(2\pi U_2), \quad Z_2 = \sqrt{-2 \log U_1} \sin(2\pi U_2)$$

are independent standard Gaussians.
Hint: Show that $\sqrt{-2 \log U_1}$ has the same distribution as R in Example 2.6 and that $2\pi U_2$ has the same distribution as Θ.

2.5. **Marginally Gaussian but not jointly Gaussian.** Let X be a standard Gaussian. We define the random variable

$$Y = \begin{cases} X & \text{if } |X| \leq 1, \\ -X & \text{if } |X| > 1. \end{cases}$$

(a) Show that Y is also standard Gaussian.
(b) Argue that $X + Y$ is not Gaussian; therefore, (X, Y) is not jointly Gaussian.

2.6. **PDF of Brownian bridge.** Let $(M_t, t \in [0, 1])$ be a Brownian bridge.
(a) Write down the PDF of $(M_{1/4}, M_{3/4})$.
(b) Write down the probability of the event $\{M_{1/4}^2 + M_{3/4}^2 \leq 1\}$ in terms of a double integral.

2.7. **The covariance matrix of a random vector is always positive semidefinite.** Let $\mathcal{C}$ be the covariance matrix of the random vector $X = (X_1, \ldots, X_n)$ (not necessarily Gaussian). Show that $\mathcal{C}$ is always positive semidefinite; i.e.,

$$\sum_{i,k=1}^{n} a_i a_j \mathcal{C}_{ij} \geq 0, \quad \text{for any } a_1, \ldots, a_n \in \mathbb{R}.$$

In particular, show that it is positive definite if and only if X is nondegenerate. *Hint: Write the left side as the variance of some random variable.*

2.8. **A linear transformation of a Gaussian vector is also Gaussian.** Let $X = (X_1, \ldots, X_n)$ be an n-dimensional Gaussian vector and M a $m \times n$ matrix.
   (a) Show that $Y = MX$ is also a Gaussian vector.
   (b) If the covariance matrix of X is $\mathcal{C}$, write the covariance matrix of Y in terms of M and $\mathcal{C}$.
   (c) If X is nondegenerate, is it always the case that Y is nondegenerate? If not, give a condition on M for Y not to be degenerate.

2.9. **IID decomposition.** Let (X, Y) be a Gaussian vector with mean 0 and covariance matrix

$$\mathcal{C} = \begin{pmatrix} 1 & \rho \\ \rho & 1 \end{pmatrix},$$

for $\rho \in (-1, 1)$. Write (X, Y) as a linear combination of IID standard Gaussians. Write down the PDF of the Gaussian vector.

2.10. **IID decomposition.** Consider the Gaussian vector $X = (X_1, X_2, X_3)$ with mean 0 and covariance matrix

$$\mathcal{C} = \begin{pmatrix} 2 & 1 & 1 \\ 1 & 2 & 1 \\ 1 & 1 & 2 \end{pmatrix}.$$

Find $Z = (Z_1, Z_2, Z_3)$ and a matrix A such that $X = AZ$. You can use the work of Example 2.17.

2.11. **IID decomposition.** Consider the Gaussian vector $X = (X_1, X_2, X_3)$ with covariance matrix

$$\mathcal{C} = \begin{pmatrix} 3 & 1 & 1 \\ 1 & 3 & -1 \\ 1 & -1 & 3 \end{pmatrix}.$$

   (a) Argue that X is nondegenerate.
   (b) Use the Gram-Schmidt procedure to find (Z_1, Z_2, Z_3) IID standard Gaussians that are linear combinations of the X's. Reverse the relations to get X in terms of the Z's.
   (c) Find the eigenvectors of $\mathcal{C}$. Use it to find (Z_1, Z_2, Z_3) IID standard Gaussians that are linear combinations of the X's. Reverse the relations to get X in terms of the Z's.
   (*They will be different from* (b)).
   (d) Explain in a few words how you would sample a point according to the distribution of X using a sample of IID standard Gaussians.

2.12. **Degenerate means linearly dependent.** Let $X = (X_1, \ldots, X_n)$ be a Gaussian vector with covariance $\mathcal{C}$. Show that $\det \mathcal{C} = 0$ if and only if there exist $c_1, \ldots, c_n$ (not all zeros) such that $c_1 X_1 + \cdots + c_n X_n = 0$. *Hint: Use $\mathcal{C} = AA^T$.*

2.13. IID decomposition for degenerate vectors. Let $X = (X_1, \ldots, X_n)$ be a degenerate Gaussian vector. Show that there exists an $n \times m$ matrix A, with $m < n$, such that

$$X = AZ,$$

where $Z = (Z_1, \ldots, Z_m)$ are IID standard Gaussian random variables. Show also that we still have $\mathcal{C} = AA^T$.

Hint: Apply the same method as in the proof of Proposition 2.15.

2.14. Brownian bridge from Brownian motion. Let $(B_t, t \in [0, 1])$ be a Brownian motion.

(a) Show that the process given by $M_t = B_t - tB_1, t \in [0, 1]$, is a Brownian bridge by arguing it is a Gaussian process with the right mean and covariance.

(b) Show that the random variable $B_t - tB_1$ is independent of B_1 for any $t \in [0, 1]$.
Hint: Compute the covariance.

See Example 4.21 for more on this.

2.15. Triangle inequality. Let X, Y be two random variables in $L^2(\Omega, \mathcal{F}, \mathbf{P})$. Prove the triangle inequality

$$(\mathbf{E}[(X + Y)^2])^{1/2} \leq (\mathbf{E}[X^2])^{1/2} + (\mathbf{E}[Y^2])^{1/2}$$

by expanding the square and by using the Cauchy-Schwarz inequality. Use the above to get the more general inequality: for $X_1, \ldots, X_n$ random variables in $L^2(\Omega, \mathcal{F}, \mathbf{P})$, we have

$$(\mathbf{E}[(X_1 + \cdots + X_n)^2])^{1/2} \leq \sum_{j=1}^{n} (\mathbf{E}[X_j^2])^{1/2}.$$

2.16. The space of integrable random variable is L^1. Recall from Definition 1.27 that a random variable defined on a probability space $(\Omega, \mathcal{F}, \mathbf{P})$ is integrable if $\mathbf{E}[|X|] < \infty$. The space $L^1(\Omega, \mathcal{F}, \mathbf{P})$ is defined to be the space of integrable random variables. Show that $L^1(\Omega, \mathcal{F}, \mathbf{P})$ is a linear space; that is, if $X, Y \in L^1$, then $aX + bY \in L^1$ for any $a, b \in \mathbb{R}$. What is the zero vector? Verify that $\|X\|_1 = \mathbf{E}[|X|]$ is a norm for L^1.

2.17. ★ Wick's formula. The moments of a Gaussian random variable were computed in Example 1.31 and Exercise 1.10. We generalize this to all the joint moments $\mathbf{E}[X_1^{k_1} \ldots X_n^{k_n}], k_1, \ldots, k_n \in \mathbb{R}$, of a Gaussian vector $(X_1, \ldots, X_n)$.

(a) Let $Z = (Z_1, \ldots, Z_n)$ be IID standard Gaussians, and let $G : \mathbb{R}^n \to \mathbb{R}$ be a smooth function for which $\mathbf{E}[G(Z)]$ and $\mathbf{E}[\frac{\partial G}{\partial x_i}(Z)]$ are well-defined for every $i \leq n$. Prove that

$$\mathbf{E}[Z_i G(Z)] = \mathbf{E}\left[\frac{\partial G}{\partial x_i}(Z)\right], \qquad i \leq n.$$

Hint: Integration by parts.

(b) Let X be a nondegenerate Gaussian vector of mean 0, and let $F : \mathbb{R}^n \to \mathbb{R}$ be a smooth function for which $\mathbf{E}[F(X)]$ and $\mathbf{E}[\frac{\partial F}{\partial x_i}(X)]$ are well-defined for every $i \leq n$. Use the IID decomposition and the previous question to show that

$$\mathbf{E}[X_i F(Z)] = \sum_{j \leq n} \mathbf{E}[X_i X_j] \mathbf{E}\left[\frac{\partial F}{\partial x_j}(X)\right], \qquad i \leq n.$$

(c) Use the above to show that for any m-tuple $(i_1, \ldots, i_m)$ where $i_k \le n$ we have that $\mathbf{E}[X_{i_1} \ldots X_{i_m}] = 0$ if m is odd, and if m is even,

$$\mathbf{E}[X_{i_1} \ldots X_{i_m}] = \sum_{\text{pairings of } (i_1, \ldots, i_m)} \prod_{\text{pairs } p = (p_1, p_2)} \mathbf{E}[X_{p_1} X_{p_2}],$$

where the last sum is over all the possible pairings of indices $(i_1, \ldots, i_m)$ and the product is over pairs $p = (p_1, p_2)$ for a given pairing.

2.6. Historical and Bibliographical Notes

The first instance of the Gaussian distribution goes back to the proof of the first *central limit theorem* by de Moivre in the eighteenth century [**dM67**]. The theorem states that if $X_1, X_2, X_3, \ldots$ is a sequence of IID random variables with $\mathbf{E}[X_1^2] < \infty$, then the distribution of the sum $S_n = X_1 + \cdots + X_n$ recentered by its mean $n \cdot \mathbf{E}[X_1]$ and divided by its standard deviation $\sqrt{n \cdot \text{Var}(X_1)}$ converges to a standard Gaussian:

$$\frac{S_n - n \mathbf{E}[X_1]}{\sqrt{n \text{ Var}(X_1)}} \to Z, \text{ a standard Gaussian random variable.}$$

This was checked numerically in Numerical Project 1.3 for IID exponential random variables. Roughly speaking, this means that the sum of IID random variables is well-approximated by $S_n \approx n\mathbf{E}[X_1] + \sqrt{n} \cdot \sqrt{\text{Var}(X_1)}Z$. This was first proved by de Moivre when X_1 is Bernoulli-distributed. In this case, the sum S_n has a binomial distribution and the proof is an application of Stirling's approximation of the factorial. The Gaussian distribution was rederived by Gauss at the beginning of the nineteenth century in his treatise on statistics of astronomical observations [**Gau11**]. Around the same time, Laplace proved a more general statement of the theorem [**Lap95**]. We had to wait until the beginning of the twentieth century to have a general proof of the above statement with notable contributions by Lévy, Chebyshev, Markov, etc. [**Fis11**].

The Cholesky decomposition was invented by André-Louis Cholesky, a French cartographer and officer killed during the First World War. It appeared in a paper in 1910 to compute the unknown in multiple regression; see [**dF19**] for an interesting account.

Chapter 3

Properties of Brownian Motion

Brownian motion is a fundamental object in probability. In this chapter, we begin an investigation of its numerous properties. After defining the process in terms of its increments, we study the elementary properties of its distribution in Section 3.1. In Section 3.2, the properties of the paths are studied. A sample of paths of Brownian motion was already shown in Figure 2.2. We could already see that Brownian paths are quite rugged. We will make this precise by computing the *quadratic variation*, a central object in stochastic calculus. We also define the Poisson process in Section 3.4, which is a helpful point of comparison, as it is another process with independent and stationary increments.

3.1. Properties of the Distribution

We start by revisiting the definition of Brownian motion as a Gaussian process given in Definition 2.25. That definition was in terms of the joint distribution of the *positions* at any choice of time $0 \leq t_1 < \cdots < t_n$:

$$(B_{t_1}, B_{t_2}, \ldots, B_{t_n}).$$

The equivalent definition below describes the joint distribution of the *difference of the positions* or *increments* between successive times:

$$(B_{t_2} - B_{t_1}, B_{t_3} - B_{t_2}, \ldots, B_{t_n} - B_{t_{n-1}}).$$

Proposition 3.1. *A process $(B_t, t \geq 0)$ defined on $(\Omega, \mathcal{F}, \mathbf{P})$ has the distribution of a standard Brownian motion on $[0, \infty)$, as in Definition 2.25, if and only if the following hold:*

(1) $B_0 = 0$.

(2) *For any $s < t$, the increment $B_t - B_s$ is Gaussian of mean 0 and variance $t - s$.*

(3) *For any $n \in \mathbb{N}$ and any choice of n times $0 \leq t_1 < t_2 < \cdots < t_n < \infty$, the increments $B_{t_2} - B_{t_1}, B_{t_3} - B_{t_2}, \ldots, B_{t_n} - B_{t_{n-1}}$, are independent.*

(4) *The path $t \mapsto B_t(\omega)$ is a continuous function for a set of ω of probability one.*

For a given $T > 0$, a standard Brownian motion on $[0, T]$ has the same definition as above with the t's restricted to $[0, T]$.

It is not hard to sample Brownian paths using this definition; see Numerical Project 3.1. The increments of Brownian motion are said to be *stationary* since their distribution depends on the difference between the times and not the times themselves. For example, the increments $B_2 - B_1$ and $B_{101} - B_{100}$ are independent and have the same distribution, namely a standard Gaussian distribution.

Proof. Both definitions have in common that $B_0 = 0$ and that the paths are continuous functions of time with probability one. For the *only if* part, consider the Gaussian vector $(B_{t_1}, \ldots, B_{t_n})$ for some n and times $0 \leq t_1 < \cdots < t_n < \infty$. Note that the vector $(B_{t_1} - 0, B_{t_2} - B_{t_1}, \ldots, B_{t_n} - B_{t_{n-1}})$ is a linear transformation of $(B_{t_1}, \ldots, B_{t_n})$. So it is also a Gaussian vector by Lemma 2.8. The mean of each coordinate is clearly 0 by linearity. For the covariance matrix, we have for $i < j$

$$\mathbf{E}[(B_{t_{j+1}} - B_{t_j})(B_{t_{i+1}} - B_{t_i})]$$
$$= \mathbf{E}[B_{t_{j+1}} B_{t_{i+1}}] - \mathbf{E}[B_{t_j} B_{t_{i+1}}] - \mathbf{E}[B_{t_{j+1}} B_{t_i}] + \mathbf{E}[B_{t_j} B_{t_i}]$$
$$= 0,$$

since $\mathbf{E}[B_t B_s] = t \wedge s$. This implies that the increments are independent by Proposition 2.10. Moreover, if $i = j$, the above is $t_{j+1} - t_j$ as claimed. The *if* part is left to the reader in Exercise 3.3. $\qquad\square$

Example 3.2 (Evaluating Brownian probabilities). Definition 2.25 and the equivalent definition given in Proposition 3.1 provide two ways to evaluate the Brownian probabilities. Let's compute the probability that $B_1 > 0$ and $B_2 > 0$. We know from Definition 2.25 that (B_1, B_2) is a Gaussian vector of mean 0 and covariance matrix

$$\mathcal{C} = \begin{pmatrix} 1 & 1 \\ 1 & 2 \end{pmatrix}.$$

The determinant of $\mathcal{C}$ is 1. Its inverse is easily checked to be

$$\begin{pmatrix} 2 & -1 \\ -1 & 1 \end{pmatrix}.$$

From Proposition 2.13, we see that the probability can be expressed as a double integral over the subset $\{(x, y) : x > 0, y > 0\}$:

$$\mathbf{P}(B_1 > 0, B_2 > 0) = \int_0^\infty \int_0^\infty \frac{1}{2\pi} \exp\left(\frac{-1}{2}(2x_1^2 - 2x_1 x_2 + x_2^2)\right) dx_1\, dx_2.$$

This integral can be evaluated using a calculator or software and is equal to 3/8.

The probability can also be computed using the independence of increments. The increments $(B_1, B_2 - B_1)$ are IID standard Gaussians. We know their joint PDF. It remains to integrate over the correct region of $\mathbb{R}^2$, which is in this case $\{(z_1, z_2) : z_1 > 0, z_1 + z_2 > 0\}$:

$$\mathbf{P}(B_1 > 0, B_2 > 0) = \iint_{\{z_2 > 0, z_1 + z_2 > 0\}} \frac{e^{\frac{-1}{2}(z_1^2 + z_2^2)}}{2\pi} \, dz_1 \, dz_2.$$

We recover the first integral by the change of variable $x_1 = z_1$ and $x_2 = z_1 + z_2$. The answer 3/8 is suspiciously simple. It turns out that the integral can be evaluated exactly. Indeed, by writing $B_1 = Z_1$ and $Z_2 = B_2 - B_1$ and splitting the probability on the event $\{Z_2 \geq 0\}$ and its complement, we have that $\mathbf{P}(B_1 \geq 0, B_2 \geq 0)$ equals

$$= \mathbf{P}(Z_1 \geq 0, Z_1 + Z_2 \geq 0, Z_2 \geq 0) + \mathbf{P}(Z_1 \geq 0, Z_1 + Z_2 \geq 0, Z_2 < 0)$$
$$= \mathbf{P}(Z_1 \geq 0, Z_2 \geq 0) + \mathbf{P}(Z_1 \geq 0, Z_1 \geq -Z_2, -Z_2 > 0)$$
$$= 1/4 + \mathbf{P}(Z_1 \geq 0, Z_1 \geq Z_2, Z_2 > 0)$$
$$= 1/4 + 1/8 = 3/8.$$

The distribution of Brownian motion enjoys many interesting symmetries. They are easily checked using the definition of the process. Some are left as exercises.

Proposition 3.3. *Let $(B_t, t \geq 0)$ be a standard Brownian motion. Then:*

(1) *Reflection at time s: The process $(-B_t, t \geq 0)$ is a Brownian motion. More generally, for any $s \geq 0$, the process $(\widetilde{B}_t, t \geq 0)$ defined by*

$$\widetilde{B}_t = \begin{cases} B_t & \text{if } t \leq s, \\ B_s - (B_t - B_s) & \text{if } t > s \end{cases}$$

is a Brownian motion.

(2) *Scaling: For any $a > 0$, the process $(\frac{1}{\sqrt{a}} B_{at}, t \geq 0)$ is a Brownian motion.*

(3) *Time reversal: The process $(B_1 - B_{1-t}, t \in [0, 1])$ is a Brownian motion on $[0, 1]$.*

It is important to keep in mind that the transformed processes above are different from the original Brownian motion. However, their distributions are the same!

Proof. The reflection at time s and the time reversal are left as exercises; see Exercises 3.4 and 3.5. For the scaling, let $X_t = \frac{1}{\sqrt{a}} B_{at}$. Then $(X_t, t \geq 0)$ is a Gaussian process as it is a linear transformation of a Gaussian process, by applying Lemma 2.23. For the covariance we easily get $\mathbf{E}[X_s X_t] = \frac{1}{a}(at \wedge as) = s \wedge t$. If $t \mapsto B_t(\omega)$ is continuous, then $t \mapsto \frac{1}{\sqrt{a}} B_{at}(\omega)$ is also continuous since it is the composition of the continuous functions $f(t) = at$ and $g(x) = \frac{1}{\sqrt{a}} x$. □

The scaling property shows that Brownian motion is *self-similar*, much like a *fractal*. To see this, suppose we zoom in on a Brownian path very close to 0, say on the interval $[0, 10^{-6}]$. If the Brownian path were smooth and differentiable, the closer we zoom in around the origin, the flatter the function would look. In the limit, we would essentially see a straight line with slope given by the derivative at 0. However, what we see with Brownian motion is very, very different. The scaling property means that for $a = 10^{-6}$,

$$(B_{10^{-6}t}, t \in [0,1]) \overset{\text{distrib.}}{=} (10^{-3}B_t, t \in [0,1]),$$

where $\overset{\text{distrib.}}{=}$ means equality of the distribution of the two processes. In words, Brownian motion on $[0, 10^{-6}]$ looks like a Brownian motion on $[0,1]$, but with its amplitude multiplied by a factor 10^{-3}. In particular, it will remain rugged as we zoom in, unlike a smooth function!

Example 3.4 (Another look at the Ornstein-Uhlenbeck process). Consider the process $(X_t, t \in \mathbb{R})$ defined by

$$X_t = \frac{e^{-2t}}{\sqrt{2}} B_{e^{4t}}, \quad t \in \mathbb{R}.$$

Here the process $(B_{e^{4t}}, t \geq 0)$ is called a *time change* of Brownian motion, since the time is now quantified by an increasing function of t, namely e^{4t}. The example $(B_{at}, t \geq 0)$ in the scaling property is another example of time change.

It turns out that $(X_t, t \in \mathbb{R})$ is a stationary Ornstein-Uhlenbeck process as in Example 2.29. (Here the index of time is $\mathbb{R}$ instead of $[0, \infty)$, but the definition also applies as the process is stationary.) The process $(X_t, t \in \mathbb{R})$ is a Gaussian process because it is a linear transformation of a Gaussian process. The mean is $\mathbf{E}[X_t] = \mathbf{E}[e^{-2t}B_{e^{4t}}/\sqrt{2}] = 0$ for all t, since the mean of Brownian motion is 0. For the covariance, we have for $s \leq t$

$$\mathbf{E}[X_t X_s] = \frac{1}{2}e^{-2t-2s}\mathbf{E}[B_{e^{4t}} B_{e^{4s}}] = \frac{1}{2}e^{-2t-2s}e^{4s} = \frac{1}{2}e^{-2(t-s)}.$$

Two Gaussian processes with the same mean and covariance have the same distribution, thereby proving the claim.

3.2. Properties of the Paths

The scaling property hints at the fact that the Brownian paths are very different from your typical smooth and differentiable function. We look at this in more detail here. What can we say about the continuous function $t \mapsto B_t(\omega)$ for a typical ω? The first striking fact is:

Proposition 3.5. *Let $(B_t, t \geq 0)$ be a standard Brownian motion. The Brownian path $t \mapsto B_t(\omega)$ is nowhere differentiable with probability one.*

We will not prove this here. The proof is not difficult, but the idea might get lost in the notations. Instead, we focus on a related property: *the quadratic variation*. To highlight the difference with differentiable functions, let's first introduce the *variation* of a function. Let f be a function on an interval $[0, t]$. We want to measure how much the function oscillates. To do this, we sum the size of the increments for time intervals $[t_j, t_{j+1}], j = 0, \ldots, n-1$, with $t_0 = 0$ and $t_n = t$. The idea is to take the partition finer as $n \to \infty$ by making the mesh go to zero with n; that is, $\max_{j \leq n} |t_{j+1} - t_j| \to 0$ as $n \to \infty$. For example, we can take the equipartition $t_j = \frac{j}{n}t$, or the dyadic partition

$t_j = \frac{j}{2^n}t$, $j \leq 2^n$. The *variation* of f is then the limit of the sum of the absolute value of the increments [1]:

$$V_t(f) = \lim_{n \to \infty} \sum_{j=0}^{n-1} |f(t_{j+1}) - f(t_j)|.$$

We say that f has *bounded variation* on $[0, t]$ if $V_t(f) < \infty$, and it has *unbounded variation* if $V_t(f)$ is infinite.

Example 3.6 (Examples of functions of bounded variations).

- A function $f : [0, t] \to \mathbb{R}$ that is increasing has bounded variation: It is easily checked that $V_t(f) = f(t) - f(0)$ in this case. Why?
- Let f be a differentiable function on $[0, t]$ with continuous derivative f'. Such functions are said to be in $\mathcal{C}^1([0, t])$. Then we have $V_t(f) = \int_0^t |f'(s)|\, ds < \infty$. This is a consequence of the mean-value theorem, which states that $f(t_{j+1}) = f(t_j) + f'(m_j)(t_{j+1} - t_j)$ where $m_j \in (t_j, t_{j+1})$. With this, the variation can be written as

$$V_t(f) = \lim_{n \to \infty} \sum_{j=0}^{n-1} |f'(m_j)|(t_{j+1} - t_j) = \int_0^t |f'(s)|\, ds.$$

The last equality comes from the definition of the Riemann integral as the limit of Riemann sums. We will go back to this in Chapter 5. Note that the variation in this case is (not surprisingly) similar to the *arc length* of the graph of f on $[0, t]$. The arc length is the length of the graph of f on $[0, t]$ (if the graph represented a rope, say). The arc length is $\ell_t(f) = \int_0^t \sqrt{1 + (f'(s))^2}\, ds$. Note that we have the inequalities

(3.1) $$V_t(f) \leq \ell_t(f) \leq t + V_t(f).$$

In other words, the variation of f is finite if and only if its arc length is.

As can be seen from the above example, there is a very close relationship between functions with bounded variation and functions for which the "classical" integral makes sense. For the Itô integral, the *quadratic variation* plays a similar role. The quadratic variation of f, denoted by $\langle f \rangle_t$, is defined by taking the square of the increments

$$\langle f \rangle_t = \lim_{n \to \infty} \sum_{j=0}^{n-1} (f(t_{j+1}) - f(t_j))^2.$$

Example 3.7 (Quadratic variation of a smooth function). To get acquainted with the concept, let's compute the quadratic variation of a smooth function $f \in \mathcal{C}^1([0, t])$. In Example 3.6, we saw that $V_t(f) = \int_0^t |f'(s)|\, ds$. Here we show $\langle f \rangle_t = 0$. Indeed, we have the obvious bound

$$\sum_{j=0}^{n-1} (f(t_{j+1}) - f(t_j))^2 \leq \max_{j \leq n-1} |f_{t_{j+1}} - f_{t_j}| \cdot \sum_{j=0}^{n-1} |f(t_{j+1}) - f(t_j)|,$$

---

[1] To be precise, the variation is defined as the supremum over all partitions. To keep the formalism simpler here, we define it as the limit of a sequence of partitions, taking for granted that this limit is independent of the choice of partition sequence. We will do the same when looking at quadratic variation.

by pulling out the worst increment from the sum. On one hand, the sum on the right-hand side converges to $V_t(f)$, which is finite. On the other hand, we also have that $\max_{j \leq n-1} |f_{t_{j+1}} - f_{t_j}| \to 0$, because f is a continuous function on $[0, t]$ and therefore is *uniformly continuous* on $[0, t]$. We conclude that $\langle f \rangle_t = 0$. It is not entirely surprising that the quadratic variation is smaller than the variation, since the square of a small number is smaller than the number itself.

Theorem 3.8 (Quadratic variation of Brownian motion). *Let $(B_t, t \geq 0)$ be a standard Brownian motion. Then for any sequence of partitions $(t_j, j \leq n)$ of $[0, t]$ we have*

$$\langle B \rangle_t = \lim_{n \to \infty} \sum_{j=0}^{n-1} (B_{t_{j+1}} - B_{t_j})^2 = t,$$

where the convergence is in the L^2-sense given in Definition 2.35.

It is reasonable here to expect to have some sort of convergence as we are dealing with a sum of independent random variables. However, the conclusion would not hold if the increments were not squared; see Numerical Project 3.5. So there is something more at play here.

Proof. Note that for any n, $\sum_{j=0}^{n-1} (B_{t_{j+1}} - B_{t_j})^2$ is a random variable in L^2. (Why?) The statement says that this random variable converges to t as the mesh of the partition goes to 0. This means that the L^2-distance between the sum of the squares and t goes to zero; that is,

$$\mathbf{E}\left[\left(\sum_{j=0}^{n-1} (B_{t_{j+1}} - B_{t_j})^2 - t\right)^2\right] \to 0.$$

To see this, since $t = \sum_{j=0}^{n-1} t_{j+1} - t_j$, the left-hand side above can be written as

$$\mathbf{E}\left[\left(\sum_{j=0}^{n-1} \{(B_{t_{j+1}} - B_{t_j})^2 - (t_{j+1} - t_j)\}\right)^2\right].$$

For simplicity, define the variables $X_j = (B_{t_{j+1}} - B_{t_j})^2 - (t_{j+1} - t_j)$, $j \leq n-1$. By writing the square as the product of two sums, we get that the above is

$$(3.2) \qquad\qquad = \sum_{i,j=0}^{n-1} \mathbf{E}[X_i X_j].$$

Note that the X_j's have mean zero and are independent, because the increments are. Therefore the above reduces to

$$\sum_{i=0}^{n-1} \mathbf{E}[X_i^2].$$

We now develop the square of X_i. Since the increments are Gaussian with variance $t_{i+1} - t_i$, we get

$$\mathbf{E}[(B_{t_{i+1}} - B_{t_i})^4] = 3(t_{i+1} - t_i)^2.$$

Remember the moments of a Gaussian in equation (1.9)! This implies

$$\mathbf{E}[X_i^2] = 3(t_{i+1} - t_i)^2 - 2(t_{i+1} - t_i)^2 + (t_{i+1} - t_i)^2.$$

Putting all this together, we finally have that

$$(3.3) \qquad \mathbf{E}\left[\left(\sum_{j=0}^{n-1}(B_{t_{j+1}} - B_{t_j})^2 - t\right)^2\right] = 2\sum_{i=0}^{n-1}(t_{i+1} - t_i)^2.$$

But as a particular case of Example 3.7, we have

$$\sum_{i=0}^{n-1}(t_{i+1} - t_i)^2 \leq \max_{i \leq n-1}(t_{i+1} - t_i) \cdot \sum_{i=0}^{n-1}(t_{i+1} - t_i) = \max_{i \leq n}(t_{i+1} - t_i) \cdot t.$$

This goes to 0 as the mesh of the partition goes to 0. □

This is the second time that we encounter L^2-convergence in action. The first time was in Example 2.36 on the law of large numbers. We take this opportunity to relate it to other types of convergence.

Definition 3.9 (Convergence in probability). A sequence of random variables $(X_n, n \geq 1)$ on $(\Omega, \mathcal{F}, \mathbf{P})$ is said to *converge in probability* to X if for any choice of $\delta > 0$, we have

$$\lim_{n \to \infty} \mathbf{P}(|X_n - X| > \delta) = 0.$$

Lemma 3.10. *Let $(X_n, n \geq 1)$ be a sequence of random variables on $(\Omega, \mathcal{F}, \mathbf{P})$ that converges to X in L^2. Then $(X_n, n \geq 1)$ also converges to X in probability.*

Proof. This is a direct consequence of Chebyshev's inequality (1.11). Indeed, we have

$$\mathbf{P}(|X_n - X| > \delta) \leq \frac{1}{\delta^2}\mathbf{E}[(X_n - X)^2].$$

Since the right side goes to 0 as n gets large by the definition of L^2-convergence, the left side also goes to 0. □

The convergence in probability is the convergence occurring in the general statement of the weak law of large numbers for integrable random variables.

Theorem 3.11 (Weak law of large numbers). *Let $(X_n, n \geq 1)$ be a sequence of IID random variables with $\mathbf{E}[|X_1|] < \infty$. Consider the sequence of empirical means $(S_N/N, N \geq 1)$ where $S_N = X_1 + \cdots + X_N$. Then S_N/N converges to $\mathbf{E}[X_1]$ in probability.*

The weak law was checked numerically in Numerical Project 1.2. There, it could be observed that the convergence in probability corresponds to a narrowing of the histograms of the empirical mean around the expectation as N got larger. Example 2.36 proved the stronger L^2-convergence, but with the stronger assumption that variables are in L^2. Interestingly, we did not need the independence assumption there; only the fact that they are uncorrelated was used.

Convergence in probability may be compared with pointwise convergence on an event of probability one. This is known as *convergence almost surely*.

Definition 3.12 (Convergence almost surely). A sequence of random variables $(X_n, n \geq 1)$ on $(\Omega, \mathcal{F}, \mathbf{P})$ is said to *converge almost surely* to X if there exists an event A with $\mathbf{P}(A) = 1$ such that

$$\lim_{n \to \infty} X_n(\omega) = X(\omega), \text{ for every } \omega \in A.$$

Note that for a fixed ω, $X_n(\omega)$ is just a number, so we are asking for convergence of that sequence of numbers to a number $X(\omega)$, for every ω in a set of probability one. This is the convergence occurring in the strong law of large numbers, as observed in Numerical Project 1.2.

Theorem 3.13 (Strong law of large numbers). *Let $(X_n, n \geq 1)$ be a sequence of IID random variables with $\mathbf{E}[|X_1|] < \infty$. Consider the sequence of empirical means $(S_N/N, N \geq 1)$ where $S_N = X_1 + \cdots + X_N$. Then S_N/N converges to $\mathbf{E}[X_1]$ almost surely.*

It turns out that convergence almost surely always implies convergence in probability, so the strong law above implies the weak law. (This is a consequence of the dominated convergence theorem; see Remark 4.39.) But convergence in probability does not imply convergence almost surely in general. See Exercise 3.13 for an example. However, we do have convergence almost surely if the probabilities decay fast enough.

Lemma 3.14. *Let $(X_n, n \geq 1)$ be a sequence of random variables such that for any $\delta > 0$,*

$$(3.4) \qquad \sum_{n \geq 1} \mathbf{P}(|X_n - X| > \delta) < \infty.$$

Then the sequence converges almost surely; that is, there exists an event of probability one on which $\lim_{n \to \infty} X_n(\omega) = X(\omega)$.

Definition 3.9 states that $\mathbf{P}(|X_n - X| > \delta) \to 0$ as $n \to \infty$. Equation (3.4) is stronger as it shows that these probabilities decay to 0 fast enough to be summable.

Proof. Recall the definition of the indicator function $\mathbf{1}_A$ for an event $A \in \mathcal{F}$ in Example 1.11. Take the event $A = \{\omega : |X_n(\omega) - X(\omega)| > \delta\}$. We saw in equation (1.6) that $\mathbf{E}[\mathbf{1}_{\{|X_n - X| > \delta\}}] = \mathbf{P}(|X_n - X| > \delta)$. With this in mind, we have

$$\sum_{n \geq 1} \mathbf{P}(|X_n - X| > \delta) = \sum_{n \geq 1} \mathbf{E}\left[\mathbf{1}_{\{|X_n - X| > \delta\}}\right] = \mathbf{E}\left[\sum_{n \geq 1} \mathbf{1}_{\{|X_n - X| > \delta\}}\right].$$

Since the sum of the probabilities is finite by assumption, we must have that the expectation is finite. Therefore, we have

$$(3.5) \qquad \sum_{n \geq 1} \mathbf{1}_{\{|X_n - X| > \delta\}}(\omega) < \infty, \quad \text{for } \omega \text{ in an event of probability one.}$$

This is a consequence of Exercise 1.15. But the sum is a sum of 0's and 1's! For the sum to be finite, it must be that for n large enough, depending on δ, the indicator is 0 for all subsequent n. In other words, for every $\delta > 0$, there exists $n(\delta)$ such that for $n > n(\delta)$, we must have $|X_n(\omega) - X(\omega)| \leq \delta$. But this is exactly the formal definition of convergence of the sequence of numbers $X_n(\omega)$ to the limit $X(\omega)$. This happens for ω in an event of probability one corresponding to the event on which the sum (3.5) is finite. $\qquad\square$

In the proof, we have exchanged the order between **E** and an infinite sum. Linearity of expectation stated in Theorem 1.28 is only for a finite sum. However, there is no problem in changing the order here because the terms of the sum are positive. This is an instance of the *monotone convergence theorem* of measure theory applied to the partial sums of the indicator functions above (see Exercise 3.11).

Theorem 3.15 (Monotone convergence theorem). *If $(X_n, n \geq 1)$ is a sequence of random variables that are positive, i.e., $X_n \geq 0$ for all n, and increasing, i.e., $X_n \leq X_{n+1}$ for all n, then we have $\lim_{n \to \infty} \mathbf{E}[X_n] = \mathbf{E}[\lim_{n \to \infty} X_n]$.*

With the above remarks in mind, we can show that if the sequence of partitions is fine enough (the dyadic partition will do), the convergence in Theorem 3.8 holds almost surely.

Corollary 3.16 (Quadratic variation of a Brownian path). *Let $(B_s, s \geq 0)$ be a Brownian motion. For every $n \in \mathbb{N}$, consider the dyadic partition $(t_j, j \leq 2^n)$ of $[0, t]$ where $t_j = \frac{j}{2^n} t$. Then we have that*

$$\langle B \rangle_t = \lim_{n \to \infty} \sum_{j=0}^{2^n - 1} (B_{t_{j+1}} - B_{t_j})^2 = t \quad \text{almost surely.}$$

Proof. Borrowing equation (3.3) from the proof of Theorem 3.8, we have that

$$\mathbf{E}\left[\left(\sum_{j=0}^{2^n - 1} (B_{t_{j+1}} - B_{t_j})^2 - t\right)^2\right] = \frac{2t^2}{2^n}.$$

In particular, for any $\delta > 0$, this implies by Chebyshev's inequality that

$$\sum_{n \geq 1} \mathbf{P}\left(\left|\sum_{j=0}^{2^n - 1} (B_{t_{j+1}} - B_{t_j})^2 - t\right| > \delta\right) < \infty.$$

By Lemma 3.14, this means that for this partition there is an event A of probability one on which

(3.6) $$\lim_{n \to \infty} \sum_{j=0}^{2^n - 1} (B_{t_{j+1}}(\omega) - B_{t_j}(\omega))^2 = t, \text{ for every } \omega \in A. \qquad \square$$

We are now ready to show that *every Brownian path has infinite variation*. In view of equation (3.1), this means that a Brownian path on $[0, T]$ has an infinite arc length. This also means that they cannot be differentiable with continuous derivative, as seen in Example 3.6. This is weaker than Proposition 3.5 but much simpler to prove. See Numerical Project 3.5 for a numerical verification of Corollaries 3.16 and 3.17.

Corollary 3.17 (Brownian paths have unbounded variation). *Let $(B_s, s \geq 0)$ be a Brownian motion. Then the continuous function $s \mapsto B_s(\omega)$ on the interval $[0, t]$ has unbounded variation for ω's on a set of probability one.*

Proof. Take the sequence of dyadic partitions of $[0, t]$: $t_j = \frac{j}{2^n} t$, $n \in \mathbb{N}$, $j \leq 2^n$. We have the trivial bound for every ω

$$(3.7) \quad \sum_{j=0}^{2^n-1} (B_{t_{j+1}}(\omega) - B_{t_j}(\omega))^2 \leq \max_{j \leq 2^n} |B_{t_{j+1}}(\omega) - B_{t_j}(\omega)| \cdot \sum_{j=0}^{2^n-1} |B_{t_{j+1}}(\omega) - B_{t_j}(\omega)|.$$

Now suppose there were an event A' with $\mathbf{P}(A') > 0$ on which the Brownian path $s \mapsto B_s(\omega)$ would have finite variation for $\omega \in A'$. By definition of the variation, this would imply that $\lim_{n \to \infty} \sum_{j=0}^{2^n-1} |B_{t_{j+1}}(\omega) - B_{t_j}(\omega)| < \infty$ for $\omega \in A'$. Since the path is continuous, we also have $\lim_{n \to \infty} \max_{j \leq 2^n-1} |B_{t_{j+1}}(\omega) - B_{t_j}(\omega)| = 0$. (This is an application of the *uniform continuity* of a continuous function on $[0, t]$.) We must conclude then that for every $\omega \in A'$, the right-hand side of (3.7), and thus the left-hand side, converges to 0. This contradicts Corollary 3.16. Therefore, the event A' must have zero probability. $\qquad \square$

3.3. A Word on the Construction of Brownian Motion

An important question is: How do we know for sure that there exists a Gaussian process on some probability space $(\Omega, \mathcal{F}, \mathbf{P})$ with the defining properties of Brownian motion? This is not a trivial matter. If we were only looking at the distribution of Brownian motion at a finite number of times, then the distribution is simply the usual distribution of a Gaussian vector with the covariance function $s \wedge t$. However, when we consider an uncountable number of times, as we do, it is not clear that the distribution is well-defined on all times and, more importantly, that the paths $t \mapsto B_t(\omega)$ are continuous with probability one.

Remark 3.18 (Continuity is not to be taken for granted). Consider $(B_t, t \in [0, 1])$ a Brownian motion on $(\Omega, \mathcal{F}, \mathbf{P})$. Let X be a uniform random variable on the same probability space that is independent of the Brownian motion. We define the process

$$\widetilde{B}_t = \begin{cases} B_t & \text{if } t \neq X, \\ 0 & \text{if } t = X. \end{cases}$$

We will show that this process has the same finite-dimensional distribution as Brownian motion. In particular, it satisfies Definition 2.25. However, its paths are clearly not continuous, since at $t = X \neq 0$, $\widetilde{B}_t = 0$ and $B_t \neq 0$ with probability one. To see that it has the same finite-dimensional distribution, let $n \in \mathbb{N}$ and $0 \leq t_1 < \cdots < t_n$. Pick n intervals $A_1, \ldots, A_n$ of $\mathbb{R}$. Note that $\mathbf{P}(X = t_i \text{ for some } i \leq n) = 0$ since X is a continuous random variable. Therefore, we have

$$\begin{aligned} \mathbf{P}(B_{t_i} \in A_i, i \leq n) &= \mathbf{P}(B_{t_i} \in A_i, i \leq n; X = t_i \text{ for some } i \leq n) \\ &\quad + \mathbf{P}(B_{t_i} \in A_i, i \leq n; X \neq t_i \text{ for all } i \leq n) \\ &= 0 + \mathbf{P}(\widetilde{B}_{t_i} \in A_i, i \leq n; X \neq t_i \text{ for all } i \leq n) \\ &= \mathbf{P}(\widetilde{B}_{t_i} \in A_i, i \leq n; X = t_i \text{ for some } i \leq n) \\ &\quad + \mathbf{P}(\widetilde{B}_{t_i} \in A_i, i \leq n; X \neq t_i \text{ for all } i \leq n) \\ &= \mathbf{P}(\widetilde{B}_{t_i} \in A_i, i \leq n). \end{aligned}$$

Luckily, the question of existence of Brownian motion was settled in the early twentieth century by Norbert Wiener. One of the most famous constructions is due to Paul Lévy. It is a construction from IID standard Gaussian random variables. Consider $(Z_n, n \geq 0)$ IID standard Gaussians. Now, we write n in terms of powers of 2; that is, we write $n = 2^j + k$ for $j = 0, 1, 2 \ldots$ and $k = 0, \ldots, 2^j - 1$. Then the Brownian motion at time t is

$$(3.8) \qquad B_t = tZ_0 + \sum_{n=1}^{\infty} 2^{-j/2} \Lambda(2^j t - k) Z_n,$$

where

$$\Lambda(t) = \begin{cases} t & \text{if } 0 \leq t \leq 1/2, \\ 1 - t & \text{if } 1/2 \leq t \leq 1. \end{cases}$$

It turns out that the infinite sum of random variables Z_n in (3.8) converges in L^2. Moreover, for a fixed outcome of the Z_n's, the partial sums are continuous functions that can be shown to converge uniformly. The limit must then be continuous. Equation (3.8) provides another way to generate Brownian paths as follows: Sample N IID standard Gaussians $(Z_n, n \leq N)$. Then take

$$(3.9) \qquad B_t^{(N)} = tZ_0 + \sum_{n=1}^{N} 2^{j/2} \Lambda(2^j t - k) Z_n, \qquad \text{for } n = 2^j + k.$$

In other words, the approximation is a combination of the continuous functions $\Lambda(t)$. This method is implemented in Numerical Project 3.6. This method has the advantage of sampling continuous paths even for finite N.

3.4. A Point of Comparison: The Poisson Process

Like Brownian motion, the Poisson process is defined as a process with stationary and independent increments.

Definition 3.19. A process $(N_t, t \geq 0)$ defined on $(\Omega, \mathcal{F}, \mathbf{P})$ has the distribution of a Poisson process with rate $\lambda > 0$ if and only if the following hold:

(1) $N_0 = 0$.

(2) For any $s < t$, the *increment* $N_t - N_s$ is a Poisson random variable with parameter $\lambda(t - s)$.

(3) For any $n \in \mathbb{N}$ and any choice $0 \leq t_1 < t_2 < \cdots < t_n < \infty$, the increments $N_{t_2} - N_{t_1}, N_{t_3} - N_{t_2}, \ldots, N_{t_n} - N_{t_{n-1}}$ are independent.

Poisson paths can be sampled using this definition; see Numerical Project 3.2. By construction, it is not hard to see that the paths of Poisson processes are piecewise constant, integer-valued, and nondecreasing; see Figure 3.1. In particular, the paths of Poisson processes have finite variation. The Poisson paths are much simpler than the ones of Brownian motion in many ways!

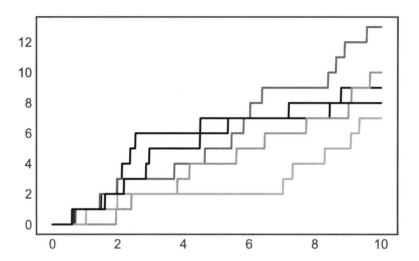

Figure 3.1. A simulation of 5 paths of the Poisson process with rate 1 on the interval [0,10].

We can construct a Poisson process as follows. Consider $(\tau_j, j \in \mathbb{N})$ IID exponential random variables with parameter $1/\lambda$. One should think of τ_j as the *waiting time before the j-th jump*. Then one defines

(3.10) $N_t = \#\{k : \tau_1 + \cdots + \tau_k \leq t\} = \#$ of jumps up to and including time t.

Using this definition, one can check that the increments of $(N_t, t \geq 0)$ are independent with the correct distribution; see Exercise 3.15.

Now here is an idea! What about defining a new process with stationary and independent increments using a given distribution other than Poisson and Gaussian? Is this even possible? The answer is yes but only if the distribution satisfies the property of being *infinitely divisible*. To see this, consider the value of the process at time 1, N_1. Then no matter in how many subintervals we chop the interval $[0, 1]$, we must have that the increments add up to N_1. In other words, we must be able to write N_1 as a sum of n IID random variables for every possible n. This is certainly true for Poisson random variables and Gaussian random variables. (Why?) Another example is the Cauchy distribution, as Numerical Project 1.4(d) demonstrates. In general, processes that can be constructed using independent, stationary increments are called *Lévy processes*.

3.5. Numerical Projects and Exercises

3.1. **Simulating Brownian motion using increments.** Use the definition of Brownian motion in Proposition 3.1 to construct a function def in Python that takes as inputs the time interval and the step size and yields as an output a Brownian path.

3.2. **Simulating the Poisson process.** Use Definition 3.19 to generate 10 paths of the Poisson process with rate 1 on the interval $[0, 10]$ with step size 0.01.

3.3. **The arcsine law.** Consider a Brownian motion on $[0,1]$, $(B_t, t \in [0,1])$. What is the proportion of time of $[0,1]$ for which a Brownian path is positive? Call this random variable X. It turns out that the CDF of X is

(3.11) $$\mathbf{P}(X \le x) = \frac{2}{\pi} \arcsin(\sqrt{x}) = \frac{1}{\pi} \int_0^x \frac{1}{\sqrt{y(1-y)}} \, dy, \quad x \le 1.$$

This will be motivated further in Example 8.29. For now, verify this law numerically by sampling 1,000 Brownian paths with time step 0.01 and plotting the histogram of the sample.

You should get a histogram resembling Figure 3.2. Can you interpret the two peaks at 0 and 1?

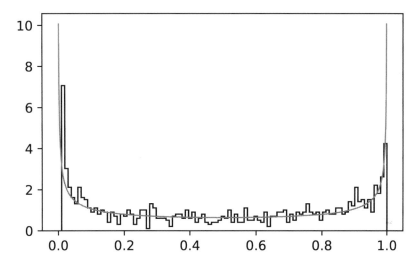

Figure 3.2. A histogram of the PDF for the arcsine law for a sample of 1,000 Brownian paths with step size 0.01. The exact PDF is in gray.

3.4. **Arcsine law for Ornstein-Uhlenbeck?** Consider a Ornstein-Uhlenbeck process on $[0,1]$, $(X_t, t \in [0,1])$ as in Numerical Project 2.3. Repeat the experiment of Project 3.3 with Ornstein-Uhlenbeck paths. What do you notice?

3.5. **Brownian variations.** In this numerical project, we observe Corollaries 3.16 and 3.17 in action. Consider $n = 1, \ldots, 20$. For each n, consider the dyadic partition of $[0,1]$ given by $t_j = j/2^n$. Sample a single Brownian path with time step 2^{-20} on $[0,1]$. For this single path:

(a) Compute for each n the variation

$$\sum_{j=0}^{2^n-1} |B_{t_{j+1}} - B_{t_j}|.$$

(b) Compute for each n the quadratic variation

$$\sum_{j=0}^{2^n-1} (B_{t_{j+1}} - B_{t_j})^2.$$

(c) Draw the graph of the variation and of the quadratic variation as a function of n, $n = 1, \ldots, 20$. What do you observe?

3.6. **Simulating Brownian motion using Lévy's construction**. Use equation (3.9) to generate 10 paths of standard Brownian motion for $N = 5$, $N = 20$, $N = 100$.

Exercises

3.1. **Brownian moments**. Use the definition of Brownian motion to compute the following moments:
  (a) $\mathbf{E}[B_t^6]$.
  (b) $\mathbf{E}[(B_{t_2} - B_{t_1})(B_{t_3} - B_{t_2})]$ if $t_1 < t_2 < t_3$.
  (c) $\mathbf{E}[B_s^2 B_t^2]$ if $s < t$.
  (d) $\mathbf{E}[B_s B_t^3]$ if $s < t$.
  (e) $\mathbf{E}[B_s^{100} B_t^{101}]$.

3.2. **Brownian probabilities**. As in Example 3.2, use one of the definitions of Brownian motion to write the following probabilities as an integral. Find the numerical value using a software:
  (a) $\mathbf{P}(B_1 > 1, B_2 > 1)$.
  (b) $\mathbf{P}(B_1 > 1, B_2 > 1, B_3 > 1)$.

3.3. **Equivalence of definition of Brownian motion**. Complete the proof of Proposition 3.1 by showing the following:
  If $(B_{t_1} - 0, B_{t_2} - B_{t_1}, \ldots, B_{t_n} - B_{t_{n-1}})$ are independent Gaussians with mean 0 and variance $t_{j+1} - t_j$, $j \leq n - 1$, then the vector $(B_{t_1}, \ldots, B_{t_n})$ is Gaussian with mean 0 and covariance $\mathbf{E}[B_t B_s] = t \wedge s$.

3.4. **Reflection at time s**. Show that the process $(\widetilde{B}_t, t \geq 0)$ defined in Proposition 3.3 is a standard Brownian motion.

3.5. **Time reversal**. Let $(B_t, t \geq 0)$ be a Brownian motion. Show that the process $(B_1 - B_{1-t}, t \in [0, 1])$ has the distribution of a standard Brownian motion on $[0, 1]$.

3.6. **Time inversion**. Let $(B_t, t \geq 0)$ be a standard Brownian motion. We consider the process

$$X_t = tB_{1/t} \text{ for } t > 0.$$

This property of Brownian motion relates the behavior for t large to the behavior for t small.
  (a) Show that $(X_t, t > 0)$ has the distribution of a Brownian motion on $t > 0$.
  (b) Argue that X_t converges to 0 as $t \to 0$ in the sense of L^2-convergence.
    It is possible to show convergence almost surely so that $(X_t, t \geq 0)$ is really a Brownian motion for $t \geq 0$, Exercise 4.20.
  (c) Use this property of Brownian motion to show the *law of large numbers* for Brownian motion

$$\lim_{t \to \infty} \frac{X_t}{t} = 0 \text{ almost surely.}$$

3.7. **Brownian motion from the bridge.** Let $(U_t, t \in [0,1])$ be a Brownian bridge.
   (a) Show that the process $B_t = (1+t)U_{\frac{t}{t+1}}$, $t \geq 0$, is a standard Brownian motion.
   (b) Use the above to get another proof that

$$\lim_{t \to \infty} \frac{B_t}{t} = 0 \text{ almost surely.}$$

3.8. **Convergence in mean or in L^1.** Consider the space $L^1(\Omega, \mathcal{F}, \mathbf{P})$ of integrable random variables; see Exercise 2.16. A sequence of random variables $(X_n, n \geq 1)$ is said to *converge in L^1 or in mean to X* if and only if $\lim_{n \to \infty} \|X_n - X\|_1 = 0$. Show that if X_n converges in L^1, then it also converges in probability.

3.9. **Fractional Brownian motion.** Recall the definition of fractional Brownian motion $(B_t^{(H)}, t \geq 0)$ with Hurst index $0 < H < 1$ in Example 2.28.
   (a) Verify that standard Brownian corresponds to $H = 1/2$.
   (b) Show that fractional Brownian motion has the following scaling property: $(B_{at}^{(H)}, t \geq 0)$ has the same distribution as $a^H(B_t^{(H)}, t \geq 0)$ for $a > 0$.
   (c) Show that the increment $B_t^{(H)} - B_s^{(H)}$ has a Gaussian distribution and is stationary (i.e., its distribution depends only on the difference $t - s$).
   (d) Show that the increments are independent only if $H = 1/2$. Show that they are negatively correlated if $H < 1/2$ and positively correlated if $H > 1/2$.

3.10. **The arcsine law on $[0, T]$.** Consider the arcsine law in equation (3.11). How does the CDF change when the Brownian motion is defined on $[0, T]$ instead of $[0, 1]$?

3.11. **An application of the monotone convergence theorem.** Let $(X_n, n \geq 1)$ be a sequence of random variables on $(\Omega, \mathcal{F}, \mathbf{P})$. Show that if $X_n \geq 0$ for all n, then

$$\mathbf{E}\Big[\sum_{n \geq 1} X_n \Big] = \sum_{n \geq 1} \mathbf{E}[X_n].$$

(The value $\mathbf{E}[X] = \infty$ is acceptable here.)
   This is an example of Fubini's theorem, which states that we can exchange two expectations/integrals/sums under suitable conditions.

3.12. **Borel-Cantelli Lemma I.** Lemma 3.14 is an example of a more general fact. Consider a sequence of events $A_1, A_2, A_3, \ldots$ defined on a probability space $(\Omega, \mathcal{F}, \mathbf{P})$ such that $\sum_n \mathbf{P}(A_n) < \infty$. Prove that the probability of the event $\{\omega \in A_n$ for infinitely many $n\}$ is 0.

3.13. ★ **Convergence in probability $\not\Rightarrow$ convergence almost surely.** Let U be a uniform random variable on $[0, 1]$. We construct a sequence of random variables as follows: $X_1 = \mathbf{1}_{[0,1/2]}(U)$; that is, X_1 is 1 if U is in $[0, 1/2]$ and 0 otherwise. Then $X_2 = \mathbf{1}_{[1/2,1]}(U)$, $X_3 = \mathbf{1}_{[0,1/4]}(U)$, $X_4 = \mathbf{1}_{[1/4,2/4]}(U)$, $X_5 = \mathbf{1}_{[2/4,3/4]}(U)$, $X_6 = \mathbf{1}_{[3/4,1]}(U)$, and so on by taking intervals of length 2^{-n}. In words, we divide $[0, 1]$ in dyadic intervals, and then the sequence is constructed by taking the indicator function of the successive dyadic intervals.
   Show that X_n converges to 0 in probability, but that $\lim_{n \to \infty} X_n$ does not exist for any outcome of U; i.e., the sequence does not converge almost surely.

3.14. ★ **But ok on a subsequence.** Use Lemma 3.14 to show that if $(X_n, n \geq 0)$ is a sequence of random variables on $(\Omega, \mathcal{F}, \mathbf{P})$ such that $X_n \to X$ in probability, then there is a subsequence $(X_{n_k}, k \geq 0)$ such that $X_{n_k} \to X$ almost surely.

3.15. ★ **Construction of the Poisson process**. Use Exercises 1.9 and 2.2 to show that the process (3.10) is a Poisson process with rate λ.

Hint: Note that $\mathbf{P}(N_t \geq k) = \mathbf{P}(\tau_1 + \cdots + \tau_k \leq t)$. Use this to prove that N_t is a Poisson random variable. For the independence, consider the conditional probability $\mathbf{P}(N_t - N_s = k | N_s = \ell)$.

3.6. Historical and Bibliographical Notes

Brownian motion has a rich history that is intimately connected with developments in modern physics. It got its name from Robert Brown, a botanist, who observed at the beginning of the nineteenth century that grains of pollen in water exhibit fast and jerky motion. The reason for this motion remained unsettled for a while until Albert Einstein in 1905 (the *Annus Mirabilis*) surmised that this motion was due to the multiple collisions of a grain of pollen with the molecules of water [**Ein05**]. Using the tools of statistical mechanics, he derived the probability density of the position of such a particle in time and space, with the proper diffusion constant. This was a huge achievement at the time as the existence of atoms was still controversial. The calculation of Einstein allowed for an experimental estimate of the Avogadro number, the order of magnitude of the number of atoms in a macroscopic quantity (such as a glass of water). The work of Einstein and the experimental confirmation are considered the first evidence of the existence of atoms. Ornstein and Uhlenbeck built on the work of Einstein to describe a Brownian particle in a medium with friction [**UO30**]. We refer to [**Nel67**] for a detailed account.

Interestingly, around the same time as Einstein, Louis Bachelier was building the foundations of modern asset pricing. In his thesis, *Théorie de la spéculation* [**Bac00**], he modeled prices using random walks and derived the equations in the limit where these random walks become Brownian motion. The first rigorous mathematical model for Brownian motion is due to Norbert Wiener [**Wie76**]. This is why the process bears his name. The construction presented in Section 3.3 is due to Lévy [**L65**]. A nice exposition of the construction is found in [**Ste01**]. Another way to construct Brownian motion is as a limit of rescaled random walks. It turns out that the limiting process, Brownian motion, is the same for a large class of distributions for the increment of the walk, a phenomenon now known as *Donsker's invariance principle*. This principle is in some ways a generalization of the central limit theorem (as the title Chapter 18 of [**Nel87**] highlights!). We refer to [**KS91**] for a full, but not easy, proof of Donsker's principle. For more on the mathematical theory of Brownian motion, we suggest [**MP10**]. Remark 3.18 is based on an example from this book. The arcsine law was proved by Lévy in Section 5 of [**L39**].

The Poisson process appeared at the beginning of the 1900s in applications in the actuarial sciences, in physics, and in telecommunications. We refer to [**Cra76**] for more details on this and on the development of probability theory in general.

Chapter 4

Martingales

Martingales form an important class of stochastic processes that are especially handy for modelling financial assets. The term martingale initially referred to the betting strategy consisting of doubling your bet at every trial.[1] This is a winning strategy if you have a generous banker! As we will see in Chapter 5, martingales can be seen as investing or gambling strategies built upon some underlying asset itself modelled by a martingale. Roughly speaking, a martingale is a stochastic process that has the property that the expected value of the process at a future time, given the information of the past, is equal to the present value of the process. (See Figure 4.1.) We will see that Brownian motion is a martingale. The martingale property of Brownian motion can be written in a compact way as follows:

$$\mathbf{E}[B_t|\mathcal{F}_s] = B_s, \quad s \leq t.$$

The objective of this chapter is to make sense of the above expectation. We will see in Sections 4.1 and 4.2 that conditional expectation has a nice geometric interpretation that was first introduced in Section 2.4. The object $\mathcal{F}_s$ represents the information of the Brownian path up to time s. Properties of martingales are given in Section 4.3. It is possible to construct a plethora of martingales using Brownian motion. In fact, a big part of Itô calculus is to construct such martingales. Martingales are particularly useful, because they facilitate many computations of probabilities and expectations. This is illustrated in Section 4.4, where we solve the gambler's ruin problem for Brownian motion.

4.1. Elementary Conditional Expectation

In elementary probability, the conditional expectation of a given variable Y given another random variable X refers to the expectation of Y under the conditional distribution of Y given X. To illustrate this, let's go through a simple example. Consider $\mathcal{B}_1, \mathcal{B}_2$

---

[1]See Exercise 4.15.

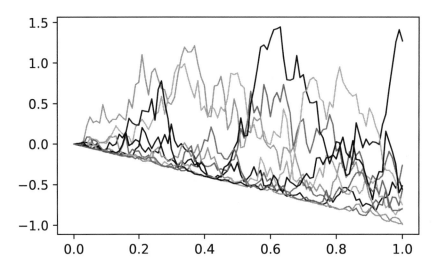

Figure 4.1. A simulation of 10 paths of the martingale $B_t^2 - t$, $t \in [0,1]$. Note that the distribution is not symmetric (as opposed to Brownian motion). However, the average is constant in time and equals 0.

be two independent Bernoulli-distributed random variables with $p = 1/2$. Then construct

$$X = \mathcal{B}_1, \qquad Y = \mathcal{B}_1 + \mathcal{B}_2.$$

It is easy to compute $\mathbf{E}[Y|X = 0]$ and $\mathbf{E}[Y|X = 1]$. By definition, it is given by

$$\mathbf{E}[Y|X = 0] = \sum_{j=0}^{2} j\mathbf{P}(Y = j|X = 0) = \sum_{j=0}^{2} j\frac{\mathbf{P}(Y = j, X = 0)}{\mathbf{P}(X = 0)} = \frac{1}{2},$$

where we use the definition of conditional probability: For two events A, B such that $\mathbf{P}(B) \neq 0$,

$$\mathbf{P}(A|B) = \frac{\mathbf{P}(A \cap B)}{\mathbf{P}(B)}.$$

The same way, one finds $\mathbf{E}[Y|X = 1] = \frac{3}{2}$. With this point of view, the conditional expectation is computed given the information that the event $\{X = 0\}$ occurred or the event $\{X = 1\}$ occurred. It is possible to regroup both conditional expectations in a single object if we think of the conditional expectation as a random variable that we denote $\mathbf{E}[Y|X]$. Namely, we take

(4.1)
$$\mathbf{E}[Y|X](\omega) = \begin{cases} \frac{1}{2} & \text{if } X(\omega) = 0, \\ \frac{3}{2} & \text{if } X(\omega) = 1. \end{cases}$$

This random variable is called the *conditional expectation of Y given X*. We make two important observations:

(A) If the value of X is known, then the value of $\mathbf{E}[Y|X]$ is determined.

(B) If we have another random variable $g(X)$ constructed from X, then we have

$$\mathbf{E}[g(X)Y] = \mathbf{E}\big[g(X)\mathbf{E}[Y|X]\big].$$

In other words, as far as X is concerned, the conditional expectation $\mathbf{E}[Y|X]$ is a proxy for Y in the expectation. We sometimes say that $\mathbf{E}[Y|X]$ is the *best estimate of Y given the information of X.*

The last observation is easy to verify since

$$\mathbf{E}[g(X)Y] = \sum_{i=0}^{1} \sum_{j=0}^{2} j\, g(i)\, \mathbf{P}(X = i, Y = j)$$

$$= \sum_{i=0}^{1} \mathbf{P}(X = i)\, g(i) \left\{ \sum_{j=0}^{2} j\, \frac{\mathbf{P}(X = i, Y = j)}{\mathbf{P}(X = i)} \right\}$$

$$= \mathbf{E}\big[g(X)\mathbf{E}[Y|X]\big].$$

We will see in the next section that the two properties above define the conditional expectation in general. For now, we just notice that the elementary definitions of conditional expectations actually possess these two properties.

Example 4.1 (Elementary definitions of conditional expectation).

(1) (X, Y) *discrete:* The treatment is very similar to the above. If a random variable X takes values $(x_i, i \geq 1)$ and Y takes values $(y_j, j \geq 1)$, we have by definition that the conditional expectation as a random variable is

$$\mathbf{E}[Y|X](\omega) = \sum_{j \geq 1} y_j \mathbf{P}(Y = y_j | X = x_i) \text{ for } \omega \text{ such that } X(\omega) = x_i.$$

(2) (X, Y) *continuous with joint PDF $f(x, y)$:* In this case, the conditional expectation is the random variable given by

$$\mathbf{E}[Y|X] = h(X), \quad \text{where } h(x) = \frac{\int_{\mathbb{R}} y f(x, y)\, dy}{\int_{\mathbb{R}} f(x, y)\, dy}.$$

The reader is invited to verify that the second example satisfies the two properties of conditional expectation above in Exercise 4.1. In the two examples above, the expectation of the random variable $\mathbf{E}[Y|X]$ is equal to $\mathbf{E}[Y]$. Indeed, in the discrete case, we have

$$\mathbf{E}[\mathbf{E}[Y|X]] = \sum_{i \leq 1} \mathbf{P}(X = x_i) \left\{ \sum_{j \geq 1} y_j \mathbf{P}(Y = y_j | X = x_i) \right\}$$

$$= \sum_{i,j \geq 1} y_j \mathbf{P}(X = x_i, Y = y_j) = \sum_{j \geq 1} y_j \mathbf{P}(Y = y_j) = \mathbf{E}[Y].$$

Example 4.2 (Conditional probability vs conditional expectation). The conditional probability of the event A given B can be recast in terms of conditional expectation using indicator functions, as defined in Example 1.11. If $0 < \mathbf{P}(B) < 1$, it is not hard to check that

$$\mathbf{P}(A|B) = \mathbf{E}[\mathbf{1}_A | \mathbf{1}_B = 1] \quad \text{and} \quad \mathbf{P}(A|B^c) = \mathbf{E}[\mathbf{1}_A | \mathbf{1}_B = 0].$$

Indeed, the random variables $\mathbf{1}_A$ and $\mathbf{1}_B$ are discrete. If we proceed as in the discrete case above, we have

$$\mathbf{E}[\mathbf{1}_A | \mathbf{1}_B = 1] = 1 \cdot \mathbf{P}(\mathbf{1}_A = 1 | \mathbf{1}_B = 1) = \frac{\mathbf{P}(\mathbf{1}_A = 1, \mathbf{1}_B = 1)}{\mathbf{P}(\mathbf{1}_B = 1)} = \frac{\mathbf{P}(A \cap B)}{\mathbf{P}(B)}.$$

A similar calculation gives $\mathbf{P}(A|B^c)$. In particular, the formula for the total probability of A is a rewriting of the expectation of the random variable $\mathbf{E}[\mathbf{1}_A|\mathbf{1}_B]$:

$$\mathbf{E}[\mathbf{E}[\mathbf{1}_A|\mathbf{1}_B]] = \mathbf{E}[\mathbf{1}_A|\mathbf{1}_B = 1]\,\mathbf{P}(\mathbf{1}_B = 1) + \mathbf{E}[\mathbf{1}_A|\mathbf{1}_B = 0]\,\mathbf{P}(\mathbf{1}_B = 0)$$
$$= \mathbf{P}(A|B)\mathbf{P}(B) + \mathbf{P}(A|B^c)\mathbf{P}(B^c) = \mathbf{P}(A).$$

In the next section, we generalize the definition of conditional expectation to give it a more precise conceptual meaning in terms of orthogonal projection.

4.2. Conditional Expectation as a Projection

Conditioning on one variable. We start by giving the definition of conditional expectation given a single variable. This relates to the two observations (A) and (B) made previously. We assume that the random variable is integrable for the expectations to be well-defined.

Definition 4.3. Let X and Y be integrable random variables on $(\Omega, \mathcal{F}, \mathbf{P})$. The conditional expectation of Y given X is the random variable denoted by $\mathbf{E}[Y|X]$ with the following two properties:

(A) There exists a function $h : \mathbb{R} \to \mathbb{R}$ such that $\mathbf{E}[Y|X] = h(X)$.

(B) For any (bounded) random variable of the form $g(X)$ for some function g [2],

(4.2) $$\mathbf{E}[g(X)Y] = \mathbf{E}[g(X)\mathbf{E}[Y|X]].$$

We can interpret the second property as follows: The conditional expectation $\mathbf{E}[Y|X]$ serves as a proxy for Y as far as X is concerned. Note that in equation (4.2) the expectation on the left can be seen as an average over the joint values of (X, Y), whereas the one on the right is an average over the values of X only! Another way to see this property is to write it as

(4.3) $$\mathbf{E}[g(X)(Y - \mathbf{E}[Y|X])] = 0.$$

In other words, in the terminology of Section 2.4,

> the random variable $Y - \mathbf{E}[Y|X]$ is orthogonal to any random variable constructed from X.

Finally, it is important to notice that if we take $g(X) = 1$, then the second property implies

(4.4) $$\mathbf{E}[Y] = \mathbf{E}[\mathbf{E}[Y|X]].$$

In other words, the expectation of the conditional expectation of Y is simply the expectation of Y.

The existence of conditional expectation $\mathbf{E}[Y|X]$ is not obvious. We know it exists in particular cases given in Example 4.1. We will show more generally that it exists and it is unique whenever Y is in $L^2(\Omega, \mathcal{F}, \mathbf{P})$. (In fact, it can be shown to exist whenever Y is integrable.) Before doing so, let's warm up by looking at the case of Gaussian vectors.

---

[2]Here, we assume that $g(X)$ is bounded to ensure that the expectation is well-defined.

Example 4.4 (Conditional expectation for Gaussian vectors. I). Let (X, Y) be a Gaussian vector of mean 0. Then

(4.5) $$\mathbf{E}[Y|X] = \frac{\mathbf{E}[XY]}{\mathbf{E}[X^2]} X.$$

This candidate satisfies the two defining properties of conditional expectation: (A) It is clearly a function of X; in fact it is a multiple of X; (B) we have that the random variable $Y - \frac{\mathbf{E}[XY]}{\mathbf{E}[X^2]} X$ is independent of X. This is a consequence of Proposition 2.10, since

$$\mathbf{E}\left[X\left(Y - \frac{\mathbf{E}[XY]}{\mathbf{E}[X^2]} X\right)\right] = 0.$$

Therefore, we have for any bounded function $g(X)$ of X

$$\mathbf{E}[g(X)(Y - \mathbf{E}[Y|X])] = \mathbf{E}[g(X)]\mathbf{E}[Y - \mathbf{E}[Y|X]] = 0,$$

since $\mathbf{E}[Y - \mathbf{E}[Y|X]] = 0$ by property (4.4).

In Exercise 4.4, it is checked that equation (4.5) corresponds to the conditional expectation calculated from the PDF, as in Example 4.1.

Example 4.5 (Brownian conditioning, I). Let $(B_t, t \geq 0)$ be a standard Brownian motion. Consider the Gaussian vector $(B_{1/2}, B_1)$. Its covariance matrix is

$$\mathcal{C} = \begin{pmatrix} 1/2 & 1/2 \\ 1/2 & 1 \end{pmatrix}.$$

Let's compute $\mathbf{E}[B_1|B_{1/2}]$ and $\mathbf{E}[B_{1/2}|B_1]$. This is easy using equation (4.5). We have

$$\mathbf{E}[B_1|B_{1/2}] = \frac{\mathbf{E}[B_1 B_{1/2}]}{\mathbf{E}[B_{1/2}^2]} B_{1/2} = \frac{1/2}{1/2} B_{1/2} = B_{1/2}.$$

In other words, the best approximation of B_1 given the information of $B_{1/2}$ is $B_{1/2}$. There is no problem in computing $\mathbf{E}[B_{1/2}|B_1]$, even though we are conditioning on a future position. Indeed, the same formula gives

$$\mathbf{E}[B_{1/2}|B_1] = \frac{\mathbf{E}[B_1 B_{1/2}]}{\mathbf{E}[B_1^2]} B_1 = \frac{1}{2} B_1.$$

This means that the best approximation of $B_{1/2}$ given the position at time 1 is $\frac{1}{2}B_1$, which makes a whole lot of sense!

In Example 4.4 for the Gaussian vector (X, Y), the conditional expectation was equal to the *orthogonal projection of Y onto X* in L^2, as defined in equation (2.12). In particular, the conditional expectation was a multiple of X. Is this always the case? Unfortunately, it is not. For example, in equation (4.1), the conditional expectation is clearly not a multiple of the random variable X. However, it is a function of X, as is always the case by Definition 4.3.

The idea to construct the conditional expectation $\mathbf{E}[Y|X]$ in general is to *project Y on the space of all random variables that can be constructed from X.* To make this precise, consider the following subspace of $L^2(\Omega, \mathcal{F}, \mathbf{P})$:

Definition 4.6. Let $(\Omega, \mathcal{F}, \mathbf{P})$ be a probability space and X a random variable defined on it. The space $L^2(\Omega, \sigma(X), \mathbf{P})$ is the linear subspace of $L^2(\Omega, \mathcal{F}, \mathbf{P})$ consisting of the square-integrable random variables of the form $g(X)$ for some function $g : \mathbb{R} \to \mathbb{R}$.

This is a *linear subspace* of $L^2(\Omega, \mathcal{F}, \mathbf{P})$: It contains the random variable 0, and any linear combination of random variables of this kind is also a function of X and must have a finite second moment. We note the following:

$L^2(\Omega, \sigma(X), \mathbf{P})$ *is a subspace of* $L^2(\Omega, \mathcal{F}, \mathbf{P})$, *very much how a plane or line (going through the origin) is a subspace of* $\mathbb{R}^3$.

In particular, as in the case of a line or a plane, we can *project* an element Y of $L^2(\Omega, \mathcal{F}, \mathbf{P})$ onto $L^2(\Omega, \sigma(X), \mathbf{P})$. The resulting projection is an element of $L^2(\Omega, \sigma(X), \mathbf{P})$, a square-integrable random variable that is a function of X. For a subspace $\mathcal{S}$ of $\mathbb{R}^3$ (e.g., a line or a plane), the projection $\text{Proj}_{\mathcal{S}}(v)$ of a vector v onto the subspace is the closest point to v lying in the subspace. Moreover, $v - \text{Proj}_{\mathcal{S}}(v)$ is orthogonal to the subspace. This picture of orthogonal projection also holds in L^2; see Figure 4.2. Let Y be a random variable in $L^2(\Omega, \mathcal{F}, \mathbf{P})$, and let $L^2(\Omega, \sigma(X), \mathbf{P})$ be the subspace of those random variables that are functions of X. We write $Y^\star$ for the random variable in $L^2(\Omega, \sigma(X), \mathbf{P})$ that is the *closest to Y*. In other words, we have (using the definition of the L^2-distance squared)

$$(4.6) \qquad \min_{Z \in L^2(\Omega, \sigma(X), \mathbf{P})} \mathbf{E}[(Y - Z)^2] = \mathbf{E}[(Y - Y^\star)^2].$$

It turns out that $Y^\star$ is the right candidate for the conditional expectation.

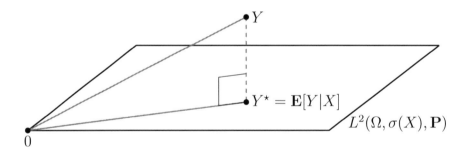

Figure 4.2. An illustration of the conditional expectation $\mathbf{E}[Y|X]$ as an orthogonal projection of Y onto the subspace $L^2(\Omega, \sigma(X), \mathbf{P})$.

Theorem 4.7 (Existence and uniqueness of conditional expectation). *Let X be a random variable on $(\Omega, \mathcal{F}, \mathbf{P})$. Let Y be a random variable in $L^2(\Omega, \mathcal{F}, \mathbf{P})$. Then the conditional expectation $\mathbf{E}[Y|X]$ is the random variable $Y^\star$ given in equation (4.6). Namely, it is the random variable in $L^2(\Omega, \sigma(X), \mathbf{P})$ that is the closest to Y in the L^2-distance. In particular, we have the following:*

- *It is the orthogonal projection of Y onto $L^2(\Omega, \sigma(X), \mathbf{P})$; i.e., $Y - Y^\star$ is orthogonal to any random variables in $L^2(\Omega, \sigma(X), \mathbf{P})$.*

- *It is unique.*

The result reinforces the meaning of the conditional expectation $\mathbf{E}[Y|X]$ as the *best estimation of Y given the information of X*: It is the closest random variable to Y among all the functions of X, in the sense of L^2.

Proof. We write for short $L^2(X)$ for the subspace $L^2(\Omega, \sigma(X), \mathbf{P})$. Let $Y^\star$ be as in (4.6). We show successively that (1) $Y - Y^\star$ is orthogonal to any element of $L^2(X)$, so it is the orthogonal projection; (2) $Y^\star$ has the properties of conditional expectation in Definition 4.3; and (3) $Y^\star$ is unique.

(1) Let $W = g(X)$ be a random variable in $L^2(X)$. We show that W is orthogonal to $Y - Y^\star$; that is, $\mathbf{E}[(Y - Y^\star)W] = 0$. This should be intuitively clear from Figure 4.2. On one hand, we have by developing the square

$$\mathbf{E}[(W - (Y - Y^\star))^2] = \mathbf{E}[W^2] - 2\mathbf{E}[W(Y - Y^\star)] + \mathbf{E}[(Y - Y^\star)^2].$$

On the other hand, since $Y^\star + W$ is in $L^2(X)$ (it is a linear combination of elements in $L^2(X)$), we must have from equation (4.6)

$$\mathbf{E}[(W - (Y - Y^\star))^2] \geq \mathbf{E}[(Y - Y^\star)^2].$$

Putting the last two equations together, we get that for any $W \in L^2(X)$,

$$\mathbf{E}[W^2] - 2\mathbf{E}[W(Y - Y^\star)] \geq 0.$$

In particular, this holds for aW for any $a > 0$, in which case we get

$$\mathbf{E}[W(Y - Y^\star)] \leq a\mathbf{E}[W^2]/2.$$

This could only be true for any $a > 0$ if $\mathbf{E}[W(Y - Y^\star)] \leq 0$. The above also holds for any $a < 0$, in which case we get the reverse inequality

$$\mathbf{E}[W(Y - Y^\star)] \geq a\mathbf{E}[W^2]/2.$$

We conclude from both cases that $\mathbf{E}[W(Y - Y^\star)] = 0$.

(2) It is clear that $Y^\star$ is a function of X by construction, since it is in $L^2(X)$. Moreover, for any $W \in L^2(X)$, we have from (1) that

$$\mathbf{E}[W(Y - Y^\star)] = 0,$$

which is the second defining property of conditional expectation.

(3) Lastly, suppose there is another element Y' that is in $L^2(X)$ that minimizes the distance to Y. Then we would get

$$\mathbf{E}[(Y - Y')^2] = \mathbf{E}[(Y - Y^\star + (Y^\star - Y'))^2] = \mathbf{E}[(Y - Y^\star)^2] + \mathbf{E}[(Y^\star - Y')^2],$$

where we used the fact that $Y^\star - Y' \in L^2(X)$ and the orthogonality of $Y - Y^\star$ with $L^2(X)$ as shown in (1). But this implies that the square-distance $\mathbf{E}[(Y - Y')^2]$ equals the minimum square-distance $\mathbf{E}[(Y - Y^\star)^2]$ if and only if $Y^\star = Y'$ with probability one. $\qquad\square$

Remark 4.8. How can we be sure that the minimum in equation (4.6) is attained? To be precise, we should have written $\inf \mathbf{E}[(Y - Z)^2]$ where inf stands for the *infimum* of $\mathbf{E}[(Y - Z)^2]$ over the set $L^2(\Omega, \sigma(X), \mathbf{P})$. The infimum $\mathcal{I}$ is the greatest lower bound over the set. The properties of the infimum imply that there will be a sequence Z_n, $n \geq 1$, of elements in the subspace $L^2(\Omega, \sigma(X), \mathbf{P})$ such that $\mathbf{E}[(Y - Z_n)^2] \to \mathcal{I}$. The triangle inequality also implies that

$$\|Z_m - Z_n\| \leq \left| \|Y - Z_m\| - \|Y - Z_n\| \right|.$$

Since $\mathbf{E}[(Y - Z_n)^2] \to \mathcal{J}$, for $\varepsilon > 0$, we can pick n large enough such that for any $m \geq n$, $\|Z_m - Z_n\| < \varepsilon$. Such a sequence $(Z_n, n \geq 1)$ is said to be Cauchy. It turns out that Cauchy sequence must converge in L^2-spaces. In particular, the sequence Z_n converges to an element $Y^\star$ in $L^2(\Omega, \sigma(X), \mathbf{P})$. So the minimum is attained. We will say more on this in Remark 5.14.

Conditioning on several random variables. We would like to generalize the conditional expectation to the case when we condition on the information of more than one variable. Taking the L^2 point of view, we should expect that the conditional expectation is the orthogonal projection of the given variable on the subspace generated by all square-integrable functions of the variables on which we condition.

It is now useful to study sigma-fields, an object we introduced briefly in Section 1.1, in more depth. This object is defined as follows.

Definition 4.9. A *sigma-field* or *sigma-algebra* $\mathcal{F}$ of a sample space Ω is a collection of events with the following properties:

(1) Ω is in $\mathcal{F}$.

(2) *Closure under complement.* If $A \in \mathcal{F}$, then A^c is in $\mathcal{F}$.

(3) *Closure under countable unions.* If $A_1, A_2, \ldots \in \mathcal{F}$, then $\bigcup_{n \geq 1} A_n$ is in $\mathcal{F}$.

Such objects play a fundamental role in the rigorous study of probability and real analysis in general. We will focus on the intuition behind them. First, let's mention some examples of sigma-fields of a given sample space Ω to get acquainted with the concept.

Example 4.10 (Examples of sigma-fields).

(i) *The trivial sigma-field.* Note that the collection of events $\{\emptyset, \Omega\}$ is a sigma-field of Ω. It is the smallest sigma-field. We will denote it by $\mathcal{F}_0$.

(ii) *The sigma-field generated by an event A.* Let A be an event that is not $\emptyset$ and not the entire Ω. Then the smallest sigma-field containing A ought to be

$$\{\emptyset, A, A^c, \Omega\}.$$

This sigma-field is denoted by $\sigma(A)$. See Exercise 4.2 for the sigma-algebra generated by two events.

(iii) *The sigma-field generated by a random variable X.* We denote by $\sigma(X)$ the smallest sigma-field containing all events pertaining to X, that is, all events of the form $\{\omega : X(\omega) \in (a, b]\}$ for some interval $(a, b] \subset \mathbb{R}$. Intuitively, we think of $\sigma(X)$ as containing all information about X.

(iv) *The sigma-field generated by a stochastic process $(X_s, s \leq t)$.* Let $(X_s, s \geq 0)$ be a stochastic process. Consider the process restricted to $[0, t]$, $(X_s, s \leq t)$. We consider the smallest sigma-field containing all events pertaining to the random variables $X_s, s \leq t$. We denote it by $\sigma(X_s, s \leq t)$ or $\mathcal{F}_t$.

(v) *Borel sets of $\mathbb{R}$.* The σ-field $\mathcal{B}(\mathbb{R})$ of Borel sets of $\mathbb{R}$ is $\mathcal{B}(\mathbb{R}) = \sigma(\{(a, b] : a < b\})$. It is the smallest sigma-field containing all intervals of the form $(a, b]$.

The sigma-fields on Ω have a natural (partial) ordering: Two sigma-fields $\mathcal{G}$ and $\mathcal{F}$ of Ω are such that $\mathcal{G} \subseteq \mathcal{F}$ if all events in $\mathcal{G}$ are in $\mathcal{F}$. For example, the trivial sigma-field $\mathcal{F}_0$ is contained in all sigma-fields of Ω. Clearly, the sigma-field $\mathcal{F}_t = \sigma(X_s, s \leq t)$ is contained in $\mathcal{F}_{t'}$ if $t \leq t'$.

If all the events pertaining to a random variable X are in the sigma-field $\mathcal{G}$, we will say that X is $\mathcal{G}$-*measurable*. This means that all information about X is contained in $\mathcal{G}$. More formally, we have

Definition 4.11. Let X be a random variable defined on $(\Omega, \mathcal{F}, \mathbf{P})$. Consider another $\mathcal{G} \subseteq \mathcal{F}$. Then X is said to be $\mathcal{G}$-*measurable* if and only if

$$\{\omega : X(\omega) \in (a, b]\} \in \mathcal{G} \text{ for all intervals } (a, b] \text{ in } \mathbb{R}.$$

Note that this definition mirrors the formal definition of random variables given in Remark 1.13.

Example 4.12 ($\mathcal{F}_0$-measurable random variables). Consider the trivial sigma-field $\mathcal{F}_0$ = $\{\emptyset, \Omega\}$. A random variable that is $\mathcal{F}_0$-measurable must be a constant. Indeed, we have that for any interval $(a, b]$, $\{\omega : X(\omega) \in (a, b]\} = \emptyset$ or $\{\omega : X(\omega) \in (a, b]\} = \Omega$. This can only hold if X takes a single value.

Example 4.13 ($\sigma(X)$-measurable random variables). Let X be a given random variable on $(\Omega, \mathcal{F}, \mathbf{P})$. Roughly speaking, a $\sigma(X)$-measurable random variable is determined by the information of X only. Here is the simplest example of a $\sigma(X)$-measurable random variable. Take the indicator function $Y = \mathbf{1}_{\{X \in B\}}$ for some event $\{X \in B\}$ pertaining to X. Then the pre-images $\{\omega : Y(\omega) \in (a, b]\}$ are either $\emptyset$, $\{X \in B\}$, $\{X \in B^c\}$, or Ω, depending if $0, 1$ are in $(a, b]$ or not. All of these events are in $\sigma(X)$. More generally, one can construct a $\sigma(X)$-measurable random variable by taking linear combinations of indicator functions of events of the form $\{X \in B\}$. It turns out that any (Borel measurable) function of X can be approximated by taking limits of such *simple* functions. Concretely, this translates into the following statement:

(4.7)   If Y is $\sigma(X)$-measurable, then $Y = g(X)$ for some function g.

In the same way, if Z is $\sigma(X, Y)$-measurable, then $Z = h(X, Y)$ for some function h. These facts can be proved rigorously using measure theory.

We are ready to give the general definition of conditional expectation.

Definition 4.14 (Conditional expectation). Let Y be an integrable random variable on $(\Omega, \mathcal{F}, \mathbf{P})$, and let $\mathcal{G} \subseteq \mathcal{F}$ be a sigma-field of Ω. The *conditional expectation of Y given $\mathcal{G}$* is the random variable denoted by $\mathbf{E}[Y|\mathcal{G}]$ such that the following hold:

(A) $\mathbf{E}[Y|\mathcal{G}]$ is $\mathcal{G}$-measurable.
    In other words, all events pertaining to the random variable $\mathbf{E}[Y|\mathcal{G}]$ are in $\mathcal{G}$.

(B) For any (bounded) random variable W that is $\mathcal{G}$-measurable,

$$\mathbf{E}[WY] = \mathbf{E}\big[W \, \mathbf{E}[Y|\mathcal{G}]\big].$$

In other words, $\mathbf{E}[Y|\mathcal{G}]$ is a proxy for Y as far as events in $\mathcal{G}$ are concerned.

Note that by taking $W = 1$ in property (B), we recover

(4.8) $$\mathbf{E}[\mathbf{E}[Y|\mathcal{G}]] = \mathbf{E}[Y].$$

Remark 4.15. Beware of the notation! If $\mathcal{G} = \sigma(X)$, then the conditional expectation $\mathbf{E}[Y|\sigma(X)]$ is usually denoted by $\mathbf{E}[Y|X]$ for short. However, one should always keep in mind that conditioning on X is in fact projecting on the linear subspace *generated by all variables constructed from X* and not on the linear space generated by X alone. In the same way, the conditional expectation $\mathbf{E}[Z|\sigma(X,Y)]$ is often written $\mathbf{E}[Z|X,Y]$ for short.

As expected, if Y is in $L^2(\Omega, \mathcal{F}, \mathbf{P})$, then $\mathbf{E}[Y|\mathcal{G}]$ is given by the orthogonal projection of Y onto the subspace $L^2(\Omega, \mathcal{G}, \mathbf{P})$, the subspace of square-integrable random variables that are $\mathcal{G}$-measurable. We write $Y^\star$ for the random variable in $L^2(\Omega, \mathcal{G}, \mathbf{P})$ that is the closest to Y; that is,

(4.9) $$\min_{Z \in L^2(\Omega, \mathcal{G}, \mathbf{P})} \mathbf{E}[(Y - Z)^2] = \mathbf{E}[(Y - Y^\star)^2].$$

Theorem 4.16 (Existence and uniqueness of conditional expectation). *Let $\mathcal{G} \subseteq \mathcal{F}$ be a sigma-field of Ω. Let Y be a random variable in $L^2(\Omega, \mathcal{F}, \mathbf{P})$. Then the conditional expectation $\mathbf{E}[Y|\mathcal{G}]$ is the random variable $Y^\star$ given in equation (4.9). Namely, it is the random variable in $L^2(\Omega, \mathcal{G}, \mathbf{P})$ that is the closest to Y in the L^2-distance. In particular, we have the following:*

- *It is the orthogonal projection of Y onto $L^2(\Omega, \mathcal{G}, \mathbf{P})$; i.e., $Y - Y^\star$ is orthogonal to any random variables in $L^2(\Omega, \mathcal{G}, \mathbf{P})$.*

- *It is unique.*

Again the result should be interpreted as follows: The conditional expectation $\mathbf{E}[Y|\mathcal{G}]$ is the best approximation of Y given the information included in $\mathcal{G}$.

Proof. The proof closely follows the one of Theorem 4.7 and is left as an exercise; see Exercise 4.3.   □

Remark 4.17. The conditional expectation in fact exists and is unique for any integrable random variable Y (i.e., $Y \in L^1(\Omega, \mathcal{F}, \mathbf{P})$) as Definition 4.14 suggests. However, there is no orthogonal projection in L^1, so the intuitive geometric picture is lost.

Example 4.18 (Conditional expectation for Gaussian vectors. II). Consider the Gaussian vector $(X_1, \ldots, X_n)$. Without loss of generality, suppose it has mean 0 and is non-degenerate. What is the best approximation of X_n given the information of $X_1, \ldots, X_{n-1}$? In other words, what is

$$\mathbf{E}[X_n|\sigma(X_1, \ldots, X_{n-1})]?$$

With Example 4.13 in mind, let's write $\mathbf{E}[X_n|X_1, \ldots, X_{n-1}]$ for short. From Example 4.4, we expect that there are $a_1, \ldots, a_{n-1}$ such that

$$\mathbf{E}[X_n|X_1, \ldots, X_{n-1}] = a_1 X_1 + \cdots + a_{n-1} X_{n-1}.$$

In particular, since the conditional expectation is a linear combination of the X's, it is itself a Gaussian random variable. The best way to find the coefficient a's is to go back to the IID decomposition of Gaussian vectors (Proposition 2.15). Let $(Z_1, \ldots, Z_{n-1})$ be

IID standard Gaussians constructed from linear combinations of $(X_1, \ldots, X_{n-1})$. Then we have

$$\mathbf{E}[X_n | X_1, \ldots, X_{n-1}] = b_1 Z_1 + \cdots + b_{n-1} Z_{n-1},$$

where $b_j = \mathbf{E}[X_n Z_j]$, $j \leq n-1$. Note that $X_n - \mathbf{E}[X_n | X_1, \ldots, X_{n-1}]$ is then independent of $X_1, \ldots, X_{n-1}$. Therefore, property (B) in Definition 4.14 holds. Property (A) is clear.

See Exercises 4.5, 4.6, 4.8 for more practice on Gaussian conditioning.

Properties of conditional expectation. We now list the properties of conditional expectation that follow from the two defining properties (A), (B) in Definition 4.14. They will be very important when doing explicit computations with martingales. A good way to remember them is to understand how they relate to the interpretation of conditional expectation as an orthogonal projection onto a subspace or, equivalently, as the best approximation of the variable given the information available.

Proposition 4.19. *Let Y be an integrable random variable on $(\Omega, \mathcal{F}, \mathbf{P})$. Let $\mathcal{G} \subseteq \mathcal{F}$ be another sigma-field of Ω. Then the conditional expectation $\mathbf{E}[Y|\mathcal{G}]$ has the following properties:*

(1) *If Y is $\mathcal{G}$-measurable, then*

$$\mathbf{E}[Y|\mathcal{G}] = Y.$$

(2) *More generally, if Y is $\mathcal{G}$-measurable and X is another integrable random variable (with XY also integrable), then*

$$\mathbf{E}[XY|\mathcal{G}] = Y\,\mathbf{E}[X|\mathcal{G}].$$

This makes sense since Y is determined by $\mathcal{G}$; it can be treated as a constant for the conditional expectation.

(3) *If Y is independent of $\mathcal{G}$, that is, for any events $\{Y \in (a, b]\}$ and $A \in \mathcal{G}$,*

$$\mathbf{P}(\{Y \in I\} \cap A) = \mathbf{P}(\{Y \in I\})\,\mathbf{P}(A),$$

then

$$\mathbf{E}[Y|\mathcal{G}] = \mathbf{E}[Y].$$

In other words, if you have no information on Y, your best guess for its value is simply the plain expectation.

(4) **Linearity:** *Let X be another integrable random variable on $(\Omega, \mathcal{F}, \mathbf{P})$. Then*

$$\mathbf{E}[aX + bY|\mathcal{G}] = a\mathbf{E}[X|\mathcal{G}] + b\mathbf{E}[Y|\mathcal{G}], \quad \text{for any } a, b \in \mathbb{R}.$$

The linearity justifies the cumbersome choice of notation $\mathbf{E}[Y|\mathcal{G}]$ for the random variable.

(5) **Tower property:** *If $\mathcal{H} \subseteq \mathcal{G}$ is another sigma-field of Ω, then*

$$\mathbf{E}[Y|\mathcal{H}] = \mathbf{E}\big[\,\mathbf{E}[Y|\mathcal{G}]\,|\mathcal{H}\big].$$

Think in terms of two successive projections: first on a plane, then on a line in the plane.

(6) **Pythagoras's theorem.** *We have*

$$\mathbf{E}[Y^2] = \mathbf{E}[(\mathbf{E}[Y|\mathcal{G}])^2] + \mathbf{E}[(Y - \mathbf{E}[Y|\mathcal{G}])^2].$$

In particular,

(4.10) $$\mathbf{E}[\mathbf{E}[Y|\mathcal{G}]^2] \leq \mathbf{E}[Y^2].$$

In words, the L^2-norm of $\mathbf{E}[X|\mathcal{G}]$ is smaller than the one of X, which is clear if you think in terms of orthogonal projection.

(7) **Expectation of condition expectation.**

$$\mathbf{E}[\mathbf{E}[Y|\mathcal{G}]] = \mathbf{E}[Y].$$

Proof. The uniqueness property of conditional expectation in Theorem 4.16 might appear to be an academic curiosity. *Au contraire!* It is very practical since it ensures that if we find a candidate for the conditional expectation that has the two properties in Definition 4.14, then it must be *the* conditional expectation. To see this, let's prove property (1). Properties (2) to (5) are left for the reader to prove in Exercise 4.9. To prove property (1), it suffices to show that Y has the two defining properties (A), (B) of the conditional expectation $\mathbf{E}[Y|\mathcal{G}]$. Property (A) is clear since X is $\mathcal{G}$-measurable by assumption. Property (B) is also a triviality. Therefore, X is the right candidate.

Property (6) follows from the orthogonal decomposition

$$Y = \mathbf{E}[Y|\mathcal{G}] + (Y - \mathbf{E}[Y|\mathcal{G}]).$$

Property (B) in Definition 4.14 implies that the two terms on the right side are orthogonal. Then, as in Pythagoras's theorem, we have by developing the square

$$\mathbf{E}[Y^2] = \mathbf{E}[(\mathbf{E}[Y|\mathcal{G}])^2] + \mathbf{E}[(Y - \mathbf{E}[Y|\mathcal{G}])^2].$$

This yields the desired inequality.

Property (7) is equation (4.8). It is simply the defining property (B) of conditional expectation applied to $W = 1$. $\qquad\square$

Example 4.20 (Brownian conditioning. II). We continue Example 4.5. Let's now compute the conditional expectations $\mathbf{E}[e^{aB_1}|B_{1/2}]$ and $\mathbf{E}[e^{aB_{1/2}}|B_1]$, for some parameter a. We shall need the properties of conditional expectation in Proposition 4.19. For the first one, we use the fact that $B_{1/2}$ is independent of $B_1 - B_{1/2}$ to get

$$\mathbf{E}[e^{aB_1}|B_{1/2}] = e^{aB_{1/2}}\mathbf{E}[e^{a(B_1-B_{1/2})}|B_{1/2}] = e^{aB_{1/2}}\mathbf{E}[e^{a(B_1-B_{1/2})}] = e^{aB_{1/2}+a^2/4},$$

where we use property (2) in the first equality, property (3) in the second. The third equality is from the MGF of a Gaussian. The answer itself has the form of the MGF of a Gaussian with mean $B_{1/2}$ and variance $1/2$. In fact, this shows that the conditional distribution of B_1 given $B_{1/2}$ is Gaussian of mean $B_{1/2}$ and variance $1/2$. We will go back to conditional distributions in Chapter 8.

For the other expectation, note that $B_{1/2} - \frac{1}{2}B_1$ is independent of B_1. (Why?) Therefore, we have

$$\mathbf{E}[e^{aB_{1/2}}|B_1] = \mathbf{E}[e^{a(B_{1/2}-\frac{1}{2}B_1)+\frac{a}{2}B_1}|B_1] = e^{\frac{a}{2}B_1}e^{a^2/8},$$

because $B_{1/2} - \frac{1}{2}B_1$ is Gaussian of mean 0 and variance 1/4. Therefore the conditional distribution of $B_{1/2}$ given B_1 is Gaussian of mean $\frac{1}{2}B_1$ and variance 1/4.

Example 4.21 (Brownian bridge is conditioned Brownian motion). It was shown in Exercise 2.14 that

$$M_t = B_t - tB_1 \text{ is independent of } B_1 \text{ for } t \in [0,1].$$

We use this to show that the conditional distribution of Brownian motion given the value at the endpoint B_1 is the one of a Brownian bridge shifted by the straight line going from 0 to B_1. To see this, we compute the conditional MGF of $(B_{t_1}, \ldots, B_{t_n})$ given B_1 for some arbitrary choices of time $t_1, \ldots, t_n$ in $[0,1]$. We get the following by adding and subtracting $t_j B_1$:

$$\mathbf{E}[e^{a_1 B_{t_1} + \cdots + a_n B_{t_n}} | B_1] = e^{a_1 t_1 B_1 + \cdots + a_n t_n B_1} \mathbf{E}[e^{a_1 M_{t_1} + \cdots + a_n M_{t_n}}].$$

To get the equality, we used property (1) and property (3) (since M_t is independent of B_1 for all t). The right side is exactly the MGF of the process $(M_t + tB_1, t \in [0,1])$ (for a fixed value B_1), where $(M_t, t \in [0,1])$ is a Brownian bridge. This proves the claim.

Equation (4.10) in Proposition 4.19 is a particular case of *Jensen's inequality* with the convex function $c(x) = x^2$.

Lemma 4.22 (Jensen's inequality). *If c is a convex function on $\mathbb{R}$ and X is a random variable on $(\Omega, \mathcal{F}, \mathbf{P})$, then*

$$\mathbf{E}[c(X)] \geq c(\mathbf{E}[X]).$$

More generally, if $\mathcal{G} \subseteq \mathcal{F}$ is a sigma-field, then

$$\mathbf{E}[c(X)|\mathcal{G}] \geq c(\mathbf{E}[X|\mathcal{G}]).$$

We will not prove Jensen's inequality here, though it is not very hard. To remember which way the inequality goes, notice that the inequality is the same as taking the absolute value $c(x) = |x|$ inside a sum or an integral.

Example 4.23 (Embeddings of L^p spaces). In Section 2.4, we claimed that square-integrable random variables are in fact integrable. In other words, there is always the inclusion $L^2(\Omega, \mathcal{F}, \mathbf{P}) \subseteq L^1(\Omega, \mathcal{F}, \mathbf{P})$. In particular, square-integrable random variables always have a well-defined variance. This embedding is a simple consequence of Jensen's inequality since

$$\mathbf{E}[|X|]^2 \leq \mathbf{E}[X^2].$$

By taking the square root on both sides, we get $\|X\|_1 \leq \|X\|_2$. This also implies that if a sequence of random variables converges in L^2, then it also converge in L^1. More generally, for any $1 < p < \infty$, we can define $L^p(\Omega, \mathcal{F}, \mathbf{P})$ to be the linear space of random variables such that $\mathbf{E}[|X|^p] < \infty$. Then, for $p < q$, we get by Jensen's inequality

$$\mathbf{E}[|X|^q] \geq (\mathbf{E}[|X|^p])^{q/p},$$

since the function $c(x) = x^{q/p}$ is convex for $q/p \geq 1$. So if $X \in L^q$, then it must also be in L^p. Concretely, this means that any random variable with a finite q-moment will also have a finite p-moment, for $q > p$.

4.3. Martingales

We now have all the tools to define martingales.

Definition 4.24. A *filtration* $(\mathcal{F}_t, t \geq 0)$ of Ω is an increasing collection of sigma-fields of Ω; i.e.,

$$\mathcal{F}_s \subseteq \mathcal{F}_t, \quad s \leq t.$$

We will usually take $\mathcal{F}_0 = \{\emptyset, \Omega\}$. The canonical example of a filtration is the *natural filtration of a given process* $(M_s, s \geq 0)$. This is the filtration given by $\mathcal{F}_t = \sigma(M_s, s \leq t)$, as in Example 4.10(iv). The inclusions of the sigma-fields are then clear. For a given Brownian motion $(B_t, t \geq 0)$, the filtration $\mathcal{F}_t = \sigma(B_s, s \leq t)$ is sometimes called the *Brownian filtration*. We think of the filtration as the *flow of information of the process*.

Definition 4.25. A process $(M_t, t \geq 0)$ is a martingale for the filtration $(\mathcal{F}_t, t \geq 0)$ if the following hold:

(1) The process is *adapted*; i.e., M_t is $\mathcal{F}_t$-measurable for all $t \geq 0$.
   (The process is automatically adapted if we take the natural filtration of the process.)

(2) $\mathbf{E}[|M_t|] < \infty$ for all $t \geq 0$.
   (This ensures that the conditional expectation is well-defined.)

(3) *Martingale property*:

$$\mathbf{E}[M_t | \mathcal{F}_s] = M_s \text{ for any } s \leq t.$$

Roughly speaking, this means that the best approximation of the process at a future time t is its value at the present time s.

In particular, the martingale property implies

(4.11) $\mathbf{E}[M_t] = \mathbf{E}[M_0]$ for any time t.

Usually we take $\mathcal{F}_0$ to be the trivial sigma-field $\{\emptyset, \Omega\}$, so that M_0 is constant; see Definition 4.11. In this case, $\mathbf{E}[M_t] = M_0$ for all t. If properties (1) and (2) are satisfied but the best approximation is larger, $\mathbf{E}[M_t | \mathcal{F}_s] \geq M_s$, the process is called a *submartingale*. If it is smaller on average, $\mathbf{E}[M_t | \mathcal{F}_s] \leq M_s$, we say that it is a *supermartingale* .

We will mostly be interested in martingales that are *continuous* and *square-integrable*. Continuous martingales are martingales whose paths $t \mapsto M_t(\omega)$ are continuous on a set of outcomes ω of probability one. Square-integrable martingales are such that $\mathbf{E}[|M_t|^2] < \infty$ for all t's. This condition is stronger than $\mathbf{E}[|M_t|] < \infty$ due to Jensen's inequality.

Remark 4.26 (Martingales in discrete time). Martingales can be defined the same way if the index set of the process is discrete. For example, the filtration $(\mathcal{F}_n, n = 0, 1, 2, \dots)$ can be indexed by integers and the martingale property is then replaced by $\mathbf{E}[M_{n+1} | \mathcal{F}_n] = M_n$ as expected. The tower property then yields the martingale property $\mathbf{E}[M_{n+k} | \mathcal{F}_n] = M_n$ for $k \geq 1$.

Remark 4.27 (Continuous filtrations). Filtrations with continuous time can be tricky to handle rigorously. For example, one has to make sense of what it means for $\mathcal{F}_s$ as s approaches t from the left. Is this equal to $\mathcal{F}_t$? Or is there actually less information in $\lim_{s \to t^-} \mathcal{F}_s$ than in $\mathcal{F}_t$? This is a bit of a headache when dealing with processes with jumps, like the Poisson process. However, if the paths are continuous, the technical problems are not as heavy.

Let's look at some important examples of martingales constructed from Brownian motion. The reader is invited to sample paths of these martingales in Numerical Project 4.1.

Example 4.28 (Examples of Brownian martingales).

(i) *Standard Brownian motion.* Let $(B_t, t \geq 0)$ be a standard Brownian motion, and let $(\mathcal{F}_t, t \geq 0)$ be the *Brownian filtration*. Then $(B_t, t \geq 0)$ is a square-integrable martingale for the filtration $(\mathcal{F}_t, t \geq 0)$. Properties (1) and (2) of Definition 4.25 are obvious. As for the martingale property, note that by the properties of conditional expectation in Proposition 4.19, we have

$$\mathbf{E}[B_t|\mathcal{F}_s] = \mathbf{E}[B_s + B_t - B_s|\mathcal{F}_s] = B_s + \mathbf{E}[B_t - B_s] = B_s.$$

(ii) *Geometric Brownian motion.* Let $(B_t, t \geq 0)$ be a standard Brownian motion, and let $\mathcal{F}_t = \sigma(B_s, s \leq t)$. A *geometric Brownian motion* is a process $(S_t, t \geq 0)$ defined by

$$S_t = S_0 \exp(\sigma B_t + \mu t),$$

for some parameter $\sigma > 0$ and $\mu \in \mathbb{R}$. This is simply the exponential of Brownian motion with drift in Example 2.26. This is not a martingale for most choices of μ! In fact, one must take

$$\mu = \frac{-1}{2}\sigma^2,$$

for the process to be a martingale for the Brownian filtration. Let's verify this. Property (1) is obvious since S_t is a function of B_t for each t. Moreover, property (2) is clear: $\mathbf{E}[\exp(\sigma B_t + \mu t)] = e^{\mu t + \frac{\sigma^2}{2}t}$. As for the martingale property, note that by the properties of conditional expectation and the MGF of Gaussians, we have for $s \leq t$

$$\mathbf{E}[S_t|\mathcal{F}_s] = S_0 e^{\sigma B_s - \frac{\sigma^2}{2}t}\mathbf{E}[e^{\sigma(B_t - B_s)}] = e^{\sigma B_s - \frac{\sigma^2}{2}t + \frac{\sigma^2}{2}(t-s)} = S_s.$$

We will sometimes abuse terminology and refer to the martingale case of geometric Brownian motion simply as geometric Brownian motion when the context is clear.

(iii) *The square of Brownian motion, compensated.* It is easy to check that $(B_t^2, t \geq 0)$ is a submartingale by direct computation using increments or by Jensen's inequality: $\mathbf{E}[B_t^2|\mathcal{F}_s] > B_s^2$, $s < t$. It is nevertheless possible to *compensate* to get a martingale:

$$M_t = B_t^2 - t.$$

The reader must verify that $(M_t, t \geq 0)$ is a martingale for the Brownian filtration in Exercise 4.10.

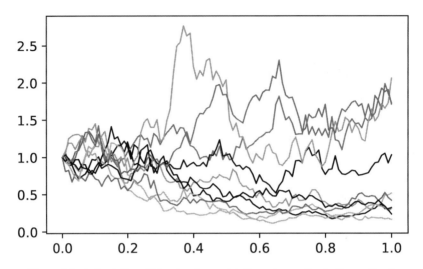

Figure 4.3. A simulation of 10 paths of geometric Brownian motion in the martingale case $\mu = \frac{-1}{2}\sigma^2$. Note that a lot of paths go to 0. This is explained in Exercise 4.13.

Example 4.29 (Other important martingales).

(i) *Symmetric random walks.* This is an example of a martingale with discrete time. Take $(X_i, i \in \mathbb{N})$ to be IID random variables with $\mathbf{E}[X_1] = 0$ and $\mathbf{E}[|X_1|] < \infty$. Take $\mathcal{F}_n = \sigma(X_i, i \leq n)$ and

$$S_n = X_1 + \cdots + X_n, \quad S_0 = 0.$$

It is easily checked that $(S_n, n \geq 0)$ is a martingale for $(\mathcal{F}_n, n \geq 0)$. A simple symmetric random walk is the particular case when the increments are distributed according to $\mathbf{P}(X_1 = 1) = \mathbf{P}(X_1 = -1) = 1/2$. Note that Brownian motion at integer times is a random walk with standard Gaussian increments.

(ii) *Compensated Poisson process.* Let $(N_t, t \geq 0)$ be a Poisson process with rate λ and $\mathcal{F}_t = \sigma(N_s, s \leq t)$. Then $(N_t, t \geq 0)$ is a submartingale for its natural filtration. Again, properties (1), (2) are easily checked. The submartingale property follows by independence of increments: For $s \leq t$,

$$\mathbf{E}[N_t | \mathcal{F}_s] = \mathbf{E}[N_s + N_t - N_s | \mathcal{F}_s] = N_s + \lambda(t - s) \geq N_s.$$

More importantly, we note that we get a martingale by slightly modifying the process. Indeed, if we subtract λt, we have that the process

$$M_t = N_t - \lambda t$$

is a martingale. This is called the *compensated Poisson process.* See Figure 4.4.

We saw in two examples that, even though a process is not itself a martingale, we can sometimes *compensate* to obtain a martingale! Itô calculus will greatly extend this perspective. We will have systematic rules that show when a function of Brownian motion is a martingale and, if not, how to modify it to get one.

For now, we observe that a convex function of a martingale is always a submartingale by Jensen's inequality.

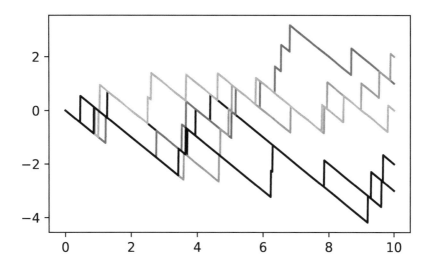

Figure 4.4. A simulation of 10 paths of the compensated Poisson process on $[0,10]$.

Corollary 4.30. *If c is a convex function on $\mathbb{R}$ and $(M_t, t \geq 0)$ is a martingale for $(\mathcal{F}_t, t \geq 0)$, then the process $(c(M_t), t \geq 0)$ is a submartingale for the same filtration, granted that $\mathbf{E}[|c(M_t)|] < \infty$.*

Proof. The fact that $c(M_t)$ is adapted to the filtration is clear since it is an explicit function of M_t. The integrability is by assumption. The submartingale property is direct from Lemma 4.22:

$$\mathbf{E}[c(M_t)|\mathcal{F}_s] \geq c(\mathbf{E}[M_t|\mathcal{F}_s]) = c(M_s). \qquad \square$$

Remark 4.31 (The Doob-Meyer decomposition). It turns out that any continuous submartingale $(X_t, t \geq 0)$ for a filtration $(\mathcal{F}_t, t \geq 0)$ can be written as a sum of two processes: a martingale $(M_t, t \geq 0)$ and an adapted increasing process $(A_t, t \geq 0)$:

$$(4.12) \qquad X_t = M_t + A_t.$$

This is called the *Doob-Meyer decomposition of the process*. In the case of $X_t = B_t^2$, the increasing process was simply the deterministic function t. Such a decomposition is important in applications as $(A_t, t \geq 0)$ is a process with bounded variation (since it is increasing), which is typically much smoother than a martingale. We will see similar decompositions in Chapters 5 and 7.

Example 4.32. Square of a Poisson process. Let $(N_t, t \geq 0)$ be a Poisson process with rate λ. We consider the compensated Poisson process $M_t = N_t - \lambda t$. By Corollary 4.30, the process $(M_t^2, t \geq 0)$ is a submartingale for the filtration $(\mathcal{F}_t, t \geq 0)$ of the Poisson process. How should we compensate M_t^2 to get a martingale? A direct computation using the properties of conditional expectation yields

$$\begin{aligned}
\mathbf{E}[M_t^2|\mathcal{F}_s] &= \mathbf{E}[(M_t - M_s + M_s)^2|\mathcal{F}_s] \\
&= M_s^2 + 2M_s\mathbf{E}[M_t - M_s|\mathcal{F}_s] + \mathbf{E}[(M_t - M_s)^2|\mathcal{F}_s] \\
&= M_s^2 + \mathbf{E}[(M_t - M_s)^2].
\end{aligned}$$

Moreover, since the variance and the mean of N_t are both λt, we have

$$\begin{aligned}
\mathbf{E}[(M_t - M_s)^2] &= \mathbf{E}[(N_t - N_s)^2] - 2\lambda(t - s)\mathbf{E}[N_t - N_s] + \lambda^2(t - s)^2 \\
&= \lambda^2(t - s)^2 + \lambda(t - s) - 2\lambda^2(t - s)^2 + \lambda^2(t - s)^2 \\
&= \lambda(t - s).
\end{aligned}$$

We conclude that the process $(M_t^2 - \lambda t, t \geq 0)$ is a martingale. The Doob-Meyer decomposition of the submartingale M_t^2 is then

$$M_t^2 = (M_t^2 - \lambda t) + \lambda t.$$

Remark 4.33 (Martingale vs Markov). We will study the Markov property in Chapter 8 . It is good to take a quick look at it now, however, to compare it to the martingale property. Some processes such as Brownian motions possess both properties, but in general the following is true:

There are martingales that are not Markov processes, and there are Markov processes that are not martingales.

Consider a process $(X_t, t \geq 0)$ and its natural filtration $(\mathcal{F}_t, t \geq 0)$. A process $(X_t, t \geq 0)$ has the *Markov property* if the conditional distribution of X_t given $\mathcal{F}_s$ depends only on X_s (and on the time s, t) for any time $s \leq t$, but not on the process prior to time s. The conditional distribution can be read off the conditional MGF $\mathbf{E}[e^{aX_t} | \mathcal{F}_s]$. With this in mind, Brownian motion has the Markov property since for any time $s \leq t$,

(4.13) $$\mathbf{E}[e^{aB_t} | \mathcal{F}_s] = e^{aB_s + \frac{a^2}{2}(t-s)}.$$

This is an explicit function of B_s and the times s, t. In fact, from the form of the MGF, we infer that the conditional distribution of B_t given $\mathcal{F}_s$ is Gaussian of mean B_s and variance $t - s$.

The Brownian motion with drift $X_t = B_t + t$, $t \geq 0$, is also a Markov process. A similar calculation to the above shows that the conditional distribution of X_t given $\mathcal{F}_s$ is Gaussian of mean $X_s + t - s$ and variance $t - s$. Again, this depends explicitly on the position X_s at time s (and the times s, t). However, it is easy to check that the process is not a martingale since $\mathbf{E}[X_t] = t$, violating equation (4.11). In Chapter 5, we will construct martingales using the Itô integral. Such martingales will not be Markov in general; see Example 5.17.

4.4. Computations with Martingales

Martingales are not only conceptually interesting, they are formidable tools to compute probabilities and expectations of processes. For example, in this section, we will solve the *gambler's ruin problem* for Brownian motion. For convenience, we introduce the notion of *stopping time* before doing so.

Definition 4.34. A random variable τ on Ω is said to be a *stopping time* for the filtration $(\mathcal{F}_t, t \geq 0)$ if

the event $\{\omega : \tau(\omega) \leq t\} \in \mathcal{F}_t$ for any $t \geq 0$.

Note that since $\mathcal{F}_t$ is a sigma-field, we must also have that $\{\omega : \tau(\omega) > t\} \in \mathcal{F}_t$.

In words, τ is a stopping time if we can decide if the events $\{\tau \leq t\}$ occurred or not based on the information available at time t.

Example 4.35 (Examples of stopping times).

(i) *First passage time.* This is the first time when a process reaches a certain value. To be precise, let $X = (X_t, t \geq 0)$ be a process and $(\mathcal{F}_t, t \geq 0)$ its natural filtration. For $a > 0$, we define the first passage time at a to be

$$\tau(\omega) = \min\{s \geq 0 : X_s(\omega) \geq a\}.$$

If the path ω never reaches a, we set $\tau(\omega) = \infty$. Now, for t fixed and for a given path $X(\omega)$, it is possible to know if $\tau(\omega) \leq t$ (the path has reached a before time t) or $\tau(\omega) > t$ (the path has not reached a before time t) with the information available at time t, since we are looking at the first time the process reaches a. We conclude that τ is a stopping time.

(ii) *Hitting time.* More generally, we can consider the first time (if ever) that the path of a process $(X_t, t \geq 0)$ enters or hits a subset B of $\mathbb{R}$,

$$\tau(\omega) = \min\{s \geq 0 : X_s \in B\}.$$

The first passage time is the particular case in which $B = [a, \infty)$.

(iii) *Minimum of two stopping times.* If τ and τ' are two stopping times for the same filtration $(\mathcal{F}_t, t \geq 0)$, then so is the minimum $\tau \wedge \tau'$ between the two, where $(\tau \wedge \tau')(\omega) = \tau(\omega) \wedge \tau'(\omega)$. This is because for any $t \geq 0$,

$$\{\omega : (\tau \wedge \tau')(\omega) \leq t\} = \{\omega : \tau(\omega) \leq t\} \cup \{\omega : \tau'(\omega) \leq t\}.$$

Since the right side is the union of two events in $\mathcal{F}_t$, it must also be in $\mathcal{F}_t$ by the properties of a sigma-field. We conclude that $\tau \wedge \tau'$ is itself a stopping time. Is it also the case that the maximum $\tau \vee \tau'$ is a stopping time?

It is good to keep the following example in mind.

Example 4.36 (Last passage time is NOT a stopping time). What if we look at the *last* time that the process reaches a, that is,

$$\rho(\omega) = \max\{t \geq 0 : X_t(\omega) \geq a\}?$$

This is a well-defined random variable, but it is not a stopping time. Based on the information available at time t, we are not able to decide whether or not $\rho(\omega) \leq t$ or not, as the path could reach a after time t.

It turns out that a martingale that is stopped when the stopping time is attained remains a martingale.

Proposition 4.37 (Stopped martingale). *If $(M_t, t \geq 0)$ is a continuous martingale for the filtration $(\mathcal{F}_t, t \geq 0)$ and τ is a stopping time for the same filtration, then the stopped process defined by*

$$M_{t \wedge \tau} = \begin{cases} M_t & \text{for } t \leq \tau, \\ M_\tau & \text{for } t \geq \tau, \end{cases}$$

is also a continuous martingale for the same filtration.

Proof. There is a simple proof of this using martingale transform; see Exercise 5.1.   □

We saw in equation (4.11) that the expectation of a martingale is the same at all times. Does this property extend to stopping time? That is, do we always have

$$\mathbf{E}[M_\tau] = \mathbf{E}[M_0] \quad \text{for a stopping time } \tau?$$

The answer cannot be yes in general, as we will see shortly in Example 4.44. If some other conditions are fulfilled, the result still holds.

Corollary 4.38 (Doob's optional stopping theorem). *If $(M_t, t \geq 0)$ is a continuous martingale for the filtration $(\mathcal{F}_t, t \geq 0)$ and τ is a stopping time such that $\tau < \infty$ and the stopped process $(M_{\tau \wedge t}, t \geq 0)$ is bounded, then*

$$\mathbf{E}[M_\tau] = \mathbf{E}[M_0].$$

Proof. Since $(M_{t \wedge \tau}, t \geq 0)$ is a martingale, we always have

$$\mathbf{E}[M_{t \wedge \tau}] = \mathbf{E}[M_0].$$

Now since $\tau(\omega) < \infty$, we must have that with probability one, $\lim_{t \to \infty} M_{t \wedge \tau} = M_\tau$. In particular, we have

$$\mathbf{E}[M_\tau] = \mathbf{E}\left[\lim_{t \to \infty} M_{t \wedge \tau}\right] = \lim_{t \to \infty} \mathbf{E}[M_{t \wedge \tau}] = \mathbf{E}[M_0]. \qquad \square$$

Remark 4.39. When can we interchange a limit and an expectation as we did above? We already saw one instance of this in Theorem 3.15. We state another set of sufficient conditions for this to hold.

Theorem 4.40 (Dominated convergence theorem). *If $(X_n, n \geq 1)$ is a sequence of random variables such that $\lim_{n \to \infty} X_n$ exists almost surely and there exists a random variable Y such $\mathbf{E}[|Y|] < \infty$ with $|X_n| \leq Y$ for all n, then $\lim_{n \to \infty} \mathbf{E}[X_n] = \mathbf{E}[\lim_{n \to \infty} X_n]$.*

An important particular case of the above, sometimes called the *bounded convergence theorem*, is when the variables X_n are uniformly bounded by a constant C, so that the dominating random variable Y can be taken to be a constant. In retrospect, we could interchange the limit and the expectation in the proof of Corollary 4.38 because the process $(M_{t \wedge \tau}, t \geq 0)$ is bounded. The theorem also shows that convergence almost surely implies convergence in probability: If $(X_n \geq 1)$ is a sequence of random variables converging to X almost surely, then for any $\delta > 0$,

$$(4.14) \qquad \lim_{n \to \infty} \mathbf{P}(|X_n - X| > \delta) = \lim_{n \to \infty} \mathbf{E}[\mathbf{1}_{\{|X_n - X| > \delta\}}] = \mathbf{E}\left[\lim_{n \to \infty} \mathbf{1}_{\{|X_n - X| > \delta\}}\right] = 0,$$

where we use the fact that the random variable $\mathbf{1}_{\{|X_n - X| > \delta\}}$ is bounded by 1. There are other variations of the optional stopping theorem with different assumptions on the process. They all reduce to conditions for taking limits inside expectations in the proof.

We are ready to dive into one of the most important computations in this book.

Example 4.41. Gambler's ruin with Brownian motion The *gambler's ruin problem* is known in different forms. Roughly speaking, it refers to the problem of computing the probability of a gambler making a series of bets reaching a certain amount before going broke. In terms of Brownian motion (and stochastic processes in general), it translates into the following questions: Let $(B_t, t \geq 0)$ be a standard Brownian motion starting at $B_0 = 0$ and $a, b > 0$.

(1) What is the probability that a Brownian path reaches a before $-b$?

(2) What is the expected waiting time for the path to reach a or $-b$?

For the first question, it is a simple computation using stopping time and martingale properties. Define the hitting time

$$\tau(\omega) = \min\{t \geq 0 : B_t(\omega) \geq a \text{ or } B_t(\omega) \leq -b\}.$$

Note that τ is the minimum between the first passage time at a and the one at $-b$.

We first show that $\tau < \infty$ with probability one. In other words, all Brownian paths reach a or $-b$ eventually. To see this, consider the event that the n-th increment exceeds $a + b$:

$$E_n = \{|B_n - B_{n-1}| > a + b\}, \quad n \geq 1.$$

Note that if E_n occurs, then we must have that the Brownian path exits the interval $[-b, a]$. Moreover, we have $\mathbf{P}(E_n) = \mathbf{P}(E_1)$ for all n. Call this probability p. Clearly, $0 < p < 1$. Since the events E_n are independent, we have

$$\mathbf{P}(E_1^c \cap \cdots \cap E_n^c) = (1 - p)^n.$$

As $n \to \infty$, we have $\lim_{n \to \infty} \mathbf{P}(E_1^c \cap \cdots \cap E_n^c) = 0$. The events $F_n = E_1^c \cap \cdots \cap E_n^c$ are decreasing in n. By the continuity of the probability (Lemma 1.4), we conclude that

$$\mathbf{P}\left(\bigcap_{n \geq 1} F_n\right) = 0.$$

Therefore, it must be that E_n occurs for some n, so all Brownian paths reach a or $-b$ eventually.

Since $\tau < \infty$ with probability one, the random variable B_τ is well-defined: $B_\tau(\omega) = B_t(\omega)$ if $\tau(\omega) = t$. It can only take two values: a or $-b$. Question (1) above translates into computing $\mathbf{P}(B_\tau = a)$. On one hand, we have

$$\mathbf{E}[B_\tau] = a\mathbf{P}(B_\tau = a) - b(1 - \mathbf{P}(B_\tau = a)).$$

On the other hand, by Corollary 4.38, we have $\mathbf{E}[B_\tau] = \mathbf{E}[B_0] = 0$. (Note that the stopped process $(B_{t \wedge \tau}, t \geq 0)$ is bounded by a above and by $-b$ below.) Putting these two observations together we get

$$\mathbf{P}(B_\tau = a) = \frac{b}{a + b}.$$

A very simple and elegant answer!

We will revisit this problem again and again. In particular, we will answer the questions above for Brownian motion with drift in Chapter 5.

Example 4.42 (Expected waiting time). Let τ be as in the last example. We now answer question (2) of the gambler's ruin problem:

$$\mathbf{E}[\tau] = ab.$$

Note that the expected waiting time is consistent with the rough heuristic that Brownian motion travels a distance $\sqrt{t}$ by time t. We now use the martingale $M_t = B_t^2 - t$ of Example 4.28. On one hand, if we apply optional stopping in Corollary 4.38, we get

$$\mathbf{E}[M_\tau] = M_0 = 0.$$

Moreover, we know the distribution of B_τ, thanks to the probability calculated in the last example. We can therefore compute $\mathbf{E}[M_\tau]$ direcly:

$$0 = \mathbf{E}[M_\tau] = \mathbf{E}[B_\tau^2 - \tau] = a^2 \cdot \frac{b}{a+b} + b^2 \cdot \frac{a}{a+b} - \mathbf{E}[\tau].$$

This implies $\mathbf{E}[\tau] = ab$.

Why can we apply optional stopping here? The random variable τ is finite with probability one as before. However, the stopped martingale $M_{t\wedge\tau}$ is not necessarily bounded: $B_{t\wedge\tau}$ is bounded but τ is not. However, the conclusion of optional stopping still holds. Indeed, we have

$$\mathbf{E}[M_{t\wedge\tau}] = \mathbf{E}[B_{t\wedge\tau}^2] - \mathbf{E}[t\wedge\tau].$$

By the bounded convergence theorem (Remark 4.39) we get $\lim_{t\to\infty} \mathbf{E}[B_{t\wedge\tau}^2] = \mathbf{E}[B_\tau^2]$. By the monotone convergence theorem (Theorem 3.15), we get $\lim_{t\to\infty} \mathbf{E}[t\wedge\tau] = \mathbf{E}[\tau]$. This is because $t\wedge\tau$ is increasing in t and $t\wedge\tau \to \tau$ almost surely as $\tau < \infty$.

Example 4.43 (First passage time of Brownian motion). We can use the previous two examples to get some very interesting information on the first passage time

$$\tau_a = \min\{t \geq 0 : B_t \geq a\}.$$

Let $\tau = \tau_a \wedge \tau_{-b}$ be as in the previous examples with $\tau_{-b} = \min\{t \geq 0 : B_t \leq -b\}$. Note that τ_{-b} is a sequence of random variables that is increasing in b. Moreover, we have $\tau_{-b} \to \infty$ almost surely as $b \to \infty$. (Why?) Moreover, the event $\{B_\tau = a\}$ is the same as $\{\tau_a < \tau_{-b}\}$. Now the events $\{\tau_a < \tau_{-b}\}$ are increasing in b, since if a path reaches a before $-b$, it will do so as well for a more negative value of $-b$. On one hand, this means by continuity of the probability (Lemma 1.4) that

$$\lim_{b\to\infty} \mathbf{P}(\tau_a < \tau_{-b}) = \mathbf{P}(\tau_a < \infty).$$

On the other hand, we have by Example 4.41

$$\lim_{b\to\infty} \mathbf{P}(\tau_a < \tau_{-b}) = \lim_{b\to\infty} \mathbf{P}(B_\tau = a) = \lim_{b\to\infty} \frac{b}{a+b} = 1.$$

We just showed that

(4.15) $$\mathbf{P}(\tau_a < \infty) = 1.$$

In other words, every Brownian path will reach a, no matter how large a is!

How long will it take to reach a on average? Well, we know from Example 4.42 that $\mathbf{E}[\tau_a \wedge \tau_{-b}] = ab$. On one hand, this means

$$\lim_{b\to\infty} \mathbf{E}[\tau_a \wedge \tau_{-b}] = \infty.$$

On the other hand, since the random variables τ_{-b} are increasing,

$$\lim_{b\to\infty} \mathbf{E}[\tau_a \wedge \tau_{-b}] = \mathbf{E}\left[\lim_{b\to\infty} \tau_a \wedge \tau_{-b}\right] = \mathbf{E}[\tau_a],$$

by Theorem 3.15. Here we also used that $\lim_{b\to\infty} \tau_{-b} = \infty$ almost surely and $\tau_a(\omega) < \infty$, from equation (4.15). We just proved that

$$\mathbf{E}[\tau_a] = \infty!$$

In other words, any Brownian path will reach a, but the expected waiting time for this to occur is infinite, no matter how small a is! What is happening here? No matter how small a is, there will always be paths that reach very large negative values before hitting a. These paths might be unlikely. However, the first passage time τ_a is so large for these paths that they affect the value of the expectation substantially. In other words, τ_a is a *heavy-tailed random variable* (flashback to Examples 1.26 and 1.34). We will describe the distribution of τ_a in more detail in the next section.

Example 4.44 (When optional stopping fails). Consider τ_a, the first passage time at $a > 0$. The random variable B_{τ_a} is well-defined since $\tau_a < \infty$. In fact, we have $B_{\tau_a} = a$ with probability one. Therefore, the following must hold:

$$\mathbf{E}[B_{\tau_a}] = a \neq B_0.$$

Corollary 4.38 does not apply here since the stopped process $(B_{t\wedge\tau_a}, t \geq 0)$ is not bounded.

4.5. Reflection Principle for Brownian Motion

It is an amazing fact that simple manipulations using stopping time yield the complete distribution of the first passage time τ_a of Brownian motion as well as the distribution for the maximum of a Brownian path on an interval of time. This is surprising since the maximum on $[0, T]$ denoted by $\max_{0\leq t\leq T} B_t$ is a random variable that depends on the whole path on $[0, T]$. This beautiful result is due to Bachelier.

Proposition 4.45 (Bachelier's formula). *Let $(B_t, t \leq T)$ be a standard Brownian motion on $[0, T]$. Then the CDF of the random variable $\max_{0\leq t\leq T} B_t$ is*

$$\mathbf{P}\left(\max_{0\leq t\leq T} B_t \leq a\right) = \mathbf{P}(|B_T| \leq a), \quad \text{for any } a \geq 0.$$

In particular, its PDF is

$$f_{\max}(a) = \frac{2}{\sqrt{2\pi T}} e^{-\frac{a^2}{2T}}.$$

In other words, the random variable $\max_{0\leq t\leq T} B_t$ has the same distribution as $|B_T|$.

Numerical Project 4.2 verifies this result empirically. Note that the paths of the random variables $\max_{0\leq s\leq t} B_s$ and $|B_t|$ are very different as t varies for a given ω: One is increasing, and the other is not. The equality holds in distribution for a fixed t. As a bonus corollary, we get the distribution of the first passage time at a.

Corollary 4.46. *Let $a \geq 0$ and $\tau_a = \min\{t \geq 0 : B_t \geq a\}$. Then*

$$\mathbf{P}(\tau_a \leq T) = \mathbf{P}\Big(\max_{0 \leq t \leq T} B_t \geq a \Big) = \int_a^\infty \frac{2}{\sqrt{2\pi T}} e^{-\frac{x^2}{2T}} \, dx.$$

In particular, the random variable τ_a has PDF

$$f_{\tau_a}(t) = \frac{a}{\sqrt{2\pi}} \frac{e^{-\frac{a^2}{2t}}}{t^{3/2}}, \quad t > 0.$$

This implies that it is heavy-tailed with $\mathbf{E}[\tau_a] = \infty$.

Proof. The maximum on $[0, T]$ is larger than or equal to a if and only if $\tau_a \leq T$. There-fore, the events $\{\max_{0 \leq t \leq T} B_t \geq a\}$ and $\{\tau_a \leq T\}$ are the same. So the CDF $\mathbf{P}(\tau_a \leq t)$ of τ_a is $\int_a^\infty \frac{2}{\sqrt{2\pi t}} e^{-\frac{x^2}{2t}} \, dx$ by Proposition 4.45. To get the PDF, it remains to differentiate the integral with respect to t. This is easy to do once we realize by a change of variable $u = x/\sqrt{t}$ that

$$\int_a^\infty \frac{2}{\sqrt{2\pi t}} e^{-\frac{x^2}{2t}} \, dx = \int_{a/\sqrt{t}}^\infty \frac{2}{\sqrt{2\pi}} e^{-\frac{u^2}{2}} \, du.$$

The derivative of the above with respect to t is $\frac{a}{\sqrt{2\pi}} \frac{e^{-\frac{a^2}{2t}}}{t^{3/2}}$. To estimate the expectation, it suffices to realize that for $t \geq 1$, $e^{-\frac{a^2}{2t}}$ is larger than $e^{-a^2/2}$. Therefore, we have

$$\mathbf{E}[\tau_a] = \int_0^\infty t \frac{a}{\sqrt{2\pi}} \frac{e^{-\frac{a^2}{2t}}}{t^{3/2}} \, dt \geq \frac{ae^{-a^2/2}}{\sqrt{2\pi}} \int_1^\infty t^{-1/2} \, dt.$$

The integral diverges like $\sqrt{t}$ and is infinite, as claimed. $\square$

To prove Proposition 4.45, we will need an important property of Brownian motion called the *reflection principle*. To motivate it, recall that the reflection symmetry of Brownian motion at time s in Proposition 3.3. It turns out that this reflection property holds if s is replaced by a stopping time. We will prove this in Chapter 8 (Theorem 8.6).

Lemma 4.47 (Reflection principle). *Let $(B_t, t \geq 0)$ be a standard Brownian motion and let τ be a stopping time for its filtration. Then the process $(\widetilde{B}_t, t \geq 0)$ defined by the reflection at time τ*

$$\widetilde{B}_t = \begin{cases} B_t & \text{if } t \leq \tau, \\ B_\tau - (B_t - B_\tau) & \text{if } t > \tau \end{cases}$$

is also a standard Brownian motion.

With this new tool, we can prove Proposition 4.45.

Proof of Proposition 4.45. Consider $\mathbf{P}(\max_{t \leq T} B_t \geq a)$. By splitting this probability over the event of the endpoint, we have

$$\mathbf{P}\Big(\max_{t \leq T} B_t \geq a \Big) = \mathbf{P}\Big(\max_{t \leq T} B_t \geq a, B_T > a \Big) + \mathbf{P}\Big(\max_{t \leq T} B_t \geq a, B_T \leq a \Big).$$

Note also that $\mathbf{P}(B_T = a) = 0$. Hence, the first probability equals $\mathbf{P}(B_T \geq a)$. As for the second, consider the time τ_a. On the event considered, we have $\tau_a \leq T$ and using Lemma 4.47 at that time, we get

$$\mathbf{P}\left(\max_{t \leq T} B_t \geq a, B_T \leq a\right) = \mathbf{P}\left(\max_{t \leq T} B_t \geq a, \widetilde{B}_T \geq a\right).$$

Observe that the event $\{\max_{t \leq T} B_t \geq a\}$ is the same as $\{\max_{t \leq T} \widetilde{B}_t \geq a\}$. (A rough picture might help here.) Therefore, the above probability is

$$\mathbf{P}\left(\max_{t \leq T} B_t \geq a, B_T \leq a\right) = \mathbf{P}\left(\max_{t \leq T} \widetilde{B}_t \geq a, \widetilde{B}_T \geq a\right) = \mathbf{P}\left(\max_{t \leq T} B_t \geq a, B_T \geq a\right),$$

where the last equality follows by the reflection principle. But as above, the last probability is equal to $\mathbf{P}(B_T \geq a)$. We conclude that

$$\mathbf{P}\left(\max_{t \leq T} B_t \geq a\right) = 2\mathbf{P}(B_T \geq a) = \mathbf{P}(|B_T| \geq a), \quad a \geq 0.$$

This implies in particular that $\mathbf{P}(\max_{t \leq T} B_t = a) = 0$. (Why ?) Thus we also have $\mathbf{P}(\max_{t \leq T} B_t \leq a) = \mathbf{P}(|B_T| \leq a)$ as claimed. $\qquad \square$

The same reasoning gives the joint distribution between the running maximum of Brownian motion and the endpoint for free,

$$\mathbf{P}\left(\max_{t \leq T} B_t > m, B_T \leq a\right) = \mathbf{P}(B_T > 2m - a),$$

for $m > 0$ and $a \leq m$; see Exercise 4.17. Direct differentiation gives the joint PDF between the two:

(4.16) $$f(m, a) = \frac{2(2m - a)}{T^{3/2}\sqrt{2\pi}} e^{\frac{-(2m-a)^2}{2T}}.$$

4.6. Numerical Projects and Exercises

4.1. **Simulating martingales.** Sample 10 paths of the following processes with a step size of 0.01:

   (a) $B_t^2 - t, t \in [0, 1]$.

   (b) Geometric Brownian motion: $S_t = \exp(B_t - t/2), t \in [0, 1]$.

   (c) Compensated Poisson process: $N_t - t, t \in [0, 10]$, where $(N_t, t \geq 0)$ is a Poisson process of rate 1.

4.2. **Maximum of Brownian motion.** Consider the maximum of Brownian motion on $[0, 1]$: $\max_{s \leq 1} B_s$.

   (a) Draw the histogram of the random variable $\max_{s \leq 1} B_s$ using 10,000 sampled Brownian paths with a step size of 0.01.

   (b) Compare this to the PDF of the random variable $|B_1|$.

4.3. **First passage time.** Let $(B_t, t \geq 0)$ be a standard Brownian motion. Consider the random variable

$$\tau = \min\{t \geq 0 : B_t \geq 1\}.$$

This is the first time that B_t reaches 1.

   (a) Draw a histogram for the distribution of $\tau \wedge 10$ on the time interval $[0, 10]$ using 10,000 Brownian paths on $[0, 10]$ with discretization 0.01.
   The notation $\tau \wedge 10$ means that if the path does not reach 1 on $[0, 10]$, then give the value 10 to the stopping time.

   (b) Estimate $\mathbf{E}[\tau \wedge 10]$.

   (c) What proportion of paths never reach 1 in the time interval $[0, 10]$?

4.4. **Gambler's ruin at the French roulette.** Consider the scenario in which you are gambling \$1 at the French roulette on the reds: You gain \$1 with probability 18/38, and you lose a dollar with probability 20/38. We estimate the probability of your fortune reaching \$200 before it reaches 0.

   (a) Write a function that samples simple random walk path from time 0 to time 5,000 with a given starting point. See Example 4.29 with $\mathbf{P}(X = +1) = 18/38$ and $\mathbf{P}(X = -1) = 20/38$.

   (b) Use the above to estimate the probability of reaching \$200 before \$0 on a sample of 100 paths if you start with \$100.

   (c) Draw the graph of the probability of your wealth reaching \$200 as a function of your starting fortune S_0, $S_0 = 1, \ldots, 199$. See Figure 4.5.

See Exercise 4.14 for the theoretical point of view on this.

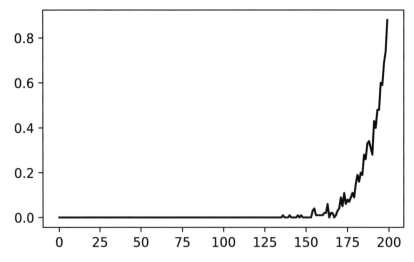

Figure 4.5. The approximate probability of reaching \$200 before \$0 as a function of the starting amount. Anybody feel like gambling anymore?

Exercises

4.1. Conditional expectation of continuous random variables. Let (X, Y) be two random variables with joint density $f(x, y)$ on $\mathbb{R}^2$. Suppose for simplicity that $\int_{\mathbb{R}} f(x, y)\, dx > 0$ for every $y \in \mathbb{R}$. Show that the conditional expectation $\mathbf{E}[Y|X]$ equals $h(X)$ where h is the function

$$(4.17) \qquad h(x) = \frac{\int_{\mathbb{R}} y f(x, y)\, dy}{\int_{\mathbb{R}} f(x, y)\, dy}.$$

In particular, verify that $\mathbf{E}[\mathbf{E}[Y|X]] = \mathbf{E}[Y]$.

Hint: To prove this, verify that the above formula satisfies both properties of the conditional expectation; then invoke uniqueness to finish it off.

4.2. Exercises on sigma-fields.

(a) Let A, B be two proper subsets of Ω such that $A \cap B \neq \emptyset$ and $A \cup B \neq \Omega$. Write down $\sigma(\{A, B\})$, the smallest sigma-field containing A and B, explicitly. What if $A \cap B = \emptyset$?

(b) The Borel sets defined in Example 4.10 form the sigma-field $\mathcal{B}(\mathbb{R})$, the smallest sigma-field containing intervals of the form $(a, b]$. Show that all singletons $\{b\}$ are in $\mathcal{B}(\mathbb{R})$ by writing $\{b\}$ as a countable intersection of intervals $(a, b]$. Conclude that all open intervals (a, b) and all closed intervals $[a, b]$ are in $\mathcal{B}(\mathbb{R})$. Is the subset $\mathbb{Q}$ of rational numbers a Borel set?

4.3. Proof of Theorem 4.16. Prove Theorem 4.16 using the proof of Theorem 4.7.

4.4. Another look at conditional expectation for Gaussians. Let (X, Y) be a Gaussian vector with mean 0 and covariance matrix

$$\mathcal{C} = \begin{pmatrix} 1 & \rho \\ \rho & 1 \end{pmatrix},$$

for $\rho \in (-1, 1)$. We verify that Example 4.4 and Exercise 4.1 yield the same conditional expectation.

(a) Use equation (4.5) to show that $\mathbf{E}[Y|X] = \rho X$.

(b) Write down the joint PDF $f(x, y)$ of (X, Y).

(c) Show that $\int_{\mathbb{R}} y f(x, y)\, dy = \rho x$ and that $\int_{\mathbb{R}} f(x, y)\, dy = 1$.

(d) Deduce that $\mathbf{E}[Y|X] = \rho X$ using equation (4.17).

4.5. Gaussian conditioning. We consider the Gaussian vector (X_1, X_2, X_3) with mean 0 and covariance matrix

$$\mathcal{C} = \begin{pmatrix} 2 & 2 & 0 \\ 2 & 4 & 0 \\ 0 & 0 & 1 \end{pmatrix}.$$

(a) Is the vector nondegenerate?

(b) Argue that X_3 is independent of X_2 and X_1.

(c) Compute $\mathbf{E}[X_2|X_1]$. Write X_2 as a linear combination of X_1 and a random variable independent of X_1.

(d) Compute $\mathbf{E}[e^{aX_2}|X_1]$ for any $a \in \mathbb{R}$. Use this to determine the conditional distribution of X_2 given X_1.

(e) Find (Z_1, Z_2, Z_3) IID standard Gaussians that are linear combinations of (X_1, X_2, X_3).

4.6. **Gaussian conditioning.** A Gaussian random vector (X, Y) has mean 0 and co-variance matrix given by

$$\mathcal{C} = \begin{pmatrix} 3/16 & 1/8 \\ 1/8 & 1/4 \end{pmatrix}.$$

(a) Argue briefly that the vector is nondegenerate.

(b) Without much calculation, argue that $\mathbf{E}[Y|X] = \frac{2}{3}X$.

(c) Find a random variable W such that $Y = W + \frac{2}{3}X$ and W is independent of X. Argue that the conditional distribution of Y given X is Gaussian with mean $\frac{2}{3}X$ and variance $\frac{1}{6}$.

(d) Find (Z_1, Z_2) independent standard Gaussians such that they are linear combinations of (X, Y).

Hint: Part (c) *helps.*

4.7. **Gaussian conditioning.** Consider the Gaussian process (X_1, X_2) of mean 0 and covariance

$$\mathcal{C} = \begin{pmatrix} 1 & -1 \\ -1 & 2 \end{pmatrix}.$$

(a) Find IID standard Gaussians (Z_1, Z_2) that are linear combinations of (X_1, X_2).

(b) Write down (X_1, X_2) in terms of (Z_1, Z_2).

(c) Compute $\mathbf{E}[X_2|X_1]$.

(d) Compute $\mathbf{E}[e^{aX_2}|X_1]$ for $a \in \mathbb{R}$. What is the conditional distribution of X_2 given X_1?

4.8. **Gaussian conditioning.** Consider the Gaussian vector (X_1, X_2, X_3) of mean 0 and covariance

$$\begin{pmatrix} 2 & 1 & 1 \\ 1 & 2 & 1 \\ 1 & 1 & 2 \end{pmatrix}.$$

Compute $\mathbf{E}[X_3|X_1, X_2]$ and $\mathbf{E}[e^{aX_3}|X_1, X_2]$. What is the conditional distribution of X_3 given X_1 and X_2?

It might be helpful to work out Exercise 2.10 first.

4.9. **Properties of conditional expectation.** Prove properties (2) to (5) in Proposition 4.19 by showing that the expression on the right has the two defining properties of the conditional expectation given on the left.

4.10. **Square of Brownian motion.** Let $(B_t, t \geq t)$ be a standard Brownian motion. Verify that $M_t = B_t^2 - t$ is a martingale for the Brownian filtration.

4.11. **Geometric Poisson process.** Let $(N_t, t \geq 0)$ be a Poisson process of intensity λ. For $\alpha > 0$, prove that the process $(e^{\alpha N_t - \lambda t(e^{\alpha} - 1)}, t \geq 0)$ is a martingale for the filtration of the Poisson process $(N_t, t \geq 0)$.

4.12. **Another Brownian martingale.** Let $(B_t, t \geq 0)$ be a standard Brownian motion. Consider for $a, b > 0$ the stopping time

$$\tau = \min_{t \geq 0}\{t : B_t \geq a \text{ or } B_t \leq -b\}.$$

In this exercise, we get information about the joint distribution of (τ, B_τ).
   (a) Show that $M_t = tB_t - \frac{1}{3}B_t^3$ is a martingale for the Brownian filtration.
   (b) Use (a) to show that

$$\mathbf{E}[\tau B_\tau] = \frac{ab}{3}(a - b).$$

   (c) Argue that for any $a > 0$ we have $\mathbf{E}[e^{aB_\tau - a^2\tau/2}] = 1$.
   (d) Find another proof of (b) using (c).
      Hint: Taylor series.

4.13. **Limit of geometric Brownian motion.** Consider a geometric Brownian motion $(S_t, t \geq 0)$ given by

$$S_t = S_0 \exp(\sigma B_t + \mu t), \quad t \geq 0,$$

where $\sigma > 0$ and $\mu < 0$.
   (a) Prove that $\lim_{t \to \infty} S_t = 0$ almost surely.
      Hint: Feel free to use the conclusion of Exercise 4.20 below.
   (b) Does it also converge to 0 in the L^2-sense? What about in the L^1-sense?

4.14. **Gambler's ruin at the French roulette.** Let $(S_n, n \geq 0)$ be a simple random walk with bias starting at $S_0 = 100$ with

$$S_n = S_0 + X_1 + \cdots + X_n,$$

where $\mathbf{P}(X_1 = +1) = p$ and $\mathbf{P}(X_1 = -1) = 1 - p = q$ with $p < 1/2$ (so S_n decreases on average).
   (a) Prove that $M_n = (q/p)^{S_n}$ is a martingale for the filtration $(\mathcal{F}_n, n \in \mathbb{N})$ where $\mathcal{F}_n = \sigma(X_m, m \leq n)$.
   (b) Define the stopping time $\tau = \min\{n \geq 0 : S_n = 200 \text{ or } S_n = 0\}$. Argue briefly that $\tau < \infty$ with probability one.
   (c) Use Corollary 4.38 to show that

$$\mathbf{P}(S_\tau = 200) = \frac{1 - (q/p)^{100}}{1 - (q/p)^{200}}.$$

   Corollary 4.38 also applies to discrete time.
   (d) Compute $\mathbf{P}(S_\tau = 200)$ for $p = 18/38$.
   (e) Estimate the value of S_0 for which the probability of reaching 200 equals the one of reaching 100.

4.15. **La martingale classique.** Consider the random walk with $S_0 = 0$ and $S_n = X_1 + \cdots + X_n$ where $(X_n, n \geq 1)$ are IID random variables that take the values ± 1 with probability 1/2. *La martingale classique* is in the betting system consisting of doubling the bet at every round. With this in mind, we consider the process given by

$$M_n = 1 \cdot (S_1 - S_0) + 2 \cdot (S_2 - S_1) + \cdots + 2^n \cdot (S_n - S_{n-1}).$$

We will see in the next chapter that $(M_n, n \geq 1)$ is a martingale transform and is also a martingale. Consider the stopping time τ, the first time m with $X_m = +1$.

  (a) Show that $\mathbf{E}[M_\tau] = 1$, yet $\mathbf{E}[M_0] = 0$.

  This shows that this betting strategy is always a winner.

  (b) Why doesn't optional stopping apply here?

  (c) What is the weakness of this strategy?

4.16. **A martingale from conditional expectation.** Let $X \in L^1(\Omega, \mathcal{F}, \mathbf{P})$, and let $(\mathcal{F}_t, t \geq 0)$ be a filtration. Prove that the process $(M_t, t \geq 0)$ defined by the conditional expectation

$$M_t = \mathbf{E}[X | \mathcal{F}_t]$$

is a martingale for the filtration.

4.17. **Joint distribution of** $(\max_{t \leq T} B_t, B_T)$. Prove equation (4.16) by revisiting the proof of Proposition 4.45.

4.18. ⋆ **Zeros of Brownian motion.** For any $t > 0$, prove that $\mathbf{P}(\max_{s \leq t} B_s > 0) = \mathbf{P}(\min_{s \leq t} B_s < 0) = 1$. Conclude from this that, for any $t \geq 0$, the continuous function $s \mapsto B_s(\omega)$ has infinitely many zeros in the interval $[0, t]$ for ω in a set of probability one. (In other words, 0 is an accumulation point for the set of zeros of the function.)

4.19. ⋆ **Doob's maximal inequalities.** We prove the following: Let $(M_k, k \geq 1)$ be a positive submartingale for the filtration $(\mathcal{F}_k, k \in \mathbb{N})$. Then for any $1 \leq p < \infty$ and $a > 0$

$$(4.18) \qquad \mathbf{P}\left(\max_{k \leq n} M_k > a\right) \leq \frac{1}{a^p} \mathbf{E}[M_n^p].$$

  (a) Use Jensen's inequality to show that if $(M_k, k \geq 1)$ is a positive submartingale, then so is $(M_k^p, k \geq 1)$ for $1 \leq p < \infty$. Conclude that it suffices to prove the statement for $p = 1$.

  (b) Consider the events

$$B_k = \bigcap_{j < k} \{\omega : M_j(\omega) \leq a\} \cap \{\omega : M_k(\omega) > a\}.$$

  Argue that the B_k's are disjoint and that $\bigcup_{k \leq n} B_k = \{\max_{k \leq n} M_k > a\} = B$.

  (c) Show that

$$\mathbf{E}[M_n] \geq \mathbf{E}[M_n \mathbf{1}_B] \geq a \sum_{k \leq n} \mathbf{P}(B_k) = a\mathbf{P}(B)$$

  by decomposing B in B_k's and by using the properties of expectations, as well as the submartingale property.

  (d) Argue that the inequality holds for continuous martingales with continuous paths by discretizing time and using convergence theorems: If $(M_t, t \geq 0)$ is a positive submartingale with continuous paths for the filtration $(\mathcal{F}_t, t \geq 0)$, then for any $1 \leq p < \infty$ and $a > 0$

$$\mathbf{P}\left(\max_{s \leq t} M_s > a\right) \leq \frac{1}{a^p} \mathbf{E}[M_t^p].$$

4.20. ⋆ **An application of Doob's maximal inequalities.** The goal is to show that for a Brownian motion $(B_t, t \geq 0)$ we have

$$\lim_{t \to \infty} \frac{B_t}{t} = 0 \quad \text{almost surely.}$$

(a) Show using the strong law of large numbers that $\lim_{n \to \infty} \frac{B_n}{n} = 0$ almost surely, where n is an integer.

(b) Use Doob's maximal inequality to show that for any $\delta > 0$,

$$\sum_{n \geq 0} \mathbf{P}\left(\max_{0 \leq s \leq 1} |B_{n+s} - B_n| > \delta n\right) < \infty.$$

(c) Deduce that

$$\lim_{n \to \infty} \max_{0 \leq s \leq 1} \frac{|B_{n+s} - B_n|}{n} = 0 \quad \text{almost surely.}$$

(d) Conclude from the above that for any increasing sequence $t_n \uparrow \infty$ we have $\lim_{n \to \infty} \frac{B_{t_n}}{t_n} = 0$ as desired.

(e) Go back to the time-inversion $X_t = tB_{1/t}$ of Brownian version in Exercise 3.6. Use the above to prove that $\lim_{t \to 0+} X_t = 0$ almost surely.

4.21. ⋆ **An example of Fubini's theorem.** Let $(X_n, n \geq 1)$ be a sequence of random variables on $(\Omega, \mathcal{F}, \mathbf{P})$. Show that if $\sum_{n \geq 1} \mathbf{E}[|X_n|] < \infty$, then

$$\mathbf{E}\left[\sum_{n \geq 1} X_n\right] = \sum_{n \geq 1} \mathbf{E}[X_n].$$

4.7. Historical and Bibliographical Notes

The term *martingale* originally referred to betting systems, e.g., *la martingale classique* and *la grande martingale*. They are often variations on the strategy consisting of doubling your bet at every round: a sure strategy for anyone with a good lender; see Exercise 4.15. It is very curious that the word martingale also refers to (according to the Merriam-Webster dictionary)

> *a device for steadying a horse's head (...) that typically consists of a strap fastened to the girth, passing between the forelegs, and bifurcating to end in two rings through which the reins pass.*

For an entertaining account of the origin of the word, see [**Man09**]. The first appearance of the word in the mathematical literature goes back to the 1930s in a thesis by Jean Ville [**BSS09**]. The term was adopted by Joseph L. Doob who developed much of the early theory. Bachelier's formula appears on page 75 of *Théorie de la spéculation* [**Bac00**]. It is astonishing that such a formula was found so early in the theory. The gambler's ruin problem is at the root of the birth of probability as a mathematical theory. The problem was posed by Blaise Pascal to Pierre de Fermat around 1656. Pascal, Fermat, and Huygens found different solutions to the problem [**Edw83**]. Numerical Project 4.4 and Exercise 4.14 are a variation on Exercise 1.69 of [**Wal12**] (acknowledging G. Slade).

The convergence theorems of Remark 4.39 played a fundamental role in the development of probability and analysis at the beginning of the twentieth century. They are due to Henri Lebesgue who developed an extension of the Riemann integral that is now known as the *Lebesgue integral*. The advantage of the Lebesgue integral is that it enlarges the class of functions that are integrable. Consider for example the function $f : [0, 1] \to \mathbb{R}$ that is 1 on the rationals and 0 elsewhere. This function does not have a well-defined Riemann integral. (Why?) However, its Lebesgue integral is easy to calculate and is 0 as it should be. Another big advantage of the Lebesgue integral is that its construction generalizes to spaces other than $\mathbb{R}$ and to *measures* of sets other than their length. The expectation on $(\Omega, \mathcal{F}, \mathbf{P})$ is in fact an integral on the space Ω under the measure or probability $\mathbf{P}$. The fact that more functions are integrable imply that the Lebesgue integral is much more robust in interchanging limits and integrals, sums and integrals, and integrals with integrals. This is the content of the convergence theorems. We refer to [**Dur96**] for more details on the theory.

Chapter 5

Itô Calculus

Itô calculus has profound applications in mathematics and mathematical finance. In this chapter, we construct the Itô integral on Brownian paths. The mathematical significance of Itô's work is to have given a rigorous meaning to an integral of random functions (like Brownian motion) whose paths do not have bounded variation; see Corollary 3.17. The construction is done in Section 5.3 as a limit of *martingale transforms*, which is the equivalent of the Riemann sum for stochastic integrals as explained in Section 5.2. One of the upshots is that the Itô integral gives a systematic way to construct Brownian martingales. We also derive Itô's formula in Section 5.4. The formula relates the Itô integral to explicit functions of Brownian motion. As such, it can be considered as the fundamental theorem of Itô calculus. As a point of comparison, it is useful to briefly go back to the integral of standard calculus: the Riemann integral.

5.1. Preliminaries

The construction of the classical Riemann integral goes as follows. Consider, for example, a continuous function g on $[0, t]$. We take a partition of $[0, t]$ in n intervals $(t_j, t_{j+1}]$ with $t_n = t$; for example $t_j = \frac{j}{n}t$. The Riemann integral is understood as the limit of Riemann sums:

$$\int_0^t g(s)\,\mathrm{d}s = \lim_{n\to\infty} \sum_{j=0}^{n-1} g(t_j)(t_{j+1} - t_j).$$

Note that the integral is a number for fixed t. The integral represents the area under the curve given by g on the interval $[0, t]$. It can also be seen as a continuous function of t as t varies on an interval. In fact, as a function of t, the integral is differentiable and its derivative is g. This is the fundamental theorem of calculus.

It is possible to modify the above definition slightly for more general increments. The construction is called the *Riemann-Stieltjes integral*. Let F be a function on $[0, t]$ of bounded variation, as in Example 3.6. It can be shown that the integral as a limit of

Riemann sums with the increments of F exists:

$$(5.1) \qquad \int_0^t g(s)\,dF(s) = \lim_{n \to \infty} \sum_{j=0}^{n-1} g(t_j)(F(t_{j+1}) - F(t_j)).$$

If F is a CDF of a random variable X, then $\int_{-\infty}^{\infty} g(s)\,dF(s)$ represents the expectation of the random variable $g(X)$; see Remark 1.25. Note that $F(t_{j+1}) - F(t_j)$ is the probability that X falls in the interval $(t_j, t_{j+1}]$.

The goal is to make sense of the above when F is replaced by a Brownian motion $(B_t, t \geq 0)$:

$$\int_0^t g(s)\,dB_s = \lim_{n \to \infty} \sum_{j=0}^{n-1} g(t_j)(B_{t_{j+1}} - B_{t_j}).$$

The major hurdle here is not the fact that the Brownian paths are random, but instead that these paths have *unbounded variation*, as proved in Corollary 3.17. This means that the classical construction does not apply for a given path. Therefore, another a priori construction is needed. The Poisson process has paths of bounded variation, as they are increasing. There is no problem in using the classical construction of the integral for Poisson paths.

Note that the sum $\sum_{j=0}^{n-1} g(t_j)(B_{t_{j+1}} - B_{t_j})$ above is a random variable. If the endpoint $t_n = t$ is varied, it can be seen as a stochastic process. Moreover, since the Brownian paths are continuous, this new stochastic process also has continuous paths. As we shall see, this stochastic process is in fact a continuous martingale like Brownian motion. It turns out that these properties remain in the limit $n \to \infty$.

What is the interpretation of the stochastic integral? If we think of $(B_t, t \geq 0)$ as modelling the price of a stock, then $\sum_{j=0}^{n-1} g(t_j)(B_{t_{j+1}} - B_{t_j})$ gives the value of a portfolio at time t that implements the following strategy: At t_j we buy $g(t_j)$ shares of the stock that we sell at time t_{j+1}. We do this for every $j \leq n - 1$. The net gain or loss of this strategy is the sum over j of $g(t_j)(B_{t_{j+1}} - B_{t_j})$. Of course, in this interpretation, the number of shares $g(t_j)$ put in play could be random and depend on the past information of the path up to time t_j.

In the next section, we take a first step towards the Itô integral by defining the *martingale transform*. The construction makes sense for any square-integrable martingale.

5.2. Martingale Transform

Let $(M_t, t \leq T)$ be a continuous square-integrable martingale on $[0, T]$ for the filtration $(\mathcal{F}_t, t \leq T)$, defined on some probability space $(\Omega, \mathcal{F}, \mathbf{P})$. The idea of the martingale transform is to modify the amplitude of each increment in such a way as to produce a martingale when these new increments are summed up. The martingale transforms are to the Itô integral what Riemann sums are for the Riemann integral.

More precisely, let $(t_j, j \leq n)$ be a sequence of partitions of $[0, T]$ with $t_0 = 0$ and $t_n = T$. For example, we can take $t_j = \frac{j}{n}T$. Consider n fixed numbers $(Y_0, Y_1, \ldots, Y_{n-1})$. It is convenient to construct a function of time X_t from these:

$$X_t = Y_j \qquad \text{if } t \in (t_j, t_{j+1}].$$

This can be written also as a sum of indicator functions:

$$(5.2) \qquad X_t = \sum_{j=0}^{n-1} Y_j \mathbf{1}_{(t_j, t_{j+1}]}(t), \quad t \le T.$$

The integral of $(X_t, t \le T)$ with respect to the martingale M on $[0, T]$, also called a *martingale transform*, is the sum of the increments of the martingale modulated by X; i.e.,

$$(5.3) \qquad I_T = Y_0(M_{t_1} - M_0) + \cdots + Y_{n-1}(M_T - M_{t_{n-1}}) = \sum_{j=0}^{n-1} Y_j(M_{t_{j+1}} - M_{t_j}).$$

This is a random variable in $L^2(\Omega, \mathcal{F}, \mathbf{P})$, since it is a linear combination of random variables in L^2. Note that we recover M_T when X_{t_j} is 1 for all intervals. We may think of $(M_t, s \le T)$ as the price of an asset, say a stock, on a time interval $[0, T]$. Then the term

$$Y_j(M_{j+1} - M_{t_j})$$

can be seen as the gain or loss in the time interval $(t_j, t_{j+1}]$ of buying Y_j units of the asset at time t_j at price M_{t_j} and selling these at time t_{j+1} at price $M_{t_{j+1}}$. Summing these terms over time gives the value I_t of implementing the *investing strategy X* on the interval $[0, T]$. It is not hard to modify the definition to obtain a stochastic process on the whole interval $[0, T]$. For $t \le T$, we simply sum the increments up to t. This can be written down as

$$(5.4) \quad I_t = Y_0(M_{t_1} - M_0) + Y_1(M_{t_2} - M_{t_1}) + \cdots + Y_j(M_t - M_{t_j}), \quad \text{if } t \in (t_j, t_{j+1}].$$

Example 5.1 (Integral of a simple process). Consider a standard Brownian motion $(B_t, t \in [0, 1])$ on the time interval $[0, 1]$. We know very well by now that it is a martingale. We look at a simple integral constructed from it. We take the following integrand:

$$X_t = \begin{cases} 10 & \text{if } t \in [0, 1/3], \\ 5 & \text{if } t \in (1/3, 2/3], \\ 2 & \text{if } t \in (2/3, 1]. \end{cases}$$

Then the integrals I_t as in equation (5.4) form a process $(I_t, t \in [0, 1])$ of the form

$$I_t = \begin{cases} 10B_t & \text{if } t \in [0, 1/3], \\ 10B_{1/3} + 5(B_t - B_{1/3}) & \text{if } t \in (1/3, 2/3], \\ 10B_{1/3} + 5(B_{2/3} - B_{1/3}) + 2(B_t - B_{2/3}) & \text{if } t \in (2/3, 1]. \end{cases}$$

We make three important observations. First, the paths of the process $(I_t, t \in [0, 1])$ are continuous, because Brownian paths are. Second, the process is a square-integrable martingale. It is easy to see that it is adapted and square-integrable, because I_t is a sum of square-integrable random variables. The martingale property is also not hard to verify. For example, we have for $t \in (2/3, 1]$,

$$\mathbf{E}[I_t | \mathcal{F}_{2/3}] = 10B_{1/3} + 5(B_{2/3} - B_{1/3}) + 2\mathbf{E}[B_t - B_{2/3} | \mathcal{F}_{2/3}] = I_{2/3},$$

since $\mathbf{E}[B_t - B_{2/3} | \mathcal{F}_{2/3}] = 0$ by the martingale property of Brownian motion. See Exercise 5.2 for more on this example.

We can generalize the integrand or investing strategy X by considering values X_{t_j} that depend on the process, hence are random, but in a *predictable way*. Namely, we can take X to be a random vector such that X_{t_j} is $\mathcal{F}_{t_j}$-measurable. In other words, X_{t_j} may be random but must only depend on the information up to time t_j. Common sense dictates that the number of shares you buy today should not depend on information in the future. With this in mind, for a given filtration, we define the space of *simple* (that is, discrete) *adapted* processes on $[0, T]$:
(5.5)
$$\mathcal{S}(T) = \left\{(X_t, t \leq T) : X_t = \sum_{j=0}^{n-1} Y_j \mathbf{1}_{(t_j, t_{j+1}]}(t), Y_j \text{ is } \mathcal{F}_{t_j}\text{-measurable}, \mathbf{E}[Y_j^2] < \infty \right\}.$$

In words, the processes in $\mathcal{S}(T)$ have paths that are piecewise constant on a finite number of intervals of $[0, T]$. The values $Y_j(\omega)$ on each time interval might vary depending on the paths ω. As random variables, the Y_j's depend only on the information up to time t_j and have finite second moment: $\mathbf{E}[Y_j^2] < \infty$. Note that $\mathcal{S}(T)$ is a linear space: If $X, X' \in \mathcal{S}(T)$, then $aX + bX' \in \mathcal{S}(T)$ for $a, b \in \mathbb{R}$. Indeed, if the paths of X, X' take a finite number of values, then so are the ones of $aX + bX'$.

Example 5.2 (An example of simple adapted process). Let $(B_t, t \leq 1)$ be a standard Brownian motion. For the interval $[0, 1]$, consider the investing strategy X in $\mathcal{S}(1)$ given by the position of the Brownian path at times $0, 1/3, 2/3$:

$$X_s = \begin{cases} 0 & \text{if } s \in [0, 1/3], \\ B_{1/3} & \text{if } s \in (1/3, 2/3], \\ B_{2/3} & \text{if } s \in (2/3, 1]. \end{cases}$$

Clearly, X is simple and adapted to the Brownian filtration. For example, the value at $s = 3/4$ is $B_{2/3}$. In particular, it depends only on the information prior to time $3/4$. See Figure 5.1.

For a simple adapted process X, the integral I_t of X with respect to the martingale $(M_t, t \leq T)$ is the same as in equation (5.4).

Definition 5.3. Let $(M_t, t \leq T)$ be a continuous square-integrable martingale for the filtration $(\mathcal{F}_t, t \leq T)$. Let $X \in \mathcal{S}(T)$ be a simple, adapted process $X = \sum_{j=0}^{n-1} Y_j \mathbf{1}_{(t_j, t_{j+1}]}$ on $[0, T]$. The *martingale transform* I_t is

$$I_t = Y_0(M_{t_1} - M_0) + Y_1(M_{t_2} - M_{t_1}) + \cdots + Y_j(M_t - M_{t_j}), \quad \text{if } t \in (t_j, t_{j+1}].$$

It defines a process $(I_t, t \leq T)$ on $[0, T]$.

Example 5.4 (Another integral of a simple process). Consider the simple process X of Example 5.2 defined on a Brownian motion. The integral of X as a process on $[0, 1]$ is

$$I_s = \begin{cases} 0 & \text{if } s \in [0, 1/3], \\ B_{1/3}(B_s - B_{1/3}) & \text{if } s \in (1/3, 2/3], \\ B_{1/3}(B_{2/3} - B_{1/3}) + B_{2/3}(B_s - B_{2/3}) & \text{if } s \in (2/3, 1]. \end{cases}$$

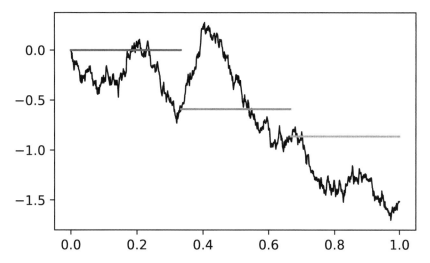

Figure 5.1. The simple process X constructed from a Brownian path in Example 5.2.
Note that the path of X is piecewise constant. However, the value on each piece is
random as it depends on the positions of the Brownian path at time $1/3$ and $2/3$.

As in Example 5.1, the paths of I_s are continuous for all $s \in [0, 1]$, since the paths of B_s
are continuous! This is also true at the integer times $s = 1/3, 2/3$, if we approached on
the left or right. The process $(I_s, s \leq 1)$ is also a martingale for the Brownian filtration.
The key here is that the value multiplying the increment on the interval $(t_j, t_{j+1}]$ is
$\mathcal{F}_{t_j}$-measurable. For example, take $t > 2/3$ and $1/3 < s < 2/3$. The properties of conditional expectation in Proposition 4.19 and the fact that Brownian motion is a martingale
give

$$
\begin{aligned}
\mathbf{E}[I_t|\mathcal{F}_s] &= \mathbf{E}[B_{1/3}(B_{2/3} - B_{1/3}) + B_{2/3}(B_t - B_{2/3})|\mathcal{F}_s] \\
&= \mathbf{E}[B_{1/3}(B_{2/3} - B_{1/3})|\mathcal{F}_s] + \mathbf{E}[B_{2/3}(B_t - B_{2/3})|\mathcal{F}_s] \\
&= B_{1/3}(B_s - B_{1/3}) + \mathbf{E}[\, \mathbf{E}[B_{2/3}(B_t - B_{2/3})|\mathcal{F}_{2/3}] \, |\mathcal{F}_s] \\
&= B_{1/3}(B_s - B_{1/3}) + \mathbf{E}[B_{2/3} \, \mathbf{E}[(B_t - B_{2/3})|\mathcal{F}_{2/3}] \, |\mathcal{F}_s] \\
&= B_{1/3}(B_s - B_{1/3}) + 0 = I_s.
\end{aligned}
$$

Note that it was crucial to use the tower property in the third equality and that we took
out what is known at $t = 2/3$ in the fourth equality.

Martingale transforms are always themselves martingales. In particular, it is not
possible in this setup to design an investing strategy whose value would be increasing
on average.

Proposition 5.5 (Martingale transforms are martingales). *Let $(M_t, t \leq T)$ be a continuous square-integrable martingale for the filtration $(\mathcal{F}_t, t \leq T)$ and let $X \in \mathcal{S}(T)$ be
a simple process as in equation (5.5). Then the martingale transform $(I_t, t \leq T)$ is a
continuous martingale on $[0, T]$ for the same filtration.*

Proof. The fact that I_t is $\mathcal{F}_t$-measurable for $t \leq T$ is clear from the construction in
equation (5.4). Indeed, the increments $M_{t_{j+1}} - M_{t_j}$ are $\mathcal{F}_t$-measurable for $t_{j+1} \leq t$ since

the martingale is adapted. The integrand X is also adapted. Moreover, I_t is integrable since

$$\mathbf{E}[|I_t|] \leq \mathbf{E}[|I_T|] \leq \sum_{j=0}^{n-1} \mathbf{E}[|Y_j||M_{t_{j+1}} - M_{t_j}|] \leq \sum_{j=0}^{n-1} (\mathbf{E}[Y_j^2])^{1/2} (\mathbf{E}[(M_{t_{j+1}} - M_{t_j})^2])^{1/2},$$

by the Cauchy-Schwarz inequality. The last term is finite by assumption on X and M. As for continuity, since $(M_t, t \leq T)$ is continuous, the only possible issue could be at the points t_j for some j. But in that case, we have for $t > t_j$ but close and any outcome ω,

$$I_t(\omega) = \sum_{i=0}^{j-1} Y_i(M_{t_{i+1}}(\omega) - M_{t_i}(\omega)) + Y_j(M_t(\omega) - M_{t_j}(\omega)),$$

which goes to I_{t_j} when $t \to t_j^+$ by continuity of $M_t(\omega)$. A similar argument holds for $t \to t_j^-$.

To prove the martingale property, consider $s < t$. We want to show that $\mathbf{E}[I_t|\mathcal{F}_s] = I_s$. Suppose $t \in (t_j, t_{j+1}]$ for some $t_j < T$. By linearity of conditional expectation in Proposition 4.19, we have

$$(5.6) \qquad \mathbf{E}[I_t|F_s] = \sum_{i=0}^{j} \mathbf{E}[Y_i(M_{t_{i+1}} - M_{t_i})|\mathcal{F}_s],$$

where it is understood that $t = t_{j+1}$ in the above to simplify notation. We can now handle each summand. There are three possibilities: $s \geq t_{i+1}$, $s \in (t_i, t_{i+1})$, and $s < t_i$. It all depends on Proposition 4.19. In the case $s \geq t_{i+1}$, we have

$$\mathbf{E}[Y_i(M_{t_{i+1}} - M_{t_i})|\mathcal{F}_s] = Y_i(M_{t_{i+1}} - M_{t_i}),$$

since the whole summand is $\mathcal{F}_s$-measurable. In the case $s \in (t_i, t_{i+1})$, we have that Y_i is $\mathcal{F}_s$-measurable; therefore

$$\mathbf{E}[Y_i(M_{t_{i+1}} - M_{t_i})|\mathcal{F}_s] = Y_i\mathbf{E}[M_{t_{i+1}} - M_{t_i}|\mathcal{F}_s] = Y_i(M_s - M_{t_i}),$$

by the martingale property. In the case $s < t_i$, we use the tower property to get

$$\mathbf{E}[Y_i(M_{t_{i+1}} - M_{t_i})|\mathcal{F}_s] = \mathbf{E}[\mathbf{E}[Y_i(M_{t_{i+1}} - M_{t_i})|\mathcal{F}_{t_i}]|\mathcal{F}_s]$$
$$= \mathbf{E}[Y_i\mathbf{E}[(M_{t_{i+1}} - M_{t_i})|\mathcal{F}_{t_i}]|\mathcal{F}_s] = 0,$$

since $\mathbf{E}[(M_{t_{i+1}} - M_{t_i})|\mathcal{F}_{t_i}] = 0$ by the martingale property. Putting all the cases together in (5.6) gives for $s \in (t_k, t_{k+1}]$, say,

$$\mathbf{E}[I_t|F_s] = Y_0(M_{t_1} - M_0) + Y_1(M_{t_2} - M_{t_1}) + \cdots + Y_k(M_s - M_{t_k}). \qquad \square$$

5.3. The Itô Integral

We now turn to martingale transforms where the underlying martingale is a standard Brownian motion $(B_t, t \geq 0)$. This gives our first definition of the Itô integral.

Definition 5.6 (Itô integral on $\mathcal{S}(T)$). Let $(B_t, t \leq T)$ be a standard Brownian motion on $[0, T]$ and let $X \in \mathcal{S}(T)$ be a simple process $X = \sum_{j=0}^{n-1} Y_j \mathbf{1}_{(t_j, t_{j+1}]}$ on $[0, T]$ adapted to the Brownian filtration. The Itô integral of X with respect to the Brownian motion is defined as the martingale transform

$$\int_0^T X_s \, dB_s = \sum_{j=0}^{n-1} Y_j (B_{t_{j+1}} - B_{t_j}),$$

and similarly for any $t \leq T$,

$$\int_0^t X_s \, dB_s = Y_0 (B_{t_1} - B_0) + Y_1 (B_{t_2} - B_{t_1}) + \cdots + Y_j (B_t - B_{t_j}), \quad \text{if } t \in (t_j, t_{j+1}].$$

Note again the similarities with Riemann sums. The interpretation of the Itô integral is as follows:

> *the value of implementing the strategy X on the underlying asset with price given by the Brownian motion.*

The martingale transform with Brownian motion has more properties than with a generic martingale as given in Definition 5.3. This is because the Brownian increments are independent. We gather the properties of the Itô integral for $X \in \mathcal{S}(T)$ in an important proposition. The same exact result will hold for continuous strategies; see Theorem 5.12.

Proposition 5.7 (Properties of the Itô integral)**.** *Let $(B_t, t \leq T)$ be a standard Brownian motion on $[0, T]$ defined on a probability space $(\Omega, \mathcal{F}, \mathbf{P})$. The Itô integral in Definition 5.6 has the following properties:*

- **Linearity:** *If $X, X' \in \mathcal{S}(T)$ and $a, b \in \mathbb{R}$, then for all $t \leq T$,*

$$\int_0^t (aX_s + bX'_s) \, dB_s = a \int_0^t X_s \, dB_s + b \int_0^t X'_s \, dB_s.$$

- **Continuous martingale:** *The process $\left(\int_0^t X_s \, dB_s, t \leq T \right)$ is a continuous martingale on $[0, T]$ for the Brownian filtration.*

- **Itô's isometry:** *The random variable $\int_0^t X_s \, dB_s$ is in $L^2(\Omega, \mathcal{F}, \mathbf{P})$ with mean 0 and variance*

$$\mathbf{E}\left[\left(\int_0^t X_s \, dB_s \right)^2 \right] = \int_0^t \mathbf{E}[X_s^2] \, ds = \mathbf{E}\left[\int_0^t X_s^2 \, ds \right], \quad t \leq T.$$

It is very important for the understanding of the theory to keep in mind that $\int_0^t X_s \, dB_s$ is a random variable. We should walk away from the temptation to use the reflexes of classical calculus to manipulate it as if it were a Riemann integral. The reason we use the integral sign to denote the random variable $\int_0^t X_s \, dB_s$ is because it shares the linearity property with the Riemann integral.

It turns out that Itô's isometry not only yields the mean and the variance of the random variable $\int_0^t X_s \, dB_s$, but also the covariances for these random variables at different times, and the covariance for two integrals built with two different strategies on the same Brownian motion; see Corollary 5.15. What about the distribution of $\int_0^t X_s \, dB_s$? It turns out that the random variable $\int_0^t X_s \, dB_s$ is not Gaussian in general. However, if the process X is not random, then it will be; see Corollary 5.18 below. For example, the process $(I_t, t \leq T)$ in Example 5.1 is Gaussian, but the one in Example 5.4 is not.

Proof of Proposition 5.7. The linearity is clear from the definition of the martingale transform. The continuity property and the martingale property follow generally from Proposition 5.5.

We now prove Itô's isometry. We will use the properties of conditional expectation in Proposition 4.19 many times, so the reader might quickly review it beforehand. To simplify notation, for fixed $t \in [0, T]$, we can suppose that the partition $(t_j, j \leq n)$ is a partition of $[0, t]$ with $t_n = t$. Since Y_j is $\mathcal{F}_{t_j}$-measurable, we have

$$\mathbf{E}[Y_j(B_{t_{j+1}} - B_{t_j})] = \mathbf{E}[\mathbf{E}[Y_j(B_{t_{j+1}} - B_{t_j})|\mathcal{F}_{t_j}]] = \mathbf{E}[Y_j \mathbf{E}[B_{t_{j+1}} - B_{t_j}|\mathcal{F}_{t_j}]] = 0,$$

since $\mathbf{E}[B_{t_{j+1}} - B_{t_j}|\mathcal{F}_{t_j}] = 0$, as Brownian motion is a martingale. Therefore, it follows that

$$\mathbf{E}\left[\int_0^t X_s \, dB_s\right] = \sum_{j=0}^{n-1} \mathbf{E}[Y_j(B_{t_{j+1}} - B_{t_j})] = 0.$$

As for the variance, we have by conditioning on $\mathcal{F}_{t_j}$ that, for $t_i < t_j$,

$$\mathbf{E}[Y_j Y_i (B_{t_{j+1}} - B_{t_j})(B_{t_{i+1}} - B_{t_i})] = \mathbf{E}[Y_j Y_i (B_{t_{i+1}} - B_{t_i}) \mathbf{E}[B_{t_{j+1}} - B_{t_j}|\mathcal{F}_{t_j}]] = 0,$$

since $\mathbf{E}[B_{t_{j+1}} - B_{t_j}|\mathcal{F}_{t_j}] = 0$ and since all factors but $B_{t_{j+1}} - B_{t_j}$ are $\mathcal{F}_{t_j}$-measurable. Thus, this yields

$$\mathbf{E}\left[\left(\int_0^t X_s \, dB_s\right)^2\right] = \sum_{i,j=0}^{n-1} \mathbf{E}[Y_j Y_i (B_{t_{j+1}} - B_{t_j})(B_{t_{i+1}} - B_{t_i})]$$

$$= \sum_{j=0}^{n-1} \mathbf{E}[Y_j^2 \, \mathbf{E}[(B_{t_{j+1}} - B_{t_j})^2|\mathcal{F}_{t_j}]],$$

by the previous equation and the fact that Y_j is $\mathcal{F}_{t_j}$-measurable. Since the increment $B_{t_{j+1}} - B_{t_j}$ is independent of $\mathcal{F}_{t_j}$, we have

$$\mathbf{E}[(B_{t_{j+1}} - B_{t_j})^2|\mathcal{F}_{t_j}] = \mathbf{E}[(B_{t_{j+1}} - B_{t_j})^2] = t_{j+1} - t_j.$$

Therefore, we conclude that

$$\mathbf{E}\left[\left(\int_0^t X_s \, dB_s\right)^2\right] = \sum_{j=0}^{n-1} \mathbf{E}[Y_j^2](t_{j+1} - t_j).$$

From the definition of X as a simple process in equation (5.2), we have $\int_0^t \mathbf{E}[X_s^2] \, ds = \sum_{j=0}^{n-1} \mathbf{E}[Y_j^2](t_{j+1} - t_j)$, since X equals Y_j on the whole interval $(t_j, t_{j+1}]$. $\square$

Example 5.8. We go back to the Itô integral in Example 5.2. The mean of I_t is 0 by Proposition 5.7 or by direct computation. It is not hard to compute the variance. For example, at $t = 1$, it is

$$\mathbf{E}[I_1^2] = \int_0^1 \mathbf{E}[X_u^2] \, du = \mathbf{E}[B_0^2] \cdot \frac{1}{3} + \mathbf{E}[B_{1/3}^2] \cdot \frac{1}{3} + \mathbf{E}[B_{2/3}^2] \cdot \frac{1}{3} = 0 + \frac{1}{9} + \frac{2}{9} = \frac{1}{3}.$$

Consider now another process Y on $[0, 1]$ defined on the same Brownian motion:

$$Y_t = B_0^2 \mathbf{1}_{(0,1/3]}(t) + B_{1/3}^2 \mathbf{1}_{(1/3,2/3]}(t) + B_{2/3}^2 \mathbf{1}_{(2/3,1]}(t).$$

Again, the Itô integral $J_t = \int_0^t Y_s^2 \, dB_s$ is well-defined as a process on $[0, 1]$:

$$J_t = \begin{cases} 0 & \text{if } t \in [0, 1/3], \\ B_{1/3}^2 (B_t - B_{1/3}) & \text{if } t \in (1/3, 2/3], \\ B_{1/3}^2 (B_{2/3} - B_{1/3}) + B_{2/3}^2 (B_t - B_{2/3}) & \text{if } t \in (2/3, 1]. \end{cases}$$

The covariance between the random variables I_1 and J_1 can be computed easily by using the independence of the increments and suitable conditioning. Indeed, we have

$$\mathbf{E}[I_1 J_1] = \sum_{i,j=0}^3 \mathbf{E}[B_{i/3} B_{j/3}^2 (B_{(i+1)/3} - B_{i/3})(B_{(j+1)/3} - B_{j/3})].$$

If $j > i$, we can condition on $\mathcal{F}_{j/3}$ in the above summand to get

$$\mathbf{E}[B_{i/3} B_{j/3}^2 (B_{(i+1)/3} - B_{i/3})(B_{(j+1)/3} - B_{j/3}) \,|\, \mathcal{F}_{j/3}]$$
$$= B_{i/3} B_{j/3}^2 (B_{(i+1)/3} - B_{i/3}) \mathbf{E}[B_{(j+1)/3} - B_{j/3} \,|\, \mathcal{F}_{j/3}] = 0.$$

The same holds for $i > j$ by conditioning on $\mathcal{F}_{i/3}$. The only remaining terms are $i = j$:

$$E[I_1 J_1] = \sum_{i=0}^3 \mathbf{E}[B_{i/3}^3 (B_{(i+1)/3} - B_{i/3})^2] = \sum_{i=0}^3 \mathbf{E}[B_{i/3}^3] \cdot \mathbf{E}[(B_{(i+1)/3} - B_{i/3})^2],$$

by independence of increments. The first factor of each term is zero (due to the nature of odd moments of a Gaussian centered at 0). Therefore, the variables I_1 and J_1 are uncorrelated. Corollary 5.15 gives a systematic way to compute covariances based on Itô's isometry.

Remark 5.9. An *isometry* is a mapping between metric spaces (i.e., with a distance) that actually preserves the distance between points. (It literally means *same measure* in Greek.) In the case of Itô's isometry, the mapping is the one that sends the integrand X to the square-integrable random variable given by the integral:

$$\mathcal{S}(T) \to L^2(\Omega, \mathcal{F}, \mathbf{P})$$

$$X \mapsto \int_0^T X_s \, dB_s.$$

The L^2-norm of $\int_0^T X_s \, dB_s$ is $(\mathbf{E}[(\int_0^T X_s \, dB_s)^2])^{1/2}$. It turns out that the space $\mathcal{S}(T)$ is also a linear space with the norm $\|X\|_{\mathcal{S}} = (\int_0^T \mathbf{E}[X_s]^2 \, ds)^{1/2}$. Itô's isometry says that these two norms (and hence the distance) are equal. In fact, this isometry extends in part to the L^2-space of functions on $\Omega \times [0, T]$, for which $\mathcal{S}(T)$ is a subspace. We will see that this isometry is central to the extension of the Itô integral in the limit $n \to \infty$.

The next goal is to extend the Itô integral to processes X other than simple processes. The integral will be defined as a limit of integrals of simple processes, very much like the Riemann integral is a limit of Riemann sums. But first, we need a good class of integrands.

Definition 5.10. For a given Brownian filtration $(\mathcal{F}_t, t \leq T)$, we consider the class of processes $\mathcal{L}_c^2(T)$ of processes $(X_t, t \leq T)$ such that the following hold:

(1) X is *adapted*; that is, X_t is $\mathcal{F}_t$-measurable for $t \leq T$.

(2) $\mathbf{E}[\int_0^T X_t^2 \, dt] = \int_0^T \mathbf{E}[X_t^2] \, dt < \infty$.

(3) X has continuous paths; that is, $t \mapsto X_t(\omega)$ is continuous on $[0, T]$ for a set of ω of probability one.

It is not hard to check that the processes $(B_t, t \leq T)$ and $(B_t^2, t \leq T)$ are in $\mathcal{L}_c^2(T)$. In fact, if f is a continuous function and $\int_0^T \mathbf{E}[f(B_t)^2] \, dt < \infty$, then the process $(f(B_t), t \leq T)$ is in $\mathcal{L}_c^2(T)$. Indeed, $f(B_t)$ is $\mathcal{F}_t$-measurable, since it is an explicit function of B_t. Moreover, the second condition is by assumption. The third holds simply because the composition of two continuous functions is continuous. Example 5.17 describes a process that is in $\mathcal{L}_c^2(T)$ but is not an explicit function of Brownian motion. See Exercise 5.7 for an example of a process of the form $(f(B_t), t \leq T)$ that is not in $\mathcal{L}_c^2(T)$. The main advantage of processes in $\mathcal{L}_c^2(T)$ is that they are easily approximated by simple adapted processes.

Lemma 5.11. *Let $X \in \mathcal{L}_c^2(T)$. Then X can be approximated by simple adapted processes in $\mathcal{S}(T)$, in the sense that there exists a sequence $X^{(n)} \in \mathcal{S}(T)$ such that*

$$\lim_{n \to \infty} \int_0^T \mathbf{E}[(X_t^{(n)} - X_t)^2] \, dt = 0.$$

Proof. For a given n, consider the partition $t_j = \frac{j}{n} T$ of $[0, T]$ and the simple adapted process given by

$$X_t^{(n)} = \sum_{j=0}^n X_{t_j} \mathbf{1}_{(t_j, t_{j+1}]}(t), \quad t \leq T.$$

In other words, we give the constant value X_{t_j} on the whole interval $(t_j, t_{j+1}]$. By continuity of the paths of X, it is clear that $X_t^{(n)}(\omega) \to X_t(\omega)$ at any $t \leq T$ and for any ω. Therefore, by Theorem 4.40, we have

$$\lim_{n \to \infty} \int_0^T \mathbf{E}[(X_t^{(n)} - X_t)^2] \, dt = 0. \qquad \square$$

We are now ready to state the most important theorem of this section.

Theorem 5.12. *Let $(B_t, t \leq T)$ be a standard Brownian motion defined on $(\Omega, \mathcal{F}, \mathbf{P})$. Let $(X_t, t \leq T)$ be a process in $\mathcal{L}_c^2(T)$. There exist random variables $\int_0^t X_s \, dB_s$, $t \leq T$, with the following properties:*

- **Linearity:** *If $X, Y \in \mathcal{L}_c^2(T)$ and $a, b \in \mathbb{R}$, then*

$$\int_0^t (aX_s + bY_s) \, dB_s = a \int_0^t X_s \, dB_s + b \int_0^t Y_s \, dB_s, \quad t \leq T.$$

- **Continuous martingale:** *The process $\left(\int_0^t X_s \, dB_s, t \leq T \right)$ is a continuous martingale for the Brownian filtration.*

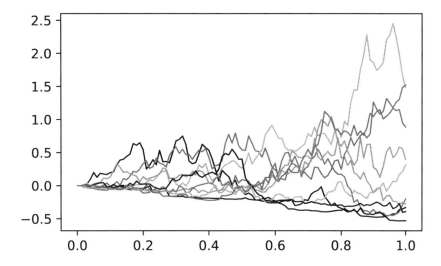

Figure 5.2. Simulation of 5 paths of the process $\int_0^t B_s \, dB_s, t \leq 1$.

- **Itô's isometry:** *The random variable $\int_0^t X_s \, dB_s$ is in $L^2(\Omega, \mathcal{F}, \mathbf{P})$ with mean 0 and variance*

$$\mathbf{E}\left[\left(\int_0^t X_s \, dB_s\right)^2\right] = \int_0^t \mathbf{E}[X_s^2] \, ds = \mathbf{E}\left[\int_0^t X_s^2 \, ds\right], \quad t \leq T.$$

Example 5.13 (Sampling Itô integrals). How can we sample paths of processes given by Itô integrals? A very simple method is to go back to the integral on simple processes. Consider the process $I_t = \int_0^t X_s \, dB_s$, $t \leq T$, constructed from $X \in \mathcal{L}_c^2(T)$ and from a standard Brownian motion $(B_t, t \geq 0)$. To simulate the paths, we fix the endpoint, say T, and a step size, say $1/n$. Then we can generate the process at every $t_j = \frac{j}{n}T$ by taking

$$I_{t_j} = \sum_{i=0}^{j-1} X_{t_i}(B_{t_{i+1}} - B_{t_i}), \quad j \leq n.$$

Here are two observations that makes this expression more palatable. First, note that the increment $B_{t_{i+1}} - B_{t_i}$ is a Gaussian random variable of mean 0 and variance $\frac{1}{n}T$ for every i. Second, we have $I_{t_j} - I_{t_{j-1}} = X_{t_{j-1}}(B_{t_j} - B_{t_{j-1}})$, so the values I_{t_j} can be computed recursively. Numerical Project 5.1 is about implementing this method.

Remark 5.14 (L^2-spaces are complete). The proof of the existence of the Itô integral is based on the completeness property of L^2-spaces: If $(X_n, n \geq 1)$ is a *Cauchy sequence* of elements in an L^2-space, then the sequence X_n converges to some element X in L^2. A sequence is Cauchy if for any choice of $\varepsilon > 0$, we can find n large enough so that

$$\|X_m - X_n\| < \varepsilon \text{ for any } m > n.$$

In other words, for an arbitrarily small distance ε, if we go further enough in the sequence the distances between increments are all smaller than ε. Another example of spaces that are complete is $\mathbb{R}$, endowed with the metric given by the absolute value.

However, the set of rational numbers $\mathbb{Q}$ as a subset of $\mathbb{R}$ is not complete, because there are sequences of rational numbers that converge to irrationals. A proof of the completeness of L^2 is outlined in Exercise 5.20.

Proof of Theorem 5.12. Consider a process $X = (X_t, t \leq T)$ in $\mathcal{L}_c^2(T)$. By Lemma 5.11, we can approximate it by a simple adapted processes $(X_t^{(n)}, t \leq T)$. In particular, this implies that the sequence is Cauchy for the metric

$$(5.7) \qquad \|X^{(n)} - X^{(m)}\| = \left(\int_0^T \mathbf{E}[(X_t^{(n)} - X_t^{(m)})^2] \, dt \right)^{1/2}.$$

The key step is the following. We know that the integral $I_t^{(n)} = \int_0^t X_s^{(n)} \, dB_s$ is well-defined as a random variable in $L^2(\Omega, \mathcal{F}, \mathbf{P})$. Moreover, we know by the Itô isometry that the L^2-distance of the processes in equation (5.7) is the same as the L^2-distance of the $I^{(n)}$'s. This means that the sequence $(I^{(n)}, n \geq 1)$ must also be Cauchy in $L^2(\Omega, \mathcal{F}, \mathbf{P})$! We conclude by completeness of the space that $I^{(n)}$ converges in L^2 to a random variable that we denote by I_t or $\int_0^t X_s \, dB_s$. Furthermore, the limit I_t does not depend on the approximating sequence $X^{(n)}$. We could have taken another sequence to approximate X and the isometry guarantees that the corresponding integrals will converge to the same random variable.

We now prove the properties:

- *Linearity*: It follows by using linearity in Proposition 5.7 for $X^{(n)}$ and $Y^{(n)}$, the two approximating processes for X and Y.

- *Isometry*: The variance follows from the following fact: If $I_t^{(n)} \to I_t$ in L^2, then $\mathbf{E}[(I_t^{(n)})^2] \to \mathbf{E}[I_t^2]$ and $\mathbf{E}[I_t^{(n)}] \to \mathbf{E}[I_t]$; see Exercise 5.3.

- *Continuous martingale*: Write $I_t = \int_0^t X_s \, dB_s$. We must show that $\mathbf{E}[I_t|\mathcal{F}_s] = I_s$ for any $t > s$. To see this, we go back to Definition 4.14. The random variable I_t is $\mathcal{F}_t$-measurable by construction. Now for a bounded random variable W that is $\mathcal{F}_s$-measurable, we need to show

$$\mathbf{E}[WI_t] = \mathbf{E}[WI_s].$$

This is clear for $I_t^{(n)}$, the approximating integrals, because $(I_t^{(n)}, t \leq T)$ is a martingale. The above then follows from the fact that $I_s^{(n)} W$ converges in L^2 to $I_s W$ (and thus the expectation converges) and the same way for t. The fact that the path $t \mapsto I_t(\omega)$ is continuous on $[0, t]$ with probability one is more involved. It uses Doob's maximal inequality; see Exercise 4.19.

$\square$

Once the conclusions of Theorem 5.12 are accepted, we are free to explore the beauty and the power of Itô calculus. As a first step, we observe that with Itô's isometry, we can compute not only variances, but also covariances between integrals. This is because an isometry also preserves the inner product in L^2-spaces.

Corollary 5.15. *Let $(B_t, t \le T)$ be a standard Brownian motion, and let $X \in \mathcal{L}_c^2(T)$. We have*

$$\mathbf{E}\left[\left(\int_0^t X_s\, dB_s\right)\left(\int_0^{t'} X_s\, dB_s\right)\right] = \int_0^{t \wedge t'} \mathbf{E}[X_s^2]\, ds, \quad t, t' \le T,$$

and for any $Y \in \mathcal{L}_c^2(T)$,

$$\mathbf{E}\left[\left(\int_0^t X_s\, dB_s\right)\left(\int_0^t Y_s\, dB_s\right)\right] = \int_0^t \mathbf{E}[X_s Y_s]\, ds, \quad t \le T.$$

Note that when X is just the constant 1, we recover from the first equation the covariance of Brownian motion.

Proof. The first assertion is Exercise 5.4. As for the second, we have on one hand by Itô's isometry

$$\mathbf{E}\left[\left(\int_0^t \{X_s + Y_s\}\, dB_s\right)^2\right] = \int_0^t \mathbf{E}[(X_s + Y_s)^2]\, ds$$

$$= \int_0^t \mathbf{E}[X_s^2]\, ds + \int_0^t \mathbf{E}[Y_s^2]\, ds + 2\int_0^t \mathbf{E}[X_s Y_s]\, ds.$$

On the other hand, by linearity of the Itô integral and of the expectation, we have

$$\mathbf{E}\left[\left(\int_0^t \{X_s + Y_s\}\, dB_s\right)^2\right] = \mathbf{E}\left[\left(\int_0^t X_s\, dB_s + \int_0^t Y_s\, dB_s\right)^2\right]$$

$$= \mathbf{E}\left[\left(\int_0^t X_s\, dB_s\right)^2\right] + \mathbf{E}\left[\left(\int_0^t Y_s\, dB_s\right)^2\right]$$

$$+ 2\mathbf{E}\left[\left(\int_0^t X_s\, dB_s\right)\left(\int_0^t Y_s\, dB_s\right)\right].$$

By combining the two equations and by using Itô's isometry, we conclude that

$$(5.8) \qquad \mathbf{E}\left[\left(\int_0^t X_s\, dB_s\right)\left(\int_0^t Y_s\, dB_s\right)\right] = \int_0^t \mathbf{E}[X_s Y_s]\, ds. \qquad \square$$

Example 5.16. Consider the processes $(B_t, t \le T)$ and $(B_t^2, t \le T)$ for a given standard Brownian motion. Note that these two processes are in $\mathcal{L}_c^2(T)$ for any $T > 0$. By Theorem 5.12, the random variables

$$I_t = \int_0^t B_s\, dB_s, \qquad J_t = \int_0^t B_s^2\, dB_s$$

exist and are in $L^2(\Omega, \mathcal{F}, \mathbf{P})$. Their mean is 0, and they have variances

$$\mathbf{E}[I_t^2] = \int_0^t \mathbf{E}[B_s^2]\, ds = \int_0^t s\, ds = \frac{t^2}{2}, \qquad \mathbf{E}[J_t^2] = \int_0^t \mathbf{E}[B_s^4]\, ds = \int_0^t 3s^2\, ds = t^3.$$

(Recall the Gaussian moments in equation (1.8).) The covariance is by Corollary 5.15:

$$\mathbf{E}[I_t J_t] = \int_0^t \mathbf{E}[B_s B_s^2]\, ds = 0.$$

The variables are uncorrelated.

Example 5.17 (A path-dependent integrand). Consider the process $X_t = \int_0^t B_s\, dB_s$ on $[0, T]$ as in Example 5.16. Note that the process $(X_t, t \le T)$ is itself in $\mathcal{L}_c^2(T)$. In particular, the integral $\int_0^t X_s\, dB_s$ is well-defined! (Note that the integrand X_t is $\mathcal{F}_t$-measurable but its value depends on the whole Brownian up to time t.) The mean of the integral is 0 and its variance is obtained by applying Itô's isometry twice:

$$\mathbf{E}\left[\left(\int_0^t X_s\, dB_s\right)^2\right] = \int_0^t \mathbf{E}[X_s^2]\, ds = \int_0^t \frac{s^2}{2}\, ds = \frac{t^3}{6}.$$

See Numerical Project 5.4.

In general, the Itô integral is not Gaussian. However, if the integrand X is not random (as in Example 5.1), the process is actually Gaussian. In this particular case, the integral is sometimes called a *Wiener integral*.

Corollary 5.18 (Wiener integral). *Let $(B_t, t \le T)$ be a standard Brownian motion and let $f : [0, T] \to \mathbb{R}$ be a function such that $\int_0^T f^2(s)\, ds < \infty$. Then the process $(\int_0^t f(s)\, dB_s, t \le T)$ is Gaussian with mean 0 and covariance*

$$\text{Cov}\left(\int_0^t f(s)\, dB_s, \int_0^{t'} f(s)\, dB_s\right) = \int_0^{t \wedge t'} f(s)^2\, ds.$$

Proof. We prove the case when f is continuous. In this case, we can use our proof of Lemma 5.11. Let $(t_j, j \le n)$ be a partition of $[0, T]$ in n intervals. The lemma shows that the sequence of simple functions

$$f^{(n)}(t) = \sum_{j=0}^{n-1} f(t_j)\mathbf{1}_{(t_j, t_{j+1}]}(t), \quad t \le T,$$

approximates f. The Itô integral of $f^{(n)}$ is

$$I_t^{(n)} = \sum_{i=0}^{j-1} f(t_j)(B_t - B_{t_j}), \quad t \in (t_j, t_{j+1}].$$

This is a Gaussian process for any n. This is because for any choice of times $s_1, \ldots, s_m$, the vector $(I_{s_1}^{(n)}, \ldots, I_{s_m}^{(n)})$ is Gaussian, since it reduces to linear combinations of Brownian motion at fixed times. Moreover, the random variable $\int_0^t f(s)\, dB_s$ is the L^2-limit of $I_t^{(n)}$ by Theorem 5.12. It remains to show that an L^2-limit of a sequence of Gaussian vectors remains Gaussian. This is sketched in Exercise 5.19. The expression of the covariances is from Corollary 5.15. $\qquad \square$

Example 5.19 (Ornstein-Uhlenbeck process as an Itô integral). Consider the function $f(s) = e^s$. The Ornstein-Uhlenbeck process starting at X_0 defined in Example 2.29 can also be written as

$$(5.9) \qquad Y_t = e^{-t} \int_0^t e^s \, dB_s, \quad t \geq 0.$$

This is tested numerically in Numerical Project 5.2. To see this mathematically, note that $(Y_t, t \geq 0)$ is a Gaussian process by Corollary 5.18. The mean is 0 and the covariance is, by Corollary 5.15,

$$\mathbf{E}[Y_t Y_s] = e^{-t-s} \int_0^s e^{2u} \, du = \frac{1}{2}(e^{-(t-s)} - e^{-(t+s)}), \quad s \leq t.$$

We can also start the process at Y_0, a Gaussian random variable of mean 0 and variance $1/2$ independent of the Brownian motion $(B_t, t \geq 0)$. The process then takes the form

$$Y_t = Y_0 e^{-t} + e^{-t} \int_0^t e^s \, dB_s.$$

Since Y_0 and the Itô integral are independent by assumption, the covariance is then

$$\mathbf{E}[Y_t Y_s] = \frac{1}{2} e^{-t-s} + \frac{1}{2}(e^{-(t-s)} - e^{-(t+s)}) = \frac{1}{2} e^{-(t-s)}, \quad s \leq t.$$

In this case, the process is stationary in the sense that $(Y_t, t \geq 0)$ has the same distribution as $(Y_{t+a}, t \geq 0)$ for any $a > 0$.

Example 5.20 (Brownian bridge as an Itô integral). The Brownian bridge $(Z_t, t \in [0, 1])$ is the stochastic process with the distribution defined in Example 2.27. Another way to construct a Brownian bridge is as follows:

$$(5.10) \qquad Z_t = (1 - t) \int_0^t \frac{1}{1 - s} \, dB_s, \quad t < 1.$$

This is tested numerically in Numerical Project 5.2. It turns out that $Z_1 = 0$. This is done in Exercise 5.21. The process Z is a Gaussian process by Corollary 5.18. The mean is 0 and the covariance is, by Corollary 5.15,

$$\mathbf{E}[Z_t Z_s] = (1 - t)(1 - s)\mathbf{E}\left[\left(\int_0^s \frac{1}{1 - u} \, dB_u\right)\left(\int_0^t \frac{1}{1 - u} \, dB_u\right)\right] = s(1 - t), \quad s \leq t.$$

The above representations of the Orstein-Uhlenbeck and the Brownian bridge implies that they are not martingales; see Exercise 5.10.

Remark 5.21 (Fubini's theorem). In Exercise 3.11 and Exercise 4.21, it was shown that we can interchange the expectation $\mathbf{E}$ and the sum $\sum$ if the random variables are positive or if $\sum_{n \geq 1} \mathbf{E}[|X_n|] < \infty$. This result holds in general when the integrands are positive or integrable. This is known as Fubini's theorem. This is applicable in particular when we calculate the variance using Itô's isometry. More precisely, we have

$$\int_0^t \mathbf{E}[X_s^2] \, ds = \mathbf{E}\left[\int_0^t X_s^2 \, ds\right].$$

Remark 5.22 (Extension to other processes). Can we define the Itô integral for processes other than the ones in $\mathcal{L}_c^2(T)$? Of course, since simple adapted processes in $\mathcal{S}(T)$ given in equation (5.5) are not continuous. In fact, the Itô construction holds whenever X is a limit of simple adapted processes. Such processes will have the property that

$$(5.11) \qquad \mathbf{E}\left[\int_0^T X_t^2\, dt\right] < \infty.$$

Theorem 5.12 is the same for these processes. In particular, they define continuous square-integrable martingales.

A further extension applies to processes such that

$$(5.12) \qquad \int_0^T X_t^2(\omega)\, dt < \infty \text{ for } \omega \text{ in a set of probability one.}$$

(Note that equation (5.11) implies the above by Exercise 1.15.) Equation (5.12) is a very weak condition, since any process $X_t = g(B_t)$ where g is continuous will satisfy it, because a continuous function is bounded on an interval. For example the process $X_t = e^{B_t^2}$ does not satisfy (5.11), but it satisfies (5.12); see Numerical Project 5.7 and Exercise 5.7. The construction of the Itô integral for such processes involves stopping times and will not be pursued here. The Itô integrals in this case are not martingales but are said to be *local martingales*; i.e., they are martingales when suitably stopped:

Definition 5.23. A process $(Y_t, t \geq 0)$ is said to be a *local martingale* for the filtration $(\mathcal{F}_t, t \geq 0)$ if there exists an increasing sequence of stopping times $(\tau_n, n \geq 1)$ for the same filtration such that $\tau_n \to +\infty$ as $n \to \infty$ almost surely, and the stopped processes $(M_{t \wedge \tau_n}, t \geq 0)$ are martingales for every $n \geq 1$.

5.4. Itô's Formula

The Itô integral was constructed in the last section in a rather abstract way. It is the limit of a sequence of random variables constructed from Brownian motion. It is good to remind ourselves that the classical Riemann integral is also very abstract! It is defined as the limit of the sequence of Riemann sums. It does not always have an explicit form. For example, the CDF of a Gaussian variable

$$\Phi(x) = \int_{-\infty}^x \frac{e^{-\frac{y^2}{2}}}{\sqrt{2\pi}}\, dy$$

is a well-defined function of x, but the integral cannot be expressed in terms of the typical elementary functions of calculus. But in some cases, a Riemann integral can be written explicitly in terms of such functions. This is the content of the *fundamental theorem of calculus*. It is useful to recall the theorem, as Itô's formula is built upon it.

Let $f : [0, T] \to \mathbb{R}$ be a function for which the derivative f' exists and is a continuous function on $[0, T]$. We will say that such a function is in $\mathcal{C}^1([0, T])$. The fundamental theorem of calculus says that we can write

$$(5.13) \qquad f(t) - f(0) = \int_0^t f'(s)\, ds, \quad t \leq T.$$

Note that we often write this result in differential form:

(5.14) $$\mathrm{d}f(t) = f'(t)\,\mathrm{d}t\,.$$

The differential form has no rigorous meaning in itself. It is simply a compact and convenient notation that encodes (5.13).

The stochastic equivalent of the fundamental theorem of calculus is Itô's formula provided below. It relates the Itô integral to an explicit function of Brownian motion. Note that the function f must be in $\mathcal{C}^2(\mathbb{R})$; i.e., f' and f'' exist and are continuous on the whole space $\mathbb{R}$.

Theorem 5.24 (Itô's formula). *Let* $(B_t, t \le T)$ *be a standard Brownian motion. Consider* $f \in \mathcal{C}^2(\mathbb{R})$. *Then, with probability one, we have*

(5.15) $$f(B_t) - f(B_0) = \int_0^t f'(B_s)\,\mathrm{d}B_s + \frac{1}{2}\int_0^t f''(B_s)\,\mathrm{d}s, \quad t \le T.$$

We will see other variations in Proposition 5.28 and in Chapters 6 and 7. Before giving the idea of the proof, we make some important observations:

(i) Equation (5.15) is an *equality of processes*, which is much stronger than equality in distribution. In other words, if you take a path of the process on the left constructed on a given Brownian motion, then this path will be the same as the path of the process on the right constructed on the same Brownian motion. See Figure 5.3. The reader should verify this in Numerical Project 5.3. The equality holds in the limit where the mesh of the partition of the interval $[0, T]$ goes to 0. See Numerical Project 5.5.

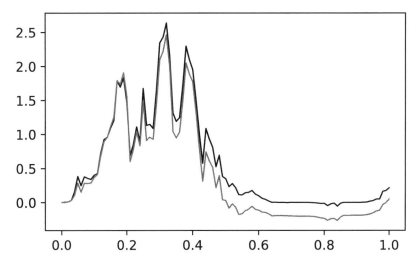

Figure 5.3. Simulation of a path of B_t^3 and of a path of $3\int_0^t B_s^2\,\mathrm{d}B_s + 3\int_0^t B_s\,\mathrm{d}s$ for a discretization of 0.01. See Numerical Project 5.5.

(ii) Note the similarity with the classical formulation in (5.13) if we replace the Riemann integral by Itô's integral. We do have the additional integral of $f''(B_s)$. As

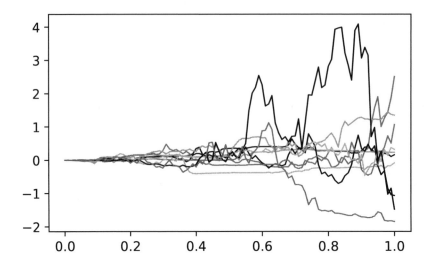

Figure 5.4. A sample of 10 paths of the martingale $B_t^3 - 3\int_0^t B_s\,ds$.

we will see in the proof, this additional term comes from the quadratic term in the Taylor approximation and from the quadratic variation of Brownian motion seen in Theorem 3.8. As in the classical case (5.14), it is very convenient to summarize the conclusion of Itô's formula in differential form:

(5.16)
$$\mathrm{d}f(B_t) = f'(B_t)\,\mathrm{d}B_t + \frac{1}{2}f''(B_t)\,\mathrm{d}t.$$

We stress that the differential form has no meaning by itself. It is a compact way to express the two integrals in Itô's formula and a powerful device for computations.

(iii) An important consequence of Itô's formula is that it provides a systematic way to construct martingales as explicit functions of Brownian motion. To make sure that $\int_0^t f'(B_s)\,\mathrm{d}B_s$, $t \le T$, defines a continuous square-integrable martingale on $[0, T]$, we might need to check that $(f'(B_t), t \le T) \in \mathcal{L}_c^2(T)$. In general the Itô integral $\int_0^t f'(B_s)\,\mathrm{d}B_s$ makes sense as a local martingale; see Remark 5.22.

Corollary 5.25 (Brownian martingales). *Let $(B_t, t \le T)$ be a standard Brownian motion. Consider $f \in \mathcal{C}^2(\mathbb{R})$ such that $\int_0^T \mathbf{E}[f'(B_s)^2]\,\mathrm{d}s < \infty$. Then the process*

$$\left(f(B_t) - \frac{1}{2}\int_0^t f''(B_s)\,\mathrm{d}s, \ t \le T\right)$$

is a martingale for the Brownian filtration.

Proof. This is straightforward from Itô's formula

$$f(B_t) - \frac{1}{2}\int_0^t f''(B_s)\,\mathrm{d}s = f(B_0) + \int_0^t f'(B_s)\,\mathrm{d}B_s.$$

The first term is a constant and the second term is a continuous martingale by Proposition 5.7.   □

The integral we subtract from $f(B_t)$ is called the *compensator*. A simple case is given by the function $f(x) = x^2$. For this function, the corollary gives that the process $B_t^2 - t, t \geq 0$, is a martingale, as we already observed in Example 4.28. The compensator was then simply t. In general, the compensator might be random.

(iv) The compensator is the Riemann integral $\int_0^t f''(B_s) \, ds$. It might seem to be a strange object at first. The function $f''(B_s)$ is random (it depends on ω), so the integral is a random variable. There is no problem in integrating the random function $f''(B_s)$ since by assumption it is a continuous function of s, since f'' and $B_s(\omega)$ are continuous. In fact, the paths of $\int_0^t f''(B_s) \, ds$ are much smoother than the ones of Brownian motion in general: The paths are differentiable everywhere (the derivative is $f''(B_t)$), and in particular, the paths have bounded variations (see Example 3.6). See Figure 5.5 for a sample of paths of the process $\int_0^t B_s \, ds$.

To sum it up, Itô's formula says that $f(B_t)$ can be expressed as a sum of two processes: one with bounded variation (the Riemann integral) and a (local) martingale with finite quadratic variation (the Itô integral). In the next chapter, we will study Itô processes in more generality, which are processes that can be expressed as the sum of a Riemann integral and an Itô integral.

Example 5.26 ($f(x) = x^3$).

In this case, Itô's formula yields

$$(5.17) \qquad B_t^3 = \int_0^t 3B_s^2 \, dB_s + \frac{1}{2} \int_0^t 6B_s \, ds = 3 \int_0^t B_s^2 \, dB_s + 3 \int_0^t B_s \, ds.$$

Figure 5.3 shows a sample of a single path of each of these two processes constructed from the same Brownian path. Note that they are almost equal (the discrepancy is only due to the discretization in the numerics)! From the above equation, we conclude that the process $B_t^3 - 3 \int_0^t B_s \, ds$ is a martingale. See Figure 5.4 for a sample of its paths. The process $(\int_0^t B_s \, ds, t \geq 0)$ is not complicated. It is a Gaussian process since the integral is the limit (almost sure and in L^2) of the Riemann sums

$$\sum_{j=0}^{n-1} B_{t_j}(t_{j+1} - t_j),$$

and each term of the sum is a Gaussian variable. (Why?) Clearly, the mean of $\int_0^t B_s \, ds$ is 0. The covariance of the process can be calculated directly by interchanging the integrals and the expectation:

$$\mathbf{E}\left[\left(\int_0^t B_s \, ds\right)\left(\int_0^{t'} B_s \, ds\right)\right] = \int_0^t \int_0^{t'} \mathbf{E}[B_s B_{s'}] \, ds \, ds' = \int_0^t \int_0^{t'} (s \wedge s') \, ds \, ds'.$$

The integral equals $\frac{t't^2}{2} - \frac{t^3}{6}$, for $t \leq t'$. In particular, the variance at time t is $\frac{t^3}{3}$. The paths of this process are very smooth as can be observed in Figure 5.5. In fact, the paths are differentiable and the derivative at time t is B_t.

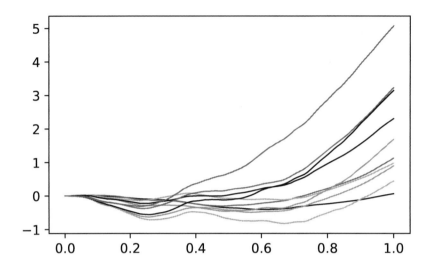

Figure 5.5. A sample of 10 paths of $3 \int_0^t B_s \, ds$.

Example 5.27 ($f(x) = \cos x$).

In this case, Itô's formula gives

$$\cos B_t - \cos 0 = \int_0^t (-\sin B_s) \, dB_s + \frac{1}{2} \int_0^t (-\cos B_s) \, ds.$$

In particular, the process

$$M_t = \cos B_t + \frac{1}{2} \int_0^t \cos B_s \, ds = 1 - \int_0^t \sin B_s \, dB_s, \quad t \geq 0,$$

is a continuous martingale starting at $M_0 = 1$. It is easy to check that the process $(\sin B_t, t \leq T)$ is in $\mathcal{L}_c^2(T)$ for any T. A sample of the paths of $(M_t, t \leq 1)$ is depicted in Figure 5.6.

Where does Itô's formula come from? It is the same idea as for the proof of the fundamental theorem of calculus. Let's start with the latter. Suppose $f \in \mathcal{C}^1(\mathbb{R})$; that is, f is differentiable with a continuous derivative. Then f admits a Taylor approximation around s of the form

(5.18) $$f(t) - f(s) = f'(s)(t - s) + \mathcal{E}(s, t).$$

(This is in the spirit of the *mean-value theorem*.) Here, $\mathcal{E}(s, t)$ is an error term that goes to 0 faster than $(t - s)$ as $s \to t$. Now, for a partition $(t_j, j \leq n)$ of $[0, t]$, say $t_j = \frac{j}{n}t$, we can trivially write for any n

$$f(t) - f(0) = \sum_{j=0}^{n} f(t_{j+1}) - f(t_j).$$

Now, we can use equation (5.18) at $s = t_j$:

$$f(t_{j+1}) - f(t_j) = f'(t_j)(t_{j+1} - t_j) + \mathcal{E}(t_j, t_{j+1}).$$

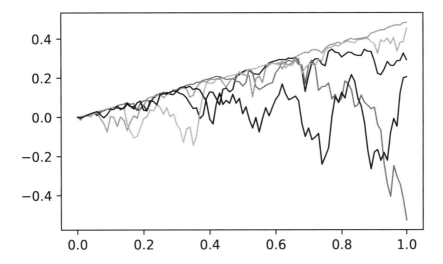

Figure 5.6. A sample of 5 paths of the martingale $\left(\cos B_t - 1 + \frac{1}{2}\int_0^t \cos B_s \, ds, t \leq 1\right)$.

Therefore, we have by taking the limit of large n

$$f(t) - f(0) = \lim_{n \to \infty} \sum_{j=0}^{n} f'(t_j)(t_{j+1} - t_j) + \sum_{j=0}^{n} \mathcal{E}(t_j, t_{j+1}) = \int_0^t f'(s) \, ds + 0.$$

The idea for Itô's formula is similar to the above with two big differences: First, we will consider a function f of *space* and not *time*. Second, we shall need a Taylor approximation to the second order around a point x: If $f \in \mathcal{C}^2(\mathbb{R})$, we have

(5.19) $$f(y) - f(x) = f'(x)(y - x) + \frac{1}{2}f''(x)(x - y)^2 + \mathcal{E}(x, y),$$

where $\mathcal{E}(x, y)$ is an error term that now goes to 0 faster than $(x - y)^2$ as $y \to x$.

Proof of Theorem 5.24. Recall that by assumption $f \in \mathcal{C}^2(\mathbb{R})$. We will prove the particular case where f is 0 outside a bounded interval. This implies that both derivatives are bounded, since they are continuous functions on a bounded interval. We first prove the formula for a fixed t. Then we generalize to processes on $[0, T]$. Consider a partition $(t_j, j \leq n)$ of $[0, t]$. From equation (5.19), we get

$$f(B_t) - f(B_0)$$

(5.20) $$= \sum_{j=0}^{n-1} f'(B_{t_j})(B_{t_{j+1}} - B_{t_j}) + \sum_{j=0}^{n-1} \frac{1}{2}f''(B_{t_j})(B_{t_{j+1}} - B_{t_j})^2 + \sum_{j=0}^{n} \mathcal{E}(B_{t_j}, B_{t_{j+1}}).$$

As $n \to \infty$, the first term converges (as a random variable in L^2) to the Itô integral. This is how we proved Proposition 5.7 using simple processes. We claim the second term converges to the Riemann integral. To see this, consider the corresponding Riemann sum

$$\sum_{j=0}^{n-1} f''(B_{t_j})(t_{j+1} - t_j).$$

This term converges almost surely to the Riemann integral $\int_0^t f''(B_s) \, ds$ since f'' is continuous. It also converges in L^2 by Theorem 4.40, since f'' is bounded by assumption. Therefore, to show the second term converges to the same limit, it suffices to show that the L^2-distance between the second term and the Riemann sum goes to 0; i.e.,

$$(5.21) \qquad \lim_{n \to \infty} \mathbf{E}\left[\left(\sum_{j=0}^{n-1} f''(B_{t_j})\{(B_{t_{j+1}} - B_{t_j})^2 - (t_{j+1} - t_j)\}\right)^2\right] = 0.$$

This is in the same spirit as the proof of the quadratic variation of Brownian motion in Theorem 3.8. To lighten notation, define the variables $X_j = (B_{t_{j+1}} - B_{t_j})^2 - (t_{j+1} - t_j)$, $j \le n-1$. We expand the square in (5.21) to get

$$\sum_{j,k=0}^{n-1} \mathbf{E}\left[f''(B_{t_j}) f''(B_{t_k}) X_j X_k\right].$$

For $j < k$, we condition on $\mathcal{F}_{t_k}$ to get that the summand is 0 by Proposition 4.19 and since $\mathbf{E}[(B_{t_{k+1}} - B_{t_k})^2] = t_{k+1} - t_k$. For $j = k$, the sum is

$$\sum_{j=0}^{n-1} \mathbf{E}\left[(f''(B_{t_j}))^2\{(B_{t_{j+1}} - B_{t_j})^2 - (t_{j+1} - t_j)\}^2\right].$$

By expanding the square again and conditioning on $\mathcal{F}_{t_j}$, we have by independence of the increments

$$\sum_j \mathbf{E}\left[(f''(B_{t_j}))^2\right]\{3(t_{j+1} - t_j)^2 + (t_{j+1} - t_j)^2 - 2(t_{j+1} - t_j)^2\}^2$$

$$= 2\sum_j \mathbf{E}\left[(f''(B_{t_j}))^2\right](t_{j+1} - t_j)^2.$$

Since f'' is bounded, this term goes to 0 exactly as in the proof of Theorem 3.8. It remains to handle the error term (5.20). This follows the same idea as for the second term and we omit it.

To extend the formula to the whole interval $[0, T]$, notice that the processes of both sides of equation (5.15) have continuous paths. Since they are equal (with probability one) at any fixed time by the above argument, they must be equal for any countable set of times with probability one; see Exercise 1.5. It suffices to consider the processes on the rational times in $[0, T]$, which are dense in $[0, T]$. Since the paths are continuous and they are equal on these times, they must be equal at all times on $[0, T]$. $\qquad \square$

Recall from equation (5.16) that Itô's formula can be conveniently written in the *differential form*:

$$\boxed{df(B_t) = f'(B_t) \, dB_t + \frac{1}{2} f''(B_t) \, dt.}$$

This notation has no meaning by itself. It is a compact way to write equation (5.15). This allows us to derive an easy and useful computational formula: If we blindly apply the classical differential to f to second order in the Taylor expansion, we formally obtain

$$(5.22) \qquad df(B_t) = f'(B_t) \, dB_t + \frac{1}{2} f''(B_t)(dB_t)^2.$$

Therefore, Itô's formula is equivalent to applying the rule $dt = dB_t \cdot dB_t$. In fact, it is counterproductive to learn Itô's formula by heart. It is much better to simply compute the differential up to the second order and apply the following simple *rules of Itô calculus*:

(5.23)

$\cdot$	dt	dB_t
dt	0	0
dB_t	0	dt

.

It is not hard to extend Itô's formula to a function $f(t, x)$ of both *time* and *space*:

$$f : [0, T] \times \mathbb{R} \to \mathbb{R}$$
(5.24)
$$(t, x) \mapsto f(t, x).$$

Such functions have partial derivatives that are themselves functions of time and space. We will use the following notation for the partial derivatives:

(5.25) $\quad \partial_0 f(t, x) = \dfrac{\partial f}{\partial t}(t, x), \qquad \partial_1 f(t, x) = \dfrac{\partial f}{\partial x}(t, x), \qquad \partial_1^2 f(t, x) = \dfrac{\partial^2 f}{\partial x^2}(t, x).$

The reason for this notation is to avoid confusion between the variable that is *being differentiated* and the value of time and space at which the derivative is *being evaluated*. It might appear strange at first, but it will avoid confusion down the road (especially when dealing with several space variables in Chapter 6). To apply Itô's formula, we will need that the partial derivative with respect to time $\partial_0 f$ exists and is continuous as a function on $[0, T] \times \mathbb{R}$ and that the first and second partial derivatives in space $\partial_1 f$ and $\partial_1^2 f$ exist and are continuous. We say that such a function f is in $\mathcal{C}^{1,2}([0, T] \times \mathbb{R})$.

Proposition 5.28 (Itô's formula). *Let $(B_t, t \leq T)$ be a standard Brownian motion on $[0, T]$. Consider a function f of time and space with $f \in \mathcal{C}^{1,2}([0, T] \times \mathbb{R})$. Then, with probability one, we have for every $t \in [0, T]$,*

$$f(t, B_t) - f(0, B_0) = \int_0^t \partial_1 f(s, B_s) \, dB_s + \int_0^t \left\{ \partial_0 f(s, B_s) + \frac{1}{2}\partial_1^2 f(s, B_s) \right\} ds.$$

Or in differential form we have

$$df(t, B_t) = \partial_1 f(t, B_t) \, dB_t + \left(\partial_0 f(t, B_t) + \frac{1}{2}\partial_1^2 f(t, B_t) \right) dt.$$

Note that the notation $\partial_0 f(t, B_t)$ stands for the function $\partial_0 f$ evaluated at the point (t, B_t), and the notation $\partial_1^2 f(t, B_t)$ stands for the function $\partial_1^2 f$ evaluated at the point (t, B_t).

Proof. The idea of the proof is similar as for a function of space only, as it depends on a Taylor approximation and on the quadratic variation. Here, however, we need to apply Taylor approximation to second order in space and to first order in time. We then get something of the following form:

$f(t, B_t) - f(0, B_0)$

$$= \sum_{j=0}^{n-1} \partial_1 f(t_j, B_{t_j})(B_{t_{j+1}} - B_{t_j}) + \partial_0 f(t_j, B_{t_j})(t_{j+1} - t_j) + \frac{1}{2}\partial_1^2 f(t_j, B_{t_j})(B_{t_{j+1}} - B_{t_j})^2$$

$$+ \partial_1 \partial_0 f(t_j, B_{t_j})(B_{t_{j+1}} - B_{t_j})(t_{j+1} - t_j) + \mathcal{E}.$$

The first line becomes the integrals in Itô's formula. We see a new animal in the second line: the mixed derivative $\partial_0\partial_1 f$ in time and space. This term is related to the limit in the *cross variation* between B_t and t given by

$$\lim_{n\to\infty} \sum_{j=0}^{n-1} (B_{t_{j+1}} - B_{t_j})(t_{j+1} - t_j).$$

It can be shown that it goes to 0 in a suitable sense; see Exercise 5.15. This is the rigorous reason for the rule $dt \cdot dB_t = 0$. Once this is known, the rest of the proof is done similarly to the one for a function of space only. We do notice though that the formula is easy to derive once we accept the rules of Itô calculus. By writing the differential to second order in space and to first order in time and applying the rules of Itô calculus, we get

$$df(t, B_t) = \partial_1 f(t, B_t)\, dB_t + \left(\partial_0 f(t, B_t) + \frac{1}{2}\partial_1^2 f(t, B_t)\right) dt. \qquad \square$$

As in the one variable case, we get a corollary to construct martingales:

Corollary 5.29 (Brownian martingales). *Let $(B_t, t \leq T)$ be a standard Brownian motion. Consider $f \in C^{1,2}([0, T] \times \mathbb{R})$ such that the process $(\partial_1 f(t, B_t), t \leq T) \in \mathcal{L}_c^2(T)$. Then the process*

$$\left(f(t, B_t) - \int_0^t \left\{\partial_0 f(s, B_s) + \frac{1}{2}\partial_1^2 f(s, B_s)\right\} ds, t \leq T\right)$$

is a martingale for the Brownian filtration. In particular, if $f(t, x)$ satisfies the partial differential equation $\partial_0 f = -\frac{1}{2}\partial_1^2 f$, then the process $(f(t, B_t), t \leq T)$ is itself a martingale.

We now catch a glimpse of a powerful connection between two fields of mathematics: *The study of martingales is closely related to the study of differential equations.* We will see this connection in action in the gambler's ruin problem in Section 5.5. This is also explored further in Chapter 8.

Example 5.30. Consider the function $f(t, x) = tx$. In this case, we have $\partial_0 f = x$, $\partial_1 f = t$, and $\partial_1^2 f = 0$. Itô's formula yields

$$d(tB_t) = t\, dB_t + B_t\, dt.$$

Therefore, the process $M_t = tB_t - \int_0^t B_s\, ds$ is a martingale for the Brownian filtration. It is also a Gaussian process by Corollary 5.18. The mean is 0 and the covariance is by Corollary 5.15

$$\mathbf{E}[M_t M_{t'}] = \int_0^{t \wedge t'} s^2\, ds = \frac{(t \wedge t')^3}{3}.$$

Example 5.31 (Geometric Brownian motion revisited). We know from Example 4.28 that geometric Brownian motion is a martingale for the choice $\mu = \frac{-1}{2}\sigma^2$. How does this translate in terms of Itô integrals? Note that $S_t = f(t, B_t)$ for the function of time and space $f(t, x) = e^{\sigma x + \mu t}$ with $S_0 = 1$. The relevant partial derivatives are $\partial_0 f = \mu f$,

$\partial_1 f = \sigma f$, and $\partial_1^2 f = \sigma^2 f$. Therefore, developing the function f to second order in space and first order in time and using the rules of Itô calculus yield

$$dS_t = df(t, B_t) = \partial_0 f(t, B_t)\, dt + \partial_1 f(t, B_t)\, dB_t + \frac{1}{2}\partial_1^2 f(t, B_t)(dB_t)^2$$

$$= \sigma f(t, B_t)\, dB_t + \left(\mu + \frac{1}{2}\sigma^2\right) f(t, B_t)\, dt.$$

In the integral notation, this is

$$S_t = 1 + \int_0^t \sigma f(s, B_s)\, dB_s + \int_0^t \left(\mu + \frac{1}{2}\sigma^2\right) f(s, B_s)\, ds.$$

We see that we have a martingale if $\mu = \frac{-1}{2}\sigma^2$ as expected. It is not hard to check that the integrand is in $\mathcal{L}_c^2(T)$ for any $T > 0$ (see Exercise 5.6).

5.5. Gambler's Ruin for Brownian Motion with Drift

We solved the gambler's ruin problem for standard Brownian motion in Example 4.41. We now deal with the case where a drift is present. Consider the Brownian motion with drift

$$X_t = \sigma B_t + \mu t,$$

where $(B_t, t \geq 0)$ is a standard Brownian motion. We assume that $\mu > 0$. Therefore, there is a bias upward. This is important!

We consider for $a, b > 0$ the first passage time of the level a or $-b$

$$\tau = \min\{t \geq 0 : X_\tau > a \text{ or } X_\tau < -b\}.$$

The problem consists of computing $\mathbf{P}(X_\tau = a)$. Recall that in the case of no drift, this probability was $b/(a+b)$. To solve the problem, we need to find a good martingale of X_t that gives us the desired probability using Doob's optional stopping theorem. It is not

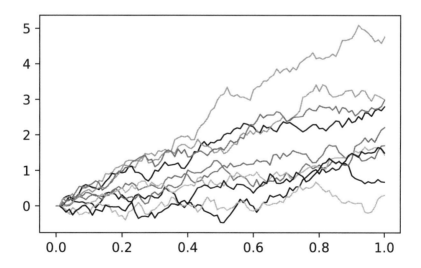

Figure 5.7. A sample of 10 paths of the process $X_t = B_t + 2t$.

hard to see here that $\tau < \infty$ (by the same argument as for standard Brownian motion). We assume the martingale is of the simplest form, that is, a function of X_t:

$$M_t = g(X_t),$$

for some function g to be found. The function needs to satisfy two properties:

- The process $(g(X_t), t \geq 0)$ is a martingale for the Brownian filtration.
- The values at $x = a$ or $x = -b$ are $g(a) = 1$ and $g(-b) = 0$.

The first condition implies by Corollary 4.38 that

$$\mathbf{E}[g(X_\tau)] = g(0).$$

The second condition is a convenient choice since we have

$$\mathbf{E}[g(X_\tau)] = g(a)\mathbf{P}(X_\tau = a) + g(-b)\mathbf{P}(X_\tau = -b) = \mathbf{P}(X_\tau = a).$$

Combining these two, we see that the ruin problem is reduced to finding $g(0)$ since

$$g(0) = \mathbf{E}[g(X_\tau)] = \mathbf{P}(X_\tau = a).$$

What are the conditions on g for $g(X_t)$ to be a martingale? Note that $g(X_t) = g(\sigma B_t + \mu t)$ is an explicit function of t and B_t: $f(t, x) = g(\sigma x + \mu t)$. By the chain rule, we have

$$\partial_0 f(t, x) = \mu g'(\sigma x + \mu t), \quad \partial_1 f(t, x) = \sigma g'(\sigma x + \mu t), \quad \partial_1^2 f(t, x) = \sigma^2 g''(\sigma x + \mu t).$$

By Corollary 5.29, for $g(X_t)$ to be martingale, we need g to satisfy the ordinary differential equation

$$\mu g' = \frac{-\sigma^2}{2} g''.$$

This is easy to solve just by integrating, and we get $g(y) = Ce^{-2\mu/\sigma^2} + C'$ for two constants C and C'. The boundary conditions $g(a) = 1$ and $g(-b) = 0$ determine those constants, and we finally have

(5.26)
$$g(y) = \frac{1 - e^{-2\mu(y+b)/\sigma^2}}{1 - e^{-2\mu(a+b)/\sigma^2}}.$$

(Notice that g is bounded, and so is the martingale $g(X_t)$. Hence, there is no problem in applying Corollary 4.38.) In particular, we get the answer to our initial question

(5.27)
$$\mathbf{P}(X_\tau = a) = g(0) = \frac{1 - e^{-2\mu b/\sigma^2}}{1 - e^{-2\mu(a+b)/\sigma^2}}.$$

This formula is tested numerically in Numerical Project 5.6. It is good to take a step back and look at what we have achieved:

- If we take the case $\mu = \sigma = 1$ and $a = b = 1$, then the probability is

$$\mathbf{P}(X_\tau = a) = \frac{1 - e^{-2}}{1 - e^{-4}} = 0.881\ldots.$$

Compare this to the case $\mu = 0$, where this probability is $1/2$!

- Notice that we reduced the problem of computing a probability to solving a *differential equation with boundary conditions*. This is amazing!

- Our answer is even more general. Had we started the process at $y \in [-b, a]$ instead of 0, then the probability would have been $g(y)$ given in equation (5.26).

- The identity

$$\mathbf{E}[g(X_\tau)] = \mathbf{P}(X_\tau = a)$$

is very intuitive. Since $g(a) = 1$ and $g(-b) = 0$, the paths that hit a (success) contribute to the expectation whereas the ones that hit $-b$ (failure) do not. Therefore, the *proportion of paths hitting* a, or in other words the probability, is given by averaging the Bernoulli variable $g(X_\tau)$ over all paths.

- Let's look at the limiting cases. If we take $b \to \infty$, then we get

(5.28) $$\mathbf{P}(X_\tau = a) \to 1, \quad b \to \infty,$$

which makes sense since the drift is upward, and we already know that it is the case when $\mu = 0$. On the other hand, if $a \to \infty$, then we get

(5.29) $$\mathbf{P}(X_\tau = -b) \to e^{\frac{-2\mu}{\sigma^2}b}.$$

It is not 1. The formula is telling us that even when $a \to \infty$, there are some paths that will never hit $-b$, because of the upward drift, no matter how small the drift is!

5.6. Tanaka's Formula

What happens to Itô's formula when f is not in $\mathcal{C}^2$? It turns out that in some cases we can still express $f(B_t)$ as a sum of a martingale and a process with bounded variation. The most famous example is when $f(B_t) = |B_t|$. (The absolute value is continuous, but the first and second derivative do not exist at 0.) Note that in this case, one can see the paths of the process $f(B_t)$ as the paths of a Brownian motion reflected on the x-axis. In this case, one recovers some, but not all, of Itô's formula as the following theorem shows.

Theorem 5.32 (Tanaka's formula). *Let $(B_t, t \geq 0)$ be a standard Brownian motion. There exists an increasing adapted process $(L_t, t \geq 0)$, called the local time of the Brownian motion at 0, such that*

$$|B_t| = \int_0^t \text{sgn}(B_s) \, dB_s + L_t, \quad t \geq 0,$$

where $\text{sgn}(x) = 1$ *if* $x \geq 0$ *and* $\text{sgn}(x) = -1$ *if* $x < 0$.

As for the case of Itô's formula where $f \in \mathcal{C}^2(\mathbb{R})$, the function of Brownian motion is expressed as a sum of an Itô integral and of a process of bounded variation, since L_t is increasing in t. (The theorem is not surprising in view of the Doob-Meyer decomposition in Remark 4.31.) The theorem is illustrated in Figure 5.8. It turns out that the Itô integral has the distribution of Brownian motion; see Section 7.6. The integrand $\text{sgn}(B_s)$ is not in $\mathcal{L}_c^2(T)$ but it can be shown that it falls in the first case in Remark 5.22, so that it is a martingale. It is the investing strategy that equals $+1$ when the Brownian motion is positive, and -1 when it is negative. The *local time at* 0, denoted by L_t, should be interpreted as the amount of time on $[0, t]$ that the Brownian motion has spent at 0. More precisely, it is equal to

$$L_t = \lim_{\varepsilon \to 0} \frac{1}{2\varepsilon} \int_0^t \mathbf{1}_{\{s \in [0,t]:|B_s| \le \varepsilon\}} \, ds \ \text{ in } L^2.$$

The existence of the process L_t is not obvious and is a consequence of the proof. The strategy of the proof is to use Itô's formula on an approximation of the absolute value that is in $\mathcal{C}^2$. The proof is technical and will be skipped. However, the result is not hard to simulate; see Numerical Project 5.8.

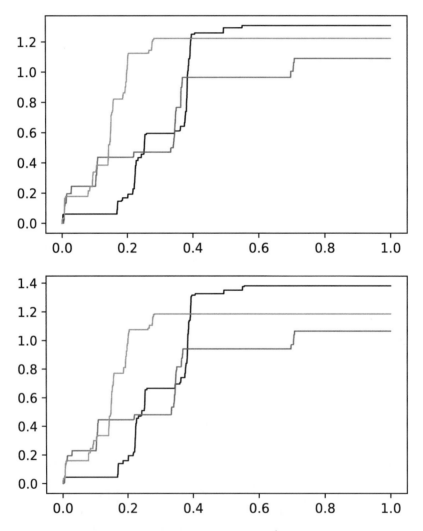

Figure 5.8. Sample of 3 paths of the processes $|B_t| - \int_0^t \text{sgn}(B_s) \, dB_s$ (top) and of L_t with the approximation $\varepsilon = 0.001$ (bottom) and step size $1/100{,}000$.

5.7. Numerical Projects and Exercises

5.1. Itô integrals. Following the procedure outlined in Example 5.13, sample 10 paths of the following three processes on $[0, 1]$ using a 0.001 discretization:

(a) $\int_0^t 4B_s^3 \, dB_s$,

(b) $\int_0^t \cos B_s \, dB_s$,

(c) $1 - \int_0^t e^{s/2} \sin B_s \, dB_s$.

5.2. Ornstein-Uhlenbeck process and Brownian bridge revisited. We saw in Examples 5.19 and 5.20 that the Ornstein-Uhlenbeck process and the Brownian bridge can be expressed in terms of Wiener integrals; see equations (5.9) and (5.10). Use these representations to sample 100 paths of each process on $[0, 1]$ with a discretization of 0.001. Compare this to the samples generated in Numerical Projects 2.3 and 2.5 using Cholesky decomposition.

5.3. Itô's formula. Sample a *single path* of the following three processes on $[0, 1]$ using a 0.01 discretization and compare to the processes in Numerical Project 5.1 constructed *on the same Brownian path*:

(a) $B_t^4 - 6 \int_0^t B_s^2 \, ds$,

(b) $\sin B_t + \frac{1}{2} \int_0^t \sin B_s \, ds$,

(c) $e^{t/2} \cos B_t$.

The command `random.seed` is useful to work on the same outcome.

5.4. A path-dependent integrand. Consider the process $(X_t, t \in [0, 1])$ with $X_t = \int_0^t B_s \, dB_s$. We construct the process $I_t = \int_0^t X_s \, dB_s$ as in Example 5.17. Following the procedure outlined in Example 5.13, sample 10 paths of this process on $[0, 1]$ using a 0.01 discretization.

5.5. Convergence of Itô's formula. Consider the two processes $(I_t, t \in [0, 1])$ and $(J_t, t \in [0, 1])$ in Example 5.26 defined by the two sides of equation (5.17) on the interval $[0, 1]$. Sample 100 paths of these two processes for each of the discretization $0.1, 0.01, 0.001, 0.0001$. Estimate $\mathbf{E}[|I_1 - J_1|]$ for each of these time steps. What do you notice?

5.6. Testing the solution to the gambler's ruin. Let's test equation (5.29).

(a) Sample 10,000 paths of Brownian paths with drift $\mu = 1$ and volatility $\sigma = 1$ on $[0, 5]$ for a step size of 0.01.

(b) Count the proportion of those paths that reach -1 on the time interval and compare with equation (5.29). Repeat the experiment for a step size of 0.001.

The experiment on $[0, 5]$ gives an approximation of the probability on $[0, \infty)$. It turns out that the probability on a finite interval can be computed exactly. See Exercise 9.5.

5.7. The integral of a process not in $\mathcal{L}_c^2(T)$.

(a) Sample 100 paths of the process $Z_t = \exp B_t^2, t \in [0, 10]$. This process in not in $\mathcal{L}_c^2(10)$ as shown in Exercise 5.7.

(b) Sample and plot 100 paths of the process $\int_0^t Z_s \, dB_s, t \in [0, 10]$. What do you notice?

5.8. **Tanaka's formula.** Generate 10 paths of Brownian motion on $[0, 1]$ using a discretization of $1/1{,}000{,}000$.

(a) Plot the paths of the process $|B_t| - \int_0^t \text{sgn}(B_s)\, dB_s$ on $[0, 1]$.

(b) Plot the paths of the process L_t^ε for $\varepsilon = 0.001$ where

$$L_t^\varepsilon = \frac{1}{2\varepsilon} |\{s \in [0, t] : |B_t| < \varepsilon\}|.$$

In other words, this is the amount of time before time t spent by Brownian motion in the interval $[-\varepsilon, \varepsilon]$ (rescaled by $1/2\varepsilon$).

Exercises

5.1. **Stopped martingales are martingales.** Let $(M_n, n = 0, 1, 2, \dots)$ be a martingale in discrete time for the filtration $(\mathcal{F}_n, n \geq 0)$. Let τ be a stopping time for the same filtration. Use the martingale transform with the process

$$X_n(\omega) = \begin{cases} +1 & \text{if } n < \tau(\omega), \\ 0 & \text{if } n \geq \tau(\omega) \end{cases}$$

to show that the stopped martingale $(M_{\tau \wedge n}, n \geq 0)$ is a martingale.

5.2. **Itô integral of a simple process.** Consider $(I_s, s \leq 1)$ the Itô integrals in Example 5.1.

(a) Argue that $(I_{1/3}, I_{2/3}, I_1)$ is a Gaussian vector.

(b) Compute the mean and the covariance matrix of $(I_{1/3}, I_{2/3}, I_1)$.

(c) Compute $\mathbf{E}[B_1 I_1]$. Are the random variables B_1 and I_1 independent? Briefly justify.

5.3. **Convergence in L^2 implies convergence of first and second moments.** Let $(X_n, n \geq 0)$ be a sequence of random variables that converge to X in $L^2(\Omega, \mathcal{F}, \mathbf{P})$.

(a) Show that $\mathbf{E}[X_n^2]$ converges to $\mathbf{E}[X^2]$.
    Hint: Write $X = (X - X_n) + X_n$. The Cauchy-Schwarz inequality might be useful.

(b) Show that $\mathbf{E}[X_n]$ converges to $\mathbf{E}[X]$.
    Hint: Write $|\mathbf{E}[X_n] - \mathbf{E}[X]|$ and use Jensen's inequality twice.

5.4. **Increments of martingales are uncorrelated.**

(a) Let $(M_t, t \geq 0)$ be a square-integrable martingale for the filtration $(\mathcal{F}_t, t \geq 0)$. Use the properties of conditional expectation to show that for $t_1 \leq t_2 \leq t_3 \leq t_4$, we have

$$\mathbf{E}[(M_{t_2} - M_{t_1})(M_{t_4} - M_{t_3})] = 0.$$

(b) Let $(B_t, t \geq 0)$ be a standard Brownian motion, and let $(X_t, t \leq T)$ be a process in $\mathcal{L}_c^2(T)$. Use part (a) to show that the covariance between integrals at different times $t < t'$ is

$$\mathbf{E}\left[\left(\int_0^t X_s\, dB_s\right)\left(\int_0^{t'} X_s\, dB_s\right)\right] = \int_0^t \mathbf{E}[X_s^2]\, ds.$$

This motivates the natural notation

$$\int_t^{t'} X_s \, dB_s = \int_0^{t'} X_s \, dB_s - \int_0^t X_s \, dB_s.$$

5.5. **Mean and variance of martingale transforms.** Let $(M_t, t \leq T)$ be a square-integrable martingale for a filtration $(\mathcal{F}_t, t \leq T)$, and let X be a simple process in $\mathcal{S}(T)$. Compute the mean and the variance of the martingale transform of X with respect to M on $[0, T]$.

5.6. **Geometric Brownian motion is in $\mathcal{L}_c^2$.** Let $M_t = \exp(\sigma B_t - \sigma^2 t/2)$ be a geometric Brownian motion. Verify that the process $(M_t, t \leq T)$ is in $\mathcal{L}_c^2(T)$ for any $T > 0$.

5.7. **A process that is not in $\mathcal{L}_c^2(T)$.** Consider the process $(e^{B_t^2}, t \leq T)$. Show that it is not in $\mathcal{L}_c^2(T)$ for $T > 1/4$.

5.8. **Practice on Itô integrals.** Consider the two processes

$$X_t = \int_0^t (1 - s) \, dB_s, \qquad Y_t = \int_0^t (1 + s) \, dB_s.$$

(a) Find the mean and the covariance of the process $(X_t, t \geq 0)$. What is its distribution?
(b) Find the mean and the covariance of the process $(Y_t, t \geq 0)$. What is its distribution?
(c) For which time t, if any, do we have that X_t and Y_t are uncorrelated? Are X_t and Y_t independent at these times?

5.9. **Practice on Itô integrals.** Consider the process $(X_t, t \geq 0)$ given by

$$X_t = \int_0^t \sin s \, dB_s.$$

(a) Argue briefly that this process is Gaussian. Find the mean and the covariance matrix.
(b) Write the covariance matrix for $(X_{\pi/2}, X_\pi)$ (i.e., the process at time $t = \pi/2$ and $t = \pi$). Write down a double integral for the probability $\mathbf{P}(X_{\pi/2} > 1, X_\pi > 1)$.
(c) On the same Brownian motion, consider the process $Y_t = \int_0^t \cos s \, dB_s$. Find for which time t the variables X_t and Y_t are independent.

5.10. **Not everything is a martingale.**
(a) Use the representation of the Ornstein-Uhlenbeck process in Example 5.19 to show that it is not a martingale for the Brownian filtration.
(b) Use the representation of the Brownian bridge in Example 5.20 to show that it is not a martingale for the Brownian filtration.
(c) Show that the process $(\int_0^t B_s \, ds, t \geq 0)$ is not a martingale for the Brownian flltration.

5.11. **Practice on Itô integrals.** Let $(B_t, t \geq 0)$ be a Brownian motion defined on $(\Omega, \mathcal{F}, \mathbf{P})$. We define for $t \geq 0$ the process

$$X_t = \int_0^t \text{sgn}(B_s) \, dB_s,$$

where $\text{sgn}(x) = -1$ if $x < 0$ and $\text{sgn}(x) = +1$ if $x \geq 0$.
The integral is well-defined even though $s \mapsto \text{sgn}(B_s)$ is not continuous.
(a) Compute the mean and the covariance of the process $(X_t, t \geq 0)$.
(b) Show that X_t and B_t are uncorrelated for all $t \geq 0$.
(c) Show that X_t and B_t are not independent. (Use $B_t^2 = 2 \int_0^t B_s \, dB_s + t$.)
It turns out that $(X_t, t \geq 0)$ is a standard Brownian motion. See Theorem 7.26.

5.12. **Integration by parts for some Itô integrals.** Let $g \in \mathcal{C}^2(\mathbb{R})$ and $(B_t, t \geq 0)$, a standard Brownian motion.
(a) Use Itô's formula to prove that for any $t \geq 0$

$$\int_0^t g(s) \, dB_s = g(t)B_t - \int_0^t B_s g'(s) \, ds.$$

(b) Use the above to show that the process given by

$$X_t = t^2 B_t - 2 \int_0^t s B_s \, ds$$

is Gaussian. Find its mean and its covariance.

5.13. **Some practice with Itô's formula.** Let $(B_t, t \geq 0)$ be a standard Brownian motion. For each of the processes $(X_t, t \leq T)$ below:
   - Determine if they are martingales for the Brownian filtration. If not, find a compensator for it.
   - Find the mean, the variance, and the covariance.
   - Is the process Gaussian? Argue briefly.
(a) $X_t = \int_0^t \cos s \, dB_s$.
(b) $X_t = B_t^4$.
(c) $X_t = e^{t/2} \cos B_t$.
   Hint: If Z is standard Gaussian, then $\mathbf{E}[\sin^2(\sigma Z)] = \frac{1 - e^{-2\sigma^2}}{2}$.
(d) $Z_t = (B_t + t) \exp(-B_t - \frac{t}{2})$.

5.14. **Gaussian moments using Itô.** Let $(B_t, t \in [0,1])$ be a Brownian motion. Use Itô's formula to show that for $k \in \mathbb{N}$

$$\mathbf{E}[B_t^k] = \frac{1}{2} k(k-1) \int_0^t \mathbf{E}[B_s^{k-2}] \, ds.$$

Conclude from this that $\mathbf{E}[B_t^4] = 3t^2$ and $\mathbf{E}[B_t^6] = 15t^3$.

5.15. **Cross-variation of t and B_t.** Let $(t_j, j \leq n)$ be a sequence of partitions of $[0, t]$ such that $\max_j |t_{j+1} - t_j| \to 0$ as $n \to \infty$. Prove that

$$\lim_{n \to \infty} \sum_{j=0}^{n-1} (t_{j+1} - t_j)(B_{t_{j+1}} - B_{t_j}) = 0 \quad \text{in } L^2.$$

This justifies the rule $dt \cdot dB_t = 0$. Can you also justify the rule $dt \cdot dt = 0$?

5.16. Exercise on Itô's formula. Consider for $t \geq 0$ the process

$$X_t = \exp(tB_t).$$

(a) Find the mean and the variance of this process.
(b) Use Itô's formula to write the process in terms of an Itô integral and a Riemann integral. Find a compensator C_t so that $X_t - C_t$ is a martingale.
(c) Argue that $(e^{tB_t}, t \leq T)$ is in $\mathcal{L}_c^2(T)$ for any $T > 0$, so that $\int_0^t e^{sB_s} dB_s$ makes sense.
(d) Show that the covariance between B_t and $\int_0^t e^{sB_s} dB_s$ is

$$\int_0^t e^{s^3/2} \, ds.$$

5.17. Itô's formula and optional stopping. Let $(B_t, t \geq 0)$ be a standard Brownian motion. Consider for $a, b > 0$ the hitting time

$$\tau = \min_{t \geq 0}\{t \, : \, B_t \geq a \text{ or } B_t \leq -b\} \,.$$

The goal of this exercise is to compute $\mathbf{E}[\tau B_\tau]$.
(a) Let $f(t, x)$ be a function of the form

$$f(t, x) = tx + g(x) \,.$$

Find an ODE for the function f for which $\partial_0 f = -\frac{1}{2}\partial_1^2 f$. Solve this ODE.
(b) Argue briefly that the process $(f(t, B_t), t \geq 0)$ is a continuous martingale.
(c) Use this to show that

$$\mathbf{E}[\tau B_\tau] = \frac{ab}{3}(a - b) \,.$$

5.18. A strange martingale. Let $(B_t, t \geq 0)$ be a standard Brownian motion. Consider the process

$$M_t = \frac{1}{\sqrt{1-t}} \exp\left(\frac{-B_t^2}{2(1-t)}\right), \quad \text{for } 0 \leq t < 1.$$

(a) Show that M_t can be represented by

$$M_t = 1 + \int_0^t \frac{-B_s M_s}{1-s} \, dB_s, \quad \text{for } 0 \leq t < 1.$$

(b) Deduce from the previous question that $(M_s, s \leq t)$ is a martingale for $t < 1$ and for the Brownian filtration.
(c) Show that $\mathbf{E}[M_t] = 1$ for all $t < 1$.
(d) Prove that $\lim_{t \to 1^-} M_t = 0$ almost surely.
(e) Argue (by contradiction) that $\mathbf{E}[\sup_{0 \leq t < 1} M_t] = +\infty$, where sup stands for the supremum.
Hint: Theorem 4.40 is useful.

5.19. $\star$ L^2-**limit of Gaussians is Gaussian.** Let $(X_n, n \geq 0)$ be a sequence of Gaussian random variables that converge to X in $L^2(\Omega, \mathcal{F}, \mathbf{P})$.

    (a) Show that X is also Gaussian.

        Hint: Use Exercise 1.14. Use also the fact that there is a subsequence that converges almost surely; see Exercise 3.14.

    (b) Find its mean and variance in terms of X.

5.20. $\star$ L^2 **is complete.** We prove that the space $L^2(\Omega, \mathcal{F}, \mathbf{P})$ is complete; that is, if $(X_n, n \geq 1)$ is a Cauchy sequence in L^2 (see Remark 5.14), then there exists $X \in L^2(\Omega, \mathcal{F}, \mathbf{P})$ such that $X_n \to X$ in L^2.

    (a) Argue from the definition of Cauchy sequence that we can find a subsequence $(X_{n_k}, k \geq 0)$ such that $\|X_m - X_{n_k}\| \leq 2^{-k}$ for all $m > n_k$, where $\| \cdot \|$ is the L^2-norm.

    (b) Consider the candidate limit $\sum_{j=0}^{\infty}(X_{n_{j+1}} - X_{n_j})$ with $X_{n_0} = 0$. Show that this sum converges almost surely (so X is well-defined) by considering

$$\sum_{j=0}^{k} \mathbf{E}[|X_{n_{j+1}} - X_{n_j}|].$$

    (c) Show that $\|X - X_{n_k}\| \to 0$ as $k \to \infty$. Conclude that $\|X\| < \infty$. (This shows the convergence in L^2 along the subsequence!)

    (d) Use again the Cauchy definition and the subsequence to show convergence of the whole sequence; i.e., $\|X_n - X\| \to 0$.

5.21. $\star$ **Another application of Doob's maximal inequality.** Let $(B_t, t \in [0, 1])$ be a Brownian motion defined on $(\Omega, \mathcal{F}, \mathbf{P})$. Recall from Example 5.20 that the process

$$Z_t = (1 - t) \int_0^t \frac{1}{1 - s} \, dB_s, \quad 0 \leq t < 1,$$

has the distribution of Brownian bridge on $[0, 1)$. In this exercise we prove $\lim_{t \to 1} Z_t = 0$ almost surely as expected.

    (a) Show that $\lim_{t \to 1} Z_t = 0$ in $L^2(\Omega, \mathcal{F}, \mathbf{P})$.

    (b) Using Doob's maximal inequality of Exercise 4.19, show that

$$\mathbf{P}\left(\max_{t \in \left[1 - \frac{1}{2^n}, 1 - \frac{1}{2^{n+1}}\right]} |Z_t| > \delta\right) \leq \frac{1}{\delta^2} \frac{1}{2^{n-1}}.$$

    (c) Deduce that $\lim_{t \to 1} Z_t = 0$ almost surely using the Borel-Cantelli lemma.

5.8. Historical and Bibliographical Notes

Stochastic integrals on Brownian motion were studied before Itô, notably by Wiener [**PW87**]. It was Kiyosi Itô who extended the definition to include integrands that were possibly dependent on the Brownian motion in a seminal paper during World War II [**Ito44**]. It is important to note that other definitions of stochastic integrals exist where the integrand is not necessarily adapted. The most famous one is arguably the

Stratonovich integral [**Str64**], for which the integrand depends symmetrically on the past and future. This definition has important applications in physics. The reader is referred to [**Øks03**] for an introduction to this integral and the comparison with the Itô integral. The proof of the continuity of the Itô integral in Theorem 5.12 is done in [**Ste01**]. Interestingly, Tanaka did not publish his formula. The first occurrence of the formula seems to have been in [**McK62**], giving credit to Tanaka.

Chapter 6

Multivariate Itô Calculus

In the last chapter, we studied functions of time and of a single standard Brownian motion. We saw how to construct martingales from such functions using Itô's formula. Here, we generalize the theory to functions of several Brownian motions. This unleashes the full power of Itô calculus. We start by defining the d-dimensional Brownian motion in Section 6.1. As in one dimension, there is an Itô formula for a function of time and of d Brownian motions as detailed in Section 6.2. Not surprisingly, there is an explicit relation between PDEs and the condition for a function of d Brownian motions to be a martingale. We demonstrate the power of the multivariate calculus by providing two important applications in Sections 6.3 and 6.4: the recurrence and transience of Brownian motion in $\mathbb{R}^d$ and the solution to the Dirichlet problem. Both applications can be seen as a generalization of the gambler's ruin problem to higher dimensions.

6.1. Multidimensional Brownian Motion

A d-dimensional Brownian motion or a Brownian motion in $\mathbb{R}^d$ is constructed by taking d independent standard Brownian motions and by considering a process $(B_t, t \geq 0)$ in $\mathbb{R}^d$ where each coordinate of B_t is a standard Brownian motion.

Definition 6.1 (Brownian motion in $\mathbb{R}^d$). Take $d \in \mathbb{N}$. Let $B^{(1)}, \dots, B^{(d)}$ be independent standard Brownian motions on $(\Omega, \mathcal{F}, \mathbf{P})$. The process $(B_t, t \geq 0)$ taking value in $\mathbb{R}^d$ defined by

$$B_t = (B_t^{(1)}, \dots, B_t^{(d)}), \quad t \geq 0,$$

is called a *d-dimensional Brownian motion* or a *Brownian motion in $\mathbb{R}^d$*.

The Brownian filtration $(\mathcal{F}_t, t \geq 0)$ is now composed of the information of all Brownian motions. In other words, it is given by the sigma-fields

$$\mathcal{F}_t = \sigma(B_s^{(i)}, i \leq d, s \leq t).$$

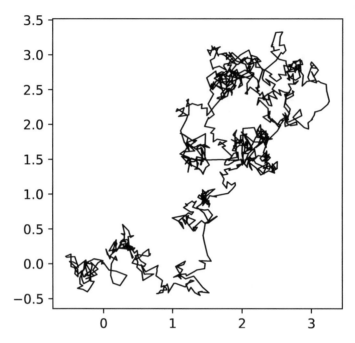

Figure 6.1. A simulation of a path of two-dimensional Brownian motion starting at $(0, 0)$ from time 0 to time 5 for a time discretization of 0.005. See Numerical Project 6.1.

For every outcome ω, the *path* or *trajectory* of a d-dimensional Brownian motion is a curve in space parametrized by the time t:

$$t \mapsto B_t(\omega) = (B_t^{(1)}(\omega), \ldots, B_t^{(d)}(\omega)).$$

Of course, this curve is continuous, since each coordinate is. Figure 6.1 gives an example of one path of a two-dimensional Brownian motion. This is a very rugged and intertwined curve! We might wonder what it does as $t \to \infty$. Does it wander around $(0, 0)$ ad infinitum or does it eventually escape to infinity? We will answer this question in Section 6.3. For doing so, we shall need a version of Itô's formula for multidimensional Brownian motion. We finish this section by noticing that it is also easy to construct Brownian motions in higher dimensions for which the coordinates are correlated.

Example 6.2 (Brownian motion with correlated coordinates). Let $(B_t, t \geq 0)$ be a two-dimensional Brownian motion. Let $-1 < \rho < 1$. We construct the two-dimensional process $(W_t, t \geq 0)$ as follows: $W_t = (W_t^{(1)}, W_t^{(2)})$ where

$$W_t^{(1)} = B_t^{(1)}, \qquad W_t^{(2)} = \rho B_t^{(1)} + \sqrt{1 - \rho^2} B_t^{(2)}.$$

Then, the coordinates of W_t are standard Brownian motions. However, the coordinates are not independent if $\rho \neq 0$ since $\mathbf{E}[W_t^{(1)} W_t^{(2)}] = \rho t$.

6.2. Itô's Formula

Let's first consider functions of *space* only. More precisely, for $d \geq 1$, consider a smooth function

$$f : \mathbb{R}^d \to \mathbb{R}$$

$$x \mapsto f(x),$$

where we write $x = (x_1, \ldots, x_d)$. By a smooth function here, we mean a function that is differentiable enough times to serve the purpose at hand. More precisely, remember from multivariable calculus that such a function has partial derivatives in each variable. There are d of them. We denote them using the following compact notation, similar to the one in equation (5.25):

$$\partial_1 f(x) = \frac{\partial f}{\partial x_1}(x) \quad , \quad \ldots \quad , \quad \partial_d f(x) = \frac{\partial f}{\partial x_d}(x).$$

The partial derivatives can be regrouped in one single object, *the gradient of f*,

$$\nabla f(x) = (\partial_1 f(x), \ldots, \partial_d f(x)).$$

We say a function $f : \mathbb{R}^d \to \mathbb{R}$ is continuously differentiable or in $\mathcal{C}^1(\mathbb{R}^d)$ if its partial derivatives are continuous functions of $\mathbb{R}^d$.

Example 6.3. The function $f(x, y) = x_1^2 + x_2^2$ is in $\mathcal{C}^1(\mathbb{R}^2)$. The partial derivatives with respect to each of the two variables are

$$\partial_1 f(x) = 2x_1, \qquad \partial_2 f(x) = 2x_2.$$

These are two continuous functions of (x_1, x_2). However, the function $g(x_1, x_2) = \sqrt{x_1^2 + x_2^2}$ is not in $\mathcal{C}^1(\mathbb{R}^2)$. The partial derivatives exist but not at the point $(0, 0)$:

$$\partial_1 g(x) = \frac{x_1}{\sqrt{x_1^2 + x_2^2}}, \qquad \partial_2 g(x) = \frac{x_2}{\sqrt{x_1^2 + x_2^2}}.$$

Note that the graph of this function is a cone. With this in mind, it is not surprising that it is not differentiable at the tip.

Similarly to functions of one variable, see equation (5.18), a function f in $\mathcal{C}^1(\mathbb{R}^d)$ admits a Taylor expansion to the first order: For a fixed $x \in \mathbb{R}^d$,

$$
\text{(6.1)} \qquad
\begin{aligned}
f(y) &= f(x) + \sum_{j=1}^{d} \partial_j f(x) \, (y_j - x_j) + \mathcal{E} \\
&= f(x) + \nabla f(x)^T (y - x) + \mathcal{E},
\end{aligned}
$$

where $\mathcal{E}$ is an error term that goes to 0 faster than the linear term as y approaches x. The second equality takes advantage of the gradient and the dot product in $\mathbb{R}^d$ (thinking of x and y as column vectors). Equation (6.1) is written in differential form as

$$\text{(6.2)} \qquad \mathrm{d}f(x) = \sum_{j=1}^{d} \partial_j f(x) \, \mathrm{d}x_j = \nabla f(x)^T \, \mathrm{d}x,$$

where $\mathrm{d}x$ stands for the column vector $(\mathrm{d}x_j, j \leq d)$.

As we learned in Chapter 5, stochastic calculus with Brownian motion often demands the existence of second-order derivatives in space to make up for the quadratic variation of Brownian motion. For a function $f : \mathbb{R}^d \to \mathbb{R}$, there are many second-order partial derivatives. (In fact, there are $d(d+1)/2$.) Again we use a compact notation to denote them:

$$\partial_i \partial_j f(x) = \frac{\partial^2 f}{\partial x_i \partial x_j}(x), \quad 1 \le i, j \le d.$$

We sometimes write ∂_i^2 for $\partial_i \partial_i$.

Definition 6.4. A function $f : \mathbb{R}^d \to \mathbb{R}$ is said to be twice continuously differentiable or in $\mathcal{C}^2(\mathbb{R}^d)$ if its second-order partial derivatives exist and are continuous functions in $\mathbb{R}^d$. More precisely, the partial derivatives

$$\partial_i \partial_j f(x), \quad 1 \le i, j \le d,$$

are continuous function in $\mathbb{R}^d$.

The second-order derivatives of a function $f \in \mathcal{C}^2(\mathbb{R}^d)$ can be gathered in a single object, *the Hessian matrix of f* denoted by $\nabla^2 f$, with entries

$$\left(\nabla^2 f(x) \right)_{ij} = \partial_i \partial_j f(x), \quad 1 \le i, j \le d.$$

This is a $d \times d$ symmetric matrix, because the mixed derivatives of a smooth function must be equal.

Example 6.5. The function $f(x) = e^{x_1} \cos x_2$ for $x = (x_1, x_2)$ is in $\mathcal{C}^2(\mathbb{R}^2)$. The second-order partial derivatives are

$$\partial_1^2 f(x) = e^{x_1} \cos x_2, \quad \partial_2^2 f(x) = -e^{x_1} \cos x_2, \quad \partial_1 \partial_2 f(x) = -e^{x_1} \sin x_2.$$

The Hessian matrix of f at $x = (0,0)$ is

$$\nabla^2 f(0,0) = \begin{pmatrix} 1 & 0 \\ 0 & -1 \end{pmatrix}.$$

As in equation (5.19), a function f in $\mathcal{C}^2(\mathbb{R}^d)$ admits a Taylor expansion to the second order that involves all partial derivatives. Around a fixed $x \in \mathbb{R}^d$, we get

(6.3)
$$f(y) - f(x) = \sum_{i=1}^{d} \partial_i f(x)\, (y_i - x_i) + \frac{1}{2} \sum_{i,j=1}^{d} \partial_i \partial_j f(x)\, (y_i - x_i)(y_j - x_j) + \mathcal{E}$$

$$= \nabla f(x)^T (y - x) + \frac{1}{2}(y - x)^T \nabla^2 f(x)(y - x) + \mathcal{E},$$

where $\mathcal{E}$ is an error term that goes to 0 faster than the quadratic term as y approaches x. We rewrote the first equality in vector notation in the second line. Again, we can write this in differential form as

(6.4)
$$df(x) = \sum_{i=1}^{d} \partial_i f(x)\, dx_i + \frac{1}{2} \sum_{i,j=1}^{d} \partial_i \partial_j f(x)\, dx_i\, dx_j = \nabla f(x)^T dx + \frac{1}{2} dx^T \nabla^2 f(x)\, dx.$$

For example, the second-order Taylor expansion at $(0,0)$ of the function $f(x) = e^{x_1} \cos x_2$ in Example 6.5 is

$$f(x) = 1 + (1,0)\begin{pmatrix} x_1 \\ x_2 \end{pmatrix} + \frac{1}{2}(x_1, x_2)\begin{pmatrix} 1 & 0 \\ 0 & -1 \end{pmatrix}\begin{pmatrix} x_1 \\ x_2 \end{pmatrix} + \mathcal{E} = 1 + x_1 + \frac{1}{2}x_1^2 - \frac{1}{2}x_2^2 + \mathcal{E}.$$

We are now ready to state Itô's formula for a function of a multidimensional Brownian motion. Let $(B_t, t \geq 0)$ be a d-dimensional Brownian motion and a function $f : \mathbb{R}^d \to \mathbb{R}$ in $\mathcal{C}^2(\mathbb{R}^d)$. We consider the stochastic process

$$(f(B_t), t \geq 0).$$

Note that, at each time t, $f(B_t)$ is simply a random variable, so $(f(B_t), t \geq 0)$ is a stochastic process as seen before, where to each t we associate a random variable. For example, if we take the functions f from Examples 6.3 and 6.5, respectively, we have the stochastic processes

$$(6.5) \qquad X_t = (B_t^{(1)})^2 + (B_t^{(2)})^2 \quad \text{and} \quad Y_t = \exp(B_t^{(1)})\cos B_t^{(2)}, \quad t \geq 0.$$

Clearly, the paths of these processes are continuous. It turns out that these stochastic processes can be re-expressed as a sum of Itô integrals and Riemann integrals.

Theorem 6.6 (Itô's formula). *Let $(B_t, t \geq 0)$ be a d-dimensional Brownian motion. Consider $f \in \mathcal{C}^2(\mathbb{R}^d)$. Then we have with probability one that for all $t \geq 0$,*

$$(6.6) \qquad f(B_t) - f(B_0) = \sum_{i=1}^{d} \int_0^t \partial_i f(B_s) \, dB_s^{(i)} + \frac{1}{2}\int_0^t \sum_{i=1}^{d} \partial_i^2 f(B_s) \, ds.$$

We stress that, as in the one-dimensional case in Theorem 5.24, Itô's formula is an equality of processes (and not an equality in distribution). Thus the processes on both sides must agree for each path; see Figures 6.2 and 6.3. Interestingly, the mixed partial derivatives $\partial_i \partial_j f$, $i \neq j$, do not appear in the formula! We see from Itô's formula that the process $f(B_t)$ can be represented as a sum of $d + 1$ processes: d Itô integrals and one Riemann integral (which is a process of finite variation). In vector notation, the formula takes the form

$$f(B_t) - f(B_0) = \int_0^t \nabla f(B_s)^T \, dB_s + \frac{1}{2}\int_0^t \Delta f(B_s) \, ds,$$

where it is understood that the first term is the sum of the d Itô integrals in equation (6.6). The symbol Δ in the second term stands for the *Laplacian of f*, defined by

$$(6.7) \qquad \Delta f(x) = \sum_{i=1}^{d} \partial_i^2 f(x).$$

In differential form, Itô's formula becomes the very neat

$$(6.8) \qquad df(B_t) = \sum_{i=1}^{d} \partial_i f(B_s) \, dB_t^{(i)} + \sum_{i=1}^{d} \frac{1}{2}\partial_i^2 f(B_s) \, dt = \nabla f(B_t)^T \, dB_s + \frac{1}{2}\Delta f(B_s) \, ds.$$

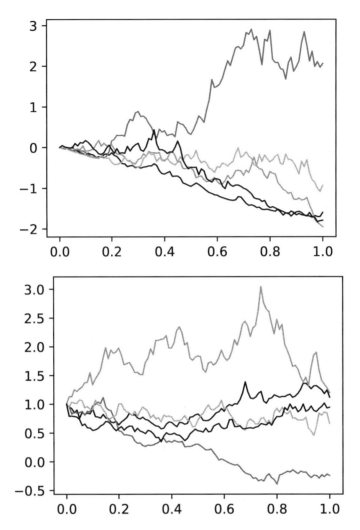

Figure 6.2. A simulation of 5 paths of the processes $(X_t - 2t, t \in [0,1])$ and $(Y_t, t \in [0,1])$ defined in equation (6.5).

Example 6.7. Consider the functions in Examples 6.3 and 6.5 and the processes $(X_t, t \geq 0)$ and $(Y_t, t \geq 0)$ in equation (6.5). If we apply Itô's formula to the first process, we have

$$X_t = \int_0^t 2B_s^{(1)} \, dB_s^{(1)} + \int_0^t 2B_s^{(2)} \, dB_s^{(2)} + 2t,$$

since $\nabla f(x) = (2x_1, 2x_2)$ and $\Delta f(x) = 4$. The second process gives

$$Y_t = 1 + \int_0^t \exp(B_s^{(1)}) \cos B_s^{(2)} \, dB_s^{(1)} - \int_0^t \exp(B_s^{(1)}) \sin B_s^{(2)} \, dB_s^{(2)},$$

since $\nabla f(x) = (e^{x_1} \cos x_2, -e^{x_1} \sin x_2)$ and $\Delta f(x) = 0$.

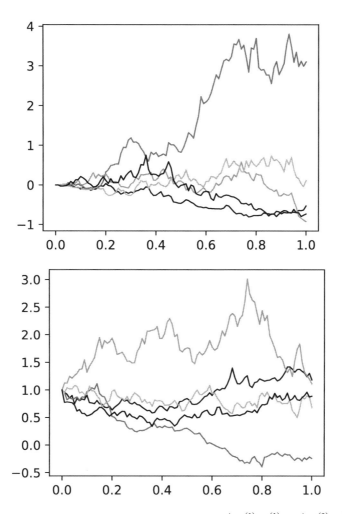

Figure 6.3. A simulation of 5 paths of the processes $\int_0^t 2B_s^{(1)} \, dB_s^{(1)} + \int_0^t 2B_s^{(2)} \, dB_s^{(2)}$ and $1 + \int_0^t \exp(B_s^{(1)}) \cos B_s^{(2)} \, dB_s^{(1)} - \int_0^t \exp(B_s^{(1)}) \sin B_s^{(2)} \, dB_s^{(2)}$. These were generated using the same Brownian motion as in Figure 6.2. Note how each path corresponds. No magic formula here, just Itô's formula. See Numerical Project 6.2.

Sketch of Proof of Theorem 6.6. The proof of the formula follows the usual recipe: Taylor's theorem together with quadratic variation and cross-variation. In this case, we do get a new cross-variation between the different Brownian motions. More precisely, consider a partition $(t_j, j \le n)$ of $[0, t]$. Then we can write

$$f(B_t) - f(B_0) = \sum_{j=0}^{n-1} f(B_{t_{j+1}}) - f(B_{t_j}).$$

We can apply the Taylor expansion (6.3) for each j to get

$$f(B_t) - f(B_0) = \sum_{j=0}^{n-1} \nabla f(B_{t_j})^T (B_{t_{j+1}} - B_{t_j})$$

$$+ \frac{1}{2} \sum_{j=0}^{n-1} (B_{t_{j+1}} - B_{t_j})^T \nabla^2 f(B_{t_j})(B_{t_{j+1}} - B_{t_j}) + \mathcal{E}.$$

We wrote the expansion using vector notation to be economical. Let's keep in mind that each term is a sum over the derivatives. The first term will converge to the d Itô integrals as in the one-dimensional case. As for the second term, it involves the *cross-variation* between the Brownian motions: For $1 \le i, k \le d$,

$$\sum_{j=0}^{n-1} (B_{t_{j+1}}^{(i)} - B_{t_j}^{(i)})(B_{t_{j+1}}^{(k)} - B_{t_j}^{(k)}).$$

It is not hard to show that this converges to 0 in L^2 when $i \ne k$; see Exercise 6.5. This explains why the mixed derivatives disappear from the formula! As for the case $i = k$, it reduces to the quadratic variation as in the one-dimensional case. This is where the Riemann integral arises, after suitable conditioning on $\mathcal{F}_{t_j}$, the sigma-field generated by $B_s, s \le t_j$. □

As in the one-dimensional case, it is not necessary to learn Itô's formula by heart. It suffices to write the differential of the function f to second order as in (6.4). We can then apply the rules of multivariate Itô calculus:

·	dt	$dB_t^{(1)}$	$dB_t^{(2)}$	...
dt	0	0	0	0
$dB_t^{(1)}$	0	dt	0	0
$dB_t^{(2)}$	0	0	dt	0
...	0	0	0	dt

Note that the rule $dB_t^{(i)} dB_t^{(j)} = 0$ for $i \ne j$ is motivated by the cross-variation result in Exercise 6.5.

How can we construct martingales using Itô's formula? Recall that an Itô integral $\int_0^t X_s \, dB_s$, $t \le T$, is a martingale whenever the integrand is in $\mathcal{L}_c^2(T)$, the space of adapted processes with continuous paths and for which

$$\int_0^T \mathbf{E}[X_s^2] \, ds < \infty.$$

The only difference here is that the integrand is a function of many Brownian motions. However, the integrands involved in the Itô integrals of Itô's formula (6.6) are clearly adapted to the filtration $(\mathcal{F}_t, t \ge 0)$ of $(B_t, t \ge 0)$, as they are functions of the Brownian motion at the time. The arguments of Proposition 5.7 and Theorem 5.12 apply verbatim if we take the definition of $\mathcal{L}_c^2(t)$ with the filtration $(\mathcal{F}_t, t \ge 0)$ of $(B_t, t \ge 0)$. With this in mind, we have the following corollary.

Corollary 6.8 (Brownian martingales). *Let $(B_t, t \geq 0)$ be a Brownian motion in $\mathbb{R}^d$. Consider $f \in \mathcal{C}^2(\mathbb{R}^d)$ such that the processes $(\partial_i f(B_t), t \leq T) \in \mathcal{L}_c^2(T)$ for every $i \leq d$. Then the process*

$$f(B_t) - \int_0^t \frac{1}{2} \Delta f(B_s) \, ds, \quad t \leq T,$$

is a martingale for the Brownian filtration.

For example, consider the processes $(X_t, t \geq 0)$ and $(Y_t, t \geq 0)$ in Example 6.7. Then the processes $X_t - 2t$ and Y_t are martingales for the Brownian filtration. In one dimension, there are no interesting martingales constructed with functions of *space* only. Indeed, $(f(B_t), t \geq 0)$ is a martingale if and only if $f''(x) = 0$ for all x. But such functions are of the form $f(x) = ax + b, a, b \in \mathbb{R}$. In other words, in one dimension, Brownian martingales of the form $f(B_t)$ are simply $aB_t + b$. Not very surprising! The situation is very different in higher dimension. Indeed, Corollary 6.8 implies that $f(B_t)$ is a martingale whenever f is a *harmonic function*:

Definition 6.9. A function $f : \mathbb{R}^d \to \mathbb{R}$ is *harmonic* in $\mathbb{R}^d$ if and only if $\Delta f(x) = 0$ for all $x \in \mathbb{R}^d$. More generally, a function $f : \mathbb{R}^d \to \mathbb{R}$ is harmonic in the region $\mathcal{O} \subset \mathbb{R}^d$ if and only if $\Delta f(x) = 0$ for all $x \in \mathcal{O}$.

Note that the function $f(x) = e^{x_1} \cos x_2$ is harmonic in $\mathbb{R}^d$. This is why the process $Y_t = \exp(B_t^{(1)}) \cos(B_t^{(2)})$ is a martingale. The distinction to a subset of $\mathbb{R}^d$ in the above definition is important since it may happen that the function is harmonic only in a subset of the space; see for example equation (6.10). It is possible to define a Brownian martingale in such cases by considering the process until it exits the region. This will be important in Section 6.3.

Itô's formula generalizes to functions of time and space as in Proposition 5.28.

Definition 6.10. A function $f : [0, \infty) \times \mathbb{R}^d \to \mathbb{R}$ is in $\mathcal{C}^{1,2}([0, T] \times \mathbb{R}^d)$ if the partial derivative in *time*

$$\partial_0 f(t, x) = \frac{\partial f}{\partial t}(t, x)$$

exists and is continuous and the second-order partial derivatives in *space*

$$\partial_i \partial_j f(t, x) = \frac{\partial^2 f}{\partial x_i \partial x_j}(t, x), \quad 1 \leq i, j \leq d,$$

exist and are continuous. Note that we use the convenient notation ∂_0 for the partial derivatives in time.

Theorem 6.11 (Itô's formula). *Let $(B_t, t \leq T)$ be a d-dimensional Brownian motion. Consider a function $f \in \mathcal{C}^{1,2}([0, T] \times \mathbb{R}^d)$. Then we have with probability one for all $t \leq T$,*

$$f(t, B_t) - f(0, B_0) = \sum_{i=1}^d \int_0^t \partial_i f(s, B_s) \, dB_s^{(i)} + \int_0^t \left(\partial_0 f(s, B_s) + \frac{1}{2} \Delta f(s, B_s) \right) ds,$$

where $\Delta = \sum_{i=1}^d \partial_{x_i}^2$.

The martingale condition is then similar to the ones in Corollary 6.8: If the processes $(\partial_i f(s, B_t), t \leq T) \in \mathcal{L}_c^2(T)$ for every $1 \leq i \leq d$, then the process

$$f(t, B_t) - \int_0^t \left\{ \partial_0 f(s, B_s) + \frac{1}{2} \Delta f(s, B_s) \right\} ds, \quad t \leq T,$$

is a martingale for the Brownian filtration. In particular, if f satisfies the partial differential equation

$$(6.9) \qquad \qquad \frac{\partial f}{\partial t} + \frac{1}{2} \Delta f = 0,$$

then the process $(f(t, B_t), t \leq T)$ is itself a martingale.

6.3. Recurrence and Transience of Brownian Motion

In dimension one, we established in Example 4.43 that every path of Brownian motion reaches any level $a \in \mathbb{R}$. More precisely, for $a > 0$, we defined the stopping time

$$\tau = \min\{t \geq 0 : B_t \geq a\}$$

and showed that $\mathbf{P}(\tau_a < \infty) = 1$. This implies in particular that every path will come back to 0 and will do so infinitely many times. (This is because $(B_{t+\tau} - B_\tau, t \geq 0)$ is a standard Brownian motion, a property we have already seen in Lemma 4.47.) This property of Brownian motion is called *recurrence*. This is to be compared with Brownian motion with drift in Section 5.5. There, we found that if the drift is negative, then there are paths that will not reach a given level $a > 0$ with positive probability; see equation (5.29) where it was established for a positive drift and a negative level. Such paths go to infinity without ever going back to 0. This property of the process is called *transience*.

We will now derive similar properties for the multidimensional Brownian motion. We will rely heavily on Corollary 6.8 and on some knowledge of harmonic functions. Harmonic functions play a very important role in mathematics, physics, and in nature in general. As we mentioned earlier, if $d = 1$, the only harmonic functions are the linear functions, since the equation $f''(x) = 0$ has the solutions $f(x) = ax + b$. However, in higher dimensions, the collection of harmonic functions is extremely rich. This gives access to a plethora of Brownian martingales. For example, the following functions are harmonic in the whole space minus the origin, $\mathbb{R}^d \setminus 0$,

$$(6.10) \qquad \qquad h(x) = \begin{cases} \log \|x\|, & d = 2, \\ \|x\|^{2-d}, & d \geq 3, \end{cases} \qquad x \in \mathbb{R}^d.$$

The reader is asked to verify this in Exercise 6.8. These examples are special as they are rotationally invariant: They only depend on the distance $\|x\| = \sqrt{x_1^2 + \cdots + x_d^2}$ of x to the origin.

Interestingly enough, the answer to the recurrence vs transience question depends on the dimension. We will show that Brownian motion is recurrent in dimension $d = 2$ in the sense that every Brownian path starting from a given x will eventually enter a disc around the origin, no matter how small the disc is. The Brownian path will then enter this disc infinitely many times, as $t \to \infty$. Note that we did not say that the path actually hits 0, but that it enters a *disc around* 0. This nuance is important as we will in fact show that a Brownian path actually never hits a given point. In dimension $d = 3$ and higher, it is proved that there are some paths starting from a given x that will never enter a given ball around the origin with positive probability. This is the transience property.

The strategy of proof is similar to the approach we took in Section 5.5 for the Brownian motion with drift. We will find a good function $h : \mathbb{R}^d \to \mathbb{R}$ for which $h(B_t)$ is a martingale. In light of Corollary 6.8, this function needs to be harmonic in a suitable region. The desired probability is then obtained by considering the right boundary values.

Theorem 6.12. *Let* $(B_t, t \geq 0)$ *be a Brownian motion in* $\mathbb{R}^d$ *starting at* $B_0 = x$. *Consider for* $r < \|x\|$ *the stopping time*

$$\tau_r = \min\{t \geq 0 : \|B_t\| \leq r\},$$

the first hitting time of the paths in a ball of a radius r *around the origin. We have*

$$\mathbf{P}(\tau_r < \infty) = \begin{cases} 1 & \text{if } d \leq 2, \\ \left(\dfrac{r}{\|x\|}\right)^{d-2} & \text{if } d \geq 3. \end{cases}$$

In particular, for $d \leq 2$, *the paths are recurrent; that is, they come back infinitely many times in a neighborhood of the origin. For* $d \geq 3$, *each path will eventually never come back to a neighborhood of the origin.*

Note that we made sure that the starting point of the Brownian motion x is outside the ball of radius r.

Proof. Consider another hitting time of a ball with a radius larger than $\|x\|$:

$$\tau'_R = \min\{t \geq 0 : \|B_t\| \geq R\}, \quad r < \|x\| < R.$$

Note that τ'_R must increase with R and that it must go to $+\infty$ as $R \to \infty$. In particular, by Lemma 1.4 or Theorem 3.15, we have

$$\mathbf{P}(\tau_r < \infty) = \lim_{R \to \infty} \mathbf{P}(\tau_r < \tau'_R).$$

If we set $\tau = \tau_r \wedge \tau'_R$, then the right side is $\lim_{R \to \infty} \mathbf{P}(\|B_\tau\| = r)$. Note that $\tau < \infty$ with probability one, since $\mathbf{P}(\|B_{n+1} - B_n\| > R) > 0$ uniformly in n, so the same argument as in Example 4.41 holds.

To compute $\mathbf{P}(\|B_\tau\| = r)$, the idea is to find a good function h with the following properties:

- **Martingale:** $(h(B_t), t \leq \tau)$ is a bounded martingale, so that the optional stopping theorem, Corollary 4.38, implies

$$\mathbf{E}[h(B_\tau)] = h(B_0) = h(x).$$

- **Boundary values:** Since B_τ is a point at $\|x\| = r$ or at $\|x\| = R$, we pick boundary conditions for h so that $\mathbf{E}[h(B_\tau)] = \mathbf{P}(\|B_\tau\| = r)$.

The second point is easy; it suffices to take the *boundary conditions* $h(x) = 0$ if $\|x\| = R$ and $h(x) = 1$ if $\|x\| = r$. For the first point, by Itô's formula, we pick h to be harmonic in the annulus $\{x \in \mathbb{R}^d : r < x < R\}$. Since the annulus and the boundary values are rotationally-invariant here, we can use the function (6.10). Of course, the functions $ah(x) + b$ remain harmonic. To satisfy the boundary conditions, we can pick a and b suitably to get

(6.11)
$$h(x) = \begin{cases} \frac{R - \|x\|}{R - r}, & d = 1, \\ \frac{\log R - \log \|x\|}{\log R - \log r}, & d = 2, \\ \frac{R^{2-d} - \|x\|^{2-d}}{R^{2-d} - r^{2-d}}, & d \geq 3. \end{cases}$$

It is now straightforward to take the limit $R \to \infty$ to get the claimed result. $\square$

Example 6.13 (A Brownian path in $d \geq 2$ never hits a given point). We get more from the above proof. Indeed, in dimension $d \geq 2$, it implies that the probability that a Brownian motion starting at $B_0 = x$ eventually hits any other point y is 0. Indeed, if we take $y = 0$ and $x \neq 0$, we have

$$\mathbf{P}(\exists t > 0 : B_t = 0) = \lim_{R \to \infty} \lim_{r \to 0} \mathbf{P}(\tau_r < \tau_R').$$

Here, we shrink the inner radius to 0 first, and then we let the outer radius go to infinity after. It remains to take the limits of the expression obtained in equation (6.11). The first limit $r \to 0$ gives 0 right away.

6.4. Dynkin's Formula and the Dirichlet Problem

In the last section, we translated the problem of recurrence and transience of Brownian in terms of a *boundary-value problem*: We needed to find a particular function h that solves a PDE (i.e, $\Delta h = 0$) in a certain region of space (the annulus) and that satisfies given boundary conditions (the function is 1 on the inner circle, and 0 on the outer circle). This is a particular case of the *Dirichlet problem* in PDE.

Consider a region $\mathcal{O} \subset \mathbb{R}^d$. For our purpose, it is useful to think of $\mathcal{O}$ as a bounded region. We write $\partial \mathcal{O}$ for the boundary of $\mathcal{O}$. It is also convenient to think of $\mathcal{O}$ as *open*; i.e., it does not contain its boundary. In the proof of Theorem 6.12, the region was the annulus $\{x : r < \|x\| < R\}$ and the boundary was given by the spheres $\{x : \|x\| = r\}$ and $\{x : \|x\| = R\}$. The Dirichlet problem in the region $\mathcal{O}$ can be stated as follows:

Let $f : \partial \mathcal{O} \to \mathbb{R}$ be a function on the boundary of $\mathcal{O}$. Can we extend the function f to O in such a way that the extension is harmonic on $\mathcal{O}$ and coincides with f on the boundary?

In the instance of the proof of Theorem 6.12, we knew the solution of the problem, thanks to equation (6.10). The only job left to do was to adjust some parameters to fit the boundary conditions. It turns out that the formalism of stochastic calculus allows us to express the solution of the Dirichlet problem as an average over Brownian paths for general regions of space and boundary conditions. To do so, we first need to introduce a twist on Itô's formula called *Dynkin's formula*.

Let x be a point in $\mathcal{O}$. Consider a Brownian motion $(B_t, t \geq 0)$ starting at x, $B_0 = x$. To emphasize the dependence on x, we write $\mathbf{P}_x$ and $\mathbf{E}_x$ for the probability and the expectation for the Brownian motion. Dynkin's formula is obtained by merging Corollary 6.8 for the Brownian martingales together with the optional stopping theorem (Corollary 4.38). More precisely, consider $f \in \mathcal{C}^2(\mathbb{R}^d)$ that satisfies the assumption of Corollary 6.8. We then have that the process

$$M_t = f(B_t) - \frac{1}{2} \int_0^t \Delta f(B_s) \, ds, \quad t \geq 0,$$

is a Brownian martingale. Consider also the stopping time $\tau = \min\{t \geq 0 : B_t \in \mathcal{O}^c\}$. This is the first exit time of $\mathcal{O}$ of the Brownian path. Then if the assumptions of the optional stopping theorem are fulfilled, we get *Dynkin's formula*

$$(6.12) \qquad \mathbf{E}[M_\tau] = \mathbf{E}[f(B_\tau)] - \frac{1}{2}\mathbf{E}\left[\int_0^\tau \Delta f(B_s) \, ds\right] = f(B_0) = M_0.$$

When are the assumptions of optional stopping fulfilled? It is always true that the stopped martingale $(M_{t \wedge \tau}, t \geq 0)$ is a martingale by Proposition 4.37. Therefore, optional stopping holds whenever $\tau < \infty$ with probability one and we can take the limit $t \to \infty$ inside the expectation. This is usually justified by one of the convergence theorems (Theorem 3.15 or Theorem 4.40). If we assume that the region $\mathcal{O}$ is bounded, then we automatically have that $\tau < \infty$ by the same argument as in Example 4.41 on the gambler's ruin. The justification that the limit can be passed inside the expectation might depend on the problem at hand.

As an application of Dynkin's formula, let's express the solution of the Dirichlet problem as an average over Brownian motion. If f is harmonic in $\mathcal{O}$, then $\Delta f = 0$ in $\mathcal{O}$, so the process $(f(B_t), t \leq \tau)$ is itself a martingale. Note that since $\mathcal{O}$ is bounded, this is a bounded martingale! Moreover, the integrands $\partial_i f(B_s)$, $1 \leq i \leq d$, are bounded. We conclude that the solution to the Dirichlet problem can be represented as follows:

$$(6.13) \qquad\qquad f(x) = \mathbf{E}_x[f(B_\tau)], \quad x \in \mathcal{O}.$$

In other words, the value of the harmonic function of f at x is given by the average over the Brownian paths of the value of the function at the exit point of the path on the boundary of $\mathcal{O}$. Equation (6.13) does not prove the existence of the solution to the Dirichlet problem, as we assume that such a function exists. Existence of the solution to the Dirichlet problem can be shown this way, assuming the region $\mathcal{O}$ and the function on the boundary are nice enough. However, note that the averaging on the right-hand side of the formula makes sense even in the case where f is simply defined on the boundary $\partial\mathcal{O}$. We can then define a posteriori the *harmonic extension* of f inside $\mathcal{O}$ by defining its value at x as $\mathbf{E}_x[f(B_\tau)]$. In particular, the value of the harmonic extension can be evaluated numerically; see Numerical Project 6.3 and Figure 6.4.

As another application, we generalize the result obtained in Example 4.42 for the expected time before exiting a region.

Example 6.14 (Application of Dynkin's formula). Consider a two-dimensional Brownian motion $B_t = (B_t^{(1)}, B_t^{(2)})$ starting at x with $\|x\| < 1$. Consider also the stopping time $\tau = \min\{t \geq 0 : \|B_t\| \geq 1\}$, the first time the Brownian motion exits the unit disc. Let's compute the expected waiting time $\mathbf{E}[\tau]$. With analogy with Example 4.42 in dimension one, we consider the martingale of Example 6.7

$$M_t = (B_t^{(1)})^2 + (B_t^{(2)})^2 - 2t.$$

Applying the martingale property to the stopped martingale, we get

$$\mathbf{E}_x[M_{t \wedge \tau}] = M_0 = \|x\|^2.$$

The left-hand side is

$$\mathbf{E}_x[M_{t \wedge \tau}] = \mathbf{E}_x[(B_{t \wedge \tau}^{(1)})^2 + (B_{t \wedge \tau}^{(2)})^2] - 2\mathbf{E}_x[t \wedge \tau].$$

It remains to take the limit $t \to \infty$. The integrand in the first expectation is bounded by 1; therefore Theorem 4.40 applies. As for the second, we can apply Theorem 3.15. Putting all this together gives

$$\|x\|^2 = \mathbf{E}_x[(B_\tau^{(1)})^2 + (B_\tau^{(2)})^2] - 2\mathbf{E}_x[\tau] = 1 - 2\mathbf{E}_x[\tau].$$

We conclude that

$$\mathbf{E}_x[\tau] = \frac{1}{2}(1 - \|x\|^2).$$

Exercise 6.9 generalizes this result to all dimensions.

6.5. Numerical Projects and Exercises

6.1. **2D Brownian motion.**   Consider a two-dimensional Brownian motion $(B_t^{(1)}, B_t^{(2)})$ starting at $(0,0)$.
   (a) Plot one path of this Brownian motion on the plane $\mathbb{R}^2$ on the time interval $[0,5]$ using a discretization of 0.005 and 0.001.
   (b) Consider now the process $(W_t, t \geq 0)$ for $\rho = 1/2$ as in Example 6.2. Plot one path of this process on the plane $\mathbb{R}^2$ on the time interval $[0,5]$ using a discretization of 0.001.

6.2. **Brownian martingales.**   Consider a two-dimensional Brownian motion $(B_t^{(1)}, B_t^{(2)})$ starting at $(0,0)$.
   (a) Sample 10 paths of the processes

$$X_t = (B_t^{(1)})^2 + (B_t^{(2)})^2 - t \quad \text{and} \quad Y_t = \exp(B_t^{(1)}) \cos B_t^{(2)}$$

   on $[0,1]$ using a discretization of 0.01.
   (b) Using the *same* Brownian paths, sample 10 paths of the processes

$$\int_0^t 2B_s^{(1)} \, dB_s^{(1)} + \int_0^t 2B_s^{(2)} \, dB_s^{(2)}$$

   and

$$1 + \int_0^t \exp(B_s^{(1)}) \cos B_s^{(2)} \, dB_s^{(1)} - \int_0^t \exp(B_s^{(1)}) \sin B_s^{(2)} \, dB_s^{(2)}.$$

   The paths should be almost the same.

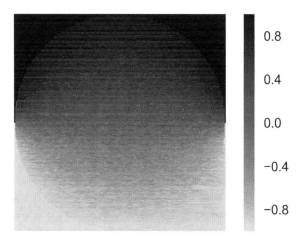

Figure 6.4. The solution to the Dirichlet problem in Numerical Project 6.3.

6.3. **Dirichlet problem.** Consider a harmonic function $h : \mathbb{R}^2 \to \mathbb{R}$ on the unit disc $\{(x, y) : x^2 + y^2 < 1\}$ with boundary values

$$h(x, y) = \begin{cases} 1 & \text{if } x^2 + y^2 = 1 \text{ and } y \geq 0, \\ -1 & \text{if } x^2 + y^2 = 1 \text{ and } y \leq 0. \end{cases}$$

(a) Estimate the value of h at the points $(0, 1/2)$ by averaging the value of h at the exit point of the disc of 1,000 Brownian paths starting at $(0, 1/2)$:

$$h(0, 1/2) \approx \frac{1}{1,000} \sum_{j=1}^{1,000} h(B_{\tau_j}^{(1)}, B_{\tau_j}^{(2)})$$

where $(B_{\tau_j}^{(1)}, B_{\tau_j}^{(2)})$ is the exit point of the j-th path on the unit circle. Use a discretization of 0.01.

(b) Repeat the above at a point (x, y), where x and y range from -1 to 1 at every 0.01. Use 100 paths per point and a discretization of 0.01. Plot your result in a heat map using seaborn.heatmap. See Figure 6.4.

Exercises

6.1. **Rotational symmetry of Brownian motion.** Let $B_t = (B_t^{(1)}, B_t^{(2)})$ be a two-dimensional Brownian motion. Let M be a rotation matrix, i.e., a matrix of the form

$$M = \begin{pmatrix} \cos\theta & \sin\theta \\ -\sin\theta & \cos\theta \end{pmatrix} \text{ for some } \theta \in [0, 2\pi).$$

Show that the process $W_t = MB_t$ is also a two-dimensional Brownian motion. *Hint: Check that the coordinates are independent standard Brownian motions.*

6.2. **Orthogonal symmetry of Brownian motion.** Let $B_t = (B_t^{(i)}, i \le d)$ be a d-dimensional Brownian motion. Let M be a $d{\times}d$ orthogonal matrix, that is, a matrix such that $M^{-1} = M^T$. Show that the process $W_t = MB_t$ is also a d-dimensional Brownian motion.

6.3. **Races between two Brownian motions.** Consider a two-dimensional Brownian motion $B_t = (B_t^{(1)}, B_t^{(2)})$. What is the probability that $B^{(1)}$ reaches 1 before $B^{(2)}$? All right, this is easy.... What about the probability that $B^{(1)}$ reaches 2 before $B^{(2)}$ reaches 1? Write you answer as an integral and evaluate it numerically.
Hint: Bachelier's formula.

6.4. **Drill.** Work out the details in Example 6.7.

6.5. **Cross-variation of $B_t^{(1)}$ and $B_t^{(2)}$.** Let $(t_j, j \le n)$ be a sequence of partitions of $[0, t]$ such that $\max_j |t_{j+1} - t_j| \to 0$ as $n \to \infty$. Prove that

$$\lim_{n \to \infty} \sum_{j=0}^{n} (B_{t_{j+1}}^{(1)} - B_{t_j}^{(1)})(B_{t_{j+1}}^{(2)} - B_{t_j}^{(2)}) = 0 \quad \text{in } L^2.$$

This justifies the rule $dB_t^{(1)} \cdot dB_t^{(2)} = 0$.
Hint: Just compute the second moment of the sum.

6.6. **A function of B_t.** Consider a two-dimensional Brownian motion $B_t = (B_t^{(1)}, B_t^{(2)})$. We consider the process

$$Z_t = B_t^{(1)} B_t^{(2)}, \quad t \ge 0.$$

Write Z_t as a sum of Itô integrals and a Riemann integral. Is this a martingale? Justify your answer.

6.7. **Another function of B_t.** Consider a two-dimensional Brownian motion $B_t = (B_t^{(1)}, B_t^{(2)})$. We consider the function $f : \mathbb{R}^2 \to \mathbb{R}$

$$f(x_1, x_2) = x_1^3 - 3x_1 x_2^2.$$

Write the process $f(B_t)$ as a sum of Itô integrals and a Riemann integral. Is this a martingale? Justify your answer.

6.8. **Harmonic functions.** Verify that the functions $h(x)$ given in equation (6.10) are harmonic in $\mathbb{R}^d \setminus 0$. Argue also that $ah(x) + b$ is harmonic for any $a, b \in \mathbb{R}$.

6.9. **Waiting time for $d > 2$.** We generalize Example 6.14. Consider a d-dimensional Brownian motion B_t starting at x with $\|x\| < 1$. Consider the exit time of the unit ball $\tau = \min\{t \ge 0 : \|B_t\| \ge 1\}$. Find $\mathbf{E}_x[\tau]$.

6.10. **The heat equation.** Recall from equation (1.14) that for any differentiable function F and random variable Z a standard Gaussian random variable, we have the Gaussian integration by parts identity

$$\mathbf{E}[ZF(Z)] = \mathbf{E}[F'(Z)],$$

whenever the expectations make sense. Use this identity to check that the solution $f(t, x)$ to the *heat equation*

$$\partial_0 f = \frac{1}{2}\Delta f \quad \text{with initial condition } f(0, x) = g(x),$$

for some function $g : \mathbb{R}^d \to \mathbb{R}$, is given by

$$f(t, x) = \mathbf{E}_x[g(B_t)],$$

where $(B_t, t \geq 0)$ is a Brownian motion in $\mathbb{R}^d$ starting at $B_0 = x$.

Note that this equation is similar to equation (6.9). *We will go back to this resemblance in Chapter* 8.

6.6. Historical and Bibliographical Notes

The question of recurrence vs transience of a stochastic process was first asked by Pólya in the context of simple random walks. Pólya credits the following anecdote in motivating the question [**DKM70**]:

> At the hotel there lived also some students with whom I usually took my meals and had friendly relations. On a certain day one of them expected the visit of his fiancee, what I knew [sic], but I did not foresee that he and his fiancee would also set out for a stroll in the woods, and then suddenly I met them there. And then I met them the same morning repeatedly. I don't remember how many times, but certainly much too often and I felt embarrassed: It looked as if I was snooping around which was, I assure you, not the case.

He proved the recurrence of the walk in $d = 2$ and the transience in $d > 2$ in [**P21**]. Paul Lévy recovered the result for Brownian motion in dimension $d = 2$ in [**L40**]. (He does make the connection with Pólya's result on p. 538.) The proof of transience of Brownian motion in higher dimension is due to Kakutani [**Kak44a**]. Durrett in [**Dur96**] attributes to Kakutani the following quote during a UCLA colloquium, which appropriately illustrates the dichotomy between dimension two and higher:

> A drunk man will eventually find his way home but a drunk bird may get lost forever.

The study of the Dirichlet problem predates stochastic calculus by more than one hundred years. It goes back to Green (the same as Green's theorem in multivariable calculus) who published *An Essay on the Application of Mathematical Analysis to the Theories of Electricity and Magnetism* in 1828. The problem played a fundamental role in electromagnetism and in potential theory in general where harmonic functions are central. The first application of stochastic calculus to the Dirichlet problem goes back to Kakutani [**Kak44b**]. For more on existence and uniqueness of the solution to the Dirichlet problem from a stochastic calculus perspective, the reader is referred to [**MP10**].

Chapter 7

Itô Processes and Stochastic Differential Equations

We saw in Chapter 5 that a function of Brownian motion can be expressed in terms of an Itô integral and a Riemann integral. This is useful since Itô integrals are martingales (when the integral is nice enough), which allow for explicit computations of expectations and probabilities, and Riemann integrals are smooth functions of time. We now extend the class of stochastic processes where similar methods apply by considering all processes constructed from an Itô integral and a Riemann integral. These are called *Itô processes*, and they are defined in detail in Section 7.1. It turns out that smooth functions of Itô processes also have an Itô's formula that follows the same rules of Itô calculus seen before; cf. Sections 7.2 and Section 7.3 for the multivariate extension. The focus of this chapter will be on a class of Itô processes called *diffusions*. As we shall see in Section 7.4, diffusions are particularly easy to sample numerically. They are also solutions of *stochastic differential equations* as explained in Section 7.5. We end the chapter in Section 7.6 with a discussion on the representation of martingales in terms of Itô integrals and with a new characterization of Brownian motion called *Lévy's characterization*.

7.1. Definition and Examples

Let's start with the definition of Itô processes.

Definition 7.1 (Itô process). Let $(B_t, t \geq 0)$ be a standard Brownian motion defined on $(\Omega, \mathcal{F}, \mathbf{P})$. An *Itô process* $(X_t, t \geq 0)$ is a process of the form

$$(7.1) \qquad X_t = X_0 + \int_0^t V_s \, dB_s + \int_0^t D_s \, ds,$$

where $(V_t, t \geq 0)$ and $(D_t, t \geq 0)$ are two adapted processes for which the integrals make sense in the sense of Itô and Riemann, respectively. We refer to $(V_t, t \geq 0)$ as the *local volatility* and to $(D_t, t \geq 0)$ as the *local drift*.

We will often denote an Itô process $(X_t, t \geq 0)$ in *differential form* as

(7.2) $$dX_t = V_t \, dB_t + D_t \, dt.$$

This form makes no rigorous sense; when we write it, we mean equation (7.1). Nevertheless, the differential notation (7.2) has two great advantages:

(1) It gives some intuition on what drives the variation of X_t. On one hand, there is a contribution of the Brownian increments which are modulated by the volatility V_t. On the other hand, there is a smoother contribution coming from the time variation which is modulated by the drift D_t.

(2) The differential notation has real computational power. In particular, evaluating Itô's formula is reduced to computing differentials, as in classical calculus, but by doing it up to the second order.

An important class of Itô processes is given by processes for which the volatility and the drift are simply functions of the position of the process.

Definition 7.2. Let $(B_t, t \geq 0)$ be a standard Brownian motion. An Itô process $(X_t, t \geq 0)$ of the form

(7.3) $$dX_t = \sigma(X_t) \, dB_t + \mu(X_t) \, dt, \quad X_0 = x,$$

where σ and μ are functions from $\mathbb{R}$ to $\mathbb{R}$, is called a *time-homogeneous diffusion*. An Itô process $(Y_t, t \geq 0)$ of the form

(7.4) $$dY_t = \sigma(t, Y_t) \, dB_t + \mu(t, Y_t) \, dt, \quad Y_0 = y,$$

where σ and μ are now functions from $[0, \infty) \times \mathbb{R}$ to $\mathbb{R}$, is called a *time-inhomogeneous diffusion*. The equations (7.3) and (7.4) are called the *stochastic differential equations* (SDE) of the respective process (X_t) and (Y_t).

In other words, a diffusion $(X_t, t \geq 0)$ is an Itô process whose local volatility V_t and local drift D_t at time t depend only on the position of the process at time t and possibly on the time t itself. It cannot depend on the path of the process before time t, or on the explicit values of the driving Brownian motion at that time (which is not the process X_t itself). The class of diffusions, and of Itô processes in general, constitutes a huge collection of stochastic processes for stochastic modelling.

Note that an SDE is a generalization of ordinary differential equations or ODEs. Indeed, if there were no randomness, i.e., no Brownian motion, the SDE (7.3) would be reduced to

$$dX_t = \mu(X_t) \, dt.$$

This can be written for $X_t = f(t)$ as

$$\frac{df}{dt} = \mu(f).$$

This is an *ordinary differential equation* (ODE). It governs the deterministic evolution of the function $X_t = f(t)$ in time. An SDE adds a random term to this evolution that is formally written as

$$\frac{dX_t}{dt} = \mu(X_t) + \sigma(X_t)\frac{dB_t}{dt}.$$

We know very well that Brownian motion is not differentiable; hence the above is not well-defined. The ill-defined term dB_t/dt is sometimes called *white noise*. However, equation (7.3) is well-defined in the sense of Itô processes. These types of equations are well-suited to model phenomena with intrinsic randomness.

Here are some examples of diffusions.

Example 7.3 (Brownian motion with drift). If we take $X_t = \sigma B_t + \mu t$ for some $\sigma > 0$ and $\mu \in \mathbb{R}$, then we can write X_t as

$$X_t = \int_0^t \sigma \, dB_t + \int_0^t \mu \, dt, \quad X_0 = 0.$$

In differential form, this becomes

$$dX_t = \sigma \, dB_t + \mu \, dt.$$

In this case, the local drift and volatility are constant.

Example 7.4 (Geometric Brownian motion). We consider the process $S_t = \exp(\sigma B_t + \mu t)$ as in Example 5.31. To see that it is a diffusion, we apply Itô's formula with $f(t,x) = \exp(\sigma x + \mu t)$. We get

$$dS_t = \sigma S_t \, dB_t + \left(\mu + \frac{1}{2}\sigma^2\right) S_t \, dt.$$

Note that the local drift and volatility are now random and proportional to the position! So the higher the S_t, the higher the volatility and the drift; see Figure 7.1.

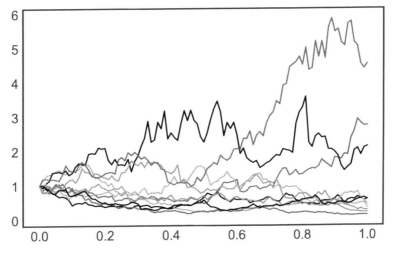

Figure 7.1. A simulation of 10 paths of geometric Brownian motion on $[0,1]$ with $\mu = -\frac{1}{2}\sigma^2$ and $\sigma = 1$.

Example 7.5 (Any smooth function of Brownian motion). Itô's formula, see equations (5.15) and (5.28), guarantees that any smooth-enough function $f(t, B_t)$ of time and a Brownian function is an Itô process with volatility $V_t = \partial_1 f(t, B_t)$ and drift $D_t = \frac{1}{2}\partial_1^2 f(t, B_t) + \partial_0 f(t, B_t)$. We will see in Theorem 7.8 that, in general, any reasonable function of an Itô process remains an Itô process.

Example 7.6 (An Itô process that is not a diffusion). Consider the process

$$X_t = \int_0^t B_s^2 \, dB_s, \quad t \geq 0.$$

This is an Itô process with local volatility $V_t = B_t^2$ and local drift $D_t = 0$. However, it is not a diffusion, because the local volatility is not an explicit function of X_t.

It turns out that Brownian bridge is a time-inhomogeneous diffusion (see Exercise 7.1) and that the Ornstein-Uhlenbeck process is a time-homogeneous diffusion (see Example 7.9). To understand these examples, we need to extend Itô's formula to Itô processes.

7.2. Itô's Formula

The first step towards a general Itô's formula is the quadratic variation of an Itô process.

Proposition 7.7 (Quadratic variation of an Itô process). *Let* $(B_t, t \geq 0)$ *be a standard Brownian motion and* $(X_t, t \geq 0)$ *an Itô process of the form* $dX_t = V_t \, dB_t + D_t \, dt$. *Then the quadratic variation of the process* $(X_t, t \geq 0)$ *is*

$$\langle X \rangle_t = \lim_{n \to \infty} \sum_{j=0}^{n-1} (X_{t_{j+1}} - X_{t_j})^2 = \int_0^t V_s^2 \, ds,$$

for any partition $(t_j, j \leq n)$ *of* $[0, t]$, *where the limit is in probability.*

Note that the quadratic variation is increasing in t, but it is not deterministic in general! The quadratic variation is a smooth stochastic process. (It is differentiable.) Observe that we recover the quadratic variation for Brownian motion for $V_t = 1$ as expected. We also notice that the formula follows easily from the rules of Itô calculus (5.23), thereby showing the consistency of the theory. Indeed, we have

(7.5) $\qquad d\langle X \rangle_t = (dX_t)^2 = (V_t \, dB_t + D_t \, dt)(V_t \, dB_t + D_t \, dt) = V_t^2 \, dt + 0.$

Proof of Proposition 7.7. The proof is involved, but it reviews some important concepts of stochastic calculus. We prove the case when the process V is in $\mathcal{L}_c^2(T)$ for some $T > 0$. Write $I_t = \int_0^t V_s \, dB_s$ and $R_t = \int_0^t D_s \, ds$, $t \leq T$. We first show that only the Itô integral contributes to the quadratic variation and that the Riemann integral does not contribute, so that

(7.6) $\qquad\qquad\qquad\qquad\qquad \langle X \rangle_t = \langle I \rangle_t.$

This was foreshadowed in (7.5) and is expected since the Riemann integral seen as a function of t has finite variation. We have that the increment square of X_t is

$$(X_{t_{j+1}} - X_{t_j})^2 = (I_{t_{j+1}} - I_{t_j})^2 + 2(I_{t_{j+1}} - I_{t_j})(R_{t_{j+1}} - R_{t_j}) + (R_{t_{j+1}} - R_{t_j})^2.$$

The Cauchy-Schwarz inequality implies

$$\sum_{j=0}^{n-1}(I_{t_{j+1}} - I_{t_j})(R_{t_{j+1}} - R_{t_j}) \le \left\{\sum_{j=0}^{n-1}(I_{t_{j+1}} - I_{t_j})^2\right\}^{1/2}\left\{\sum_{j=0}^{n-1}(R_{t_{j+1}} - R_{t_j})^2\right\}^{1/2}.$$

Therefore to prove equation (7.6), it suffices to show that $\lim_{n\to\infty}\sum_{j=0}^{n-1}(R_{t_{j+1}} - R_{t_j})^2 = 0$ almost surely. This was done in Example 3.7. Specifically, it follows from the fact that

$$\sum_{j=0}^{n-1}(R_{t_{j+1}} - R_{t_j})^2 \le \max_{j<n}\left\{\int_{t_j}^{t_{j+1}}|D_s|\,ds\right\}\int_0^t |D_s|\,ds.$$

The first term goes to 0 as $n \to \infty$ by uniform continuity of the continuous function $s \mapsto \int_0^s D_u\,du$ on $[0, t]$.

It remains to prove that $\langle I_t \rangle = \int_0^t V_s^2\,ds$. We first prove the case when $V \in \mathcal{S}(t)$ is a simple adapted process. Consider a partition $(t_j, j \le n)$ of $[0, t]$. Without loss of generality, we can suppose that V is constant on each $[t_j, t_{j+1})$ by refining the partition. We then have

$$\sum_{j=0}^{n-1}(I_{t_{j+1}} - I_{t_j})^2 = \sum_{j=0}^{n-1}V_{t_j}^2(B_{t_{j+1}} - B_{t_j})^2.$$

This converges in L^2 to $\int_0^t V_s\,ds$ as $n \to \infty$ exactly as in (5.21) by appropriate conditioning, since V_{t_j} is $\mathcal{F}_{t_j}$-measurable.

The case $V \in \mathcal{L}_c^2(T)$ is proved by approximating V by simple process in $\mathcal{S}(T)$, as in Lemma 5.11. More precisely, for any $\varepsilon > 0$, we can find a simple process $V^{(\varepsilon)}$ that is ε-close to V in the sense that

$$(7.7) \qquad \int_0^T \mathbf{E}[(V_s^{(\varepsilon)} - V_s)^2]\,ds < \varepsilon.$$

To prove the claim, we need to show that for $t \le T$,

$$\mathbf{E}\left[\left|\sum_{j=0}^{n-1}(I_{t_{j+1}} - I_{t_j})^2 - \int_0^t V_s^2\,ds\right|\right] \to 0 \text{ as } n \to \infty.$$

This implies the convergence in probability of the sequence $\sum_{j=0}^{n-1}(I_{t_{j+1}} - I_{t_j})^2$, $n \ge 1$, as desired. We now introduce the $V^{(\varepsilon)}$ approximation inside the absolute value as well as its corresponding integral $I_t^{(\varepsilon)} = \int_0^t V_s^{(\varepsilon)}\,ds$. By the triangle inequality, the left-hand side is smaller than

$$\begin{aligned}
(7.8) \qquad &\mathbf{E}\left[\left|\sum_{j=0}^{n-1}(I_{t_{j+1}} - I_{t_j})^2 - (I_{t_{j+1}}^{(\varepsilon)} - I_{t_j}^{(\varepsilon)})^2\right|\right] + \mathbf{E}\left[\left|\sum_{j=0}^{n-1}(I_{t_{j+1}}^{(\varepsilon)} - I_{t_j}^{(\varepsilon)})^2 - \int_0^t (V_s^{(\varepsilon)})^2\,ds\right|\right] \\
&+ \mathbf{E}\left[\left|\int_0^t (V_s^{(\varepsilon)})^2\,ds - \int_0^t V_s^2\,ds\right|\right].
\end{aligned}$$

We show that the first and third terms are smaller than ε uniformly in n and that the second term goes to 0 as $n \to \infty$. This proves the claim since ε is arbitrary. The second term goes to 0 as $n \to \infty$, by the argument for simple processes (note that the convergence in L^2 is stronger by Jensen's inequality). For the third term, the linearity of the integral and the Cauchy-Schwarz inequality (applied to $\mathbf{E}\int_0^t$) imply that it is

$$= \mathbf{E}\left[\left|\int_0^t (V_s^{(\varepsilon)} - V_s)(V_s^{(\varepsilon)} + V_s)\,\mathrm{d}s\right|\right]$$

$$\leq \left(\mathbf{E}\left[\int_0^t (V_s^{(\varepsilon)} - V_s)^2\,\mathrm{d}s\right]\right)^{1/2} \left(\mathbf{E}\left[\int_0^t (V_s^{(\varepsilon)} + V_s)^2\,\mathrm{d}s\right]\right)^{1/2}.$$

The first factor is smaller than the square root of ε by equation (7.7), whereas the second factor is bounded.

The first term in equation (7.8) is handled similarly. The linearity of the Itô integral and the Cauchy-Schwarz inequality give that the first term is

$$= \mathbf{E}\left[\left|\sum_{j=0}^{n-1}\left(\int_{t_j}^{t_{j+1}} (V_s^{(\varepsilon)} - V_s)\,\mathrm{d}B_s\right)\left(\int_{t_j}^{t_{j+1}} (V_s^{(\varepsilon)} + V_s)\,\mathrm{d}B_s\right)\right|\right]$$

$$\leq \left(\mathbf{E}\left[\sum_{j=0}^{n-1}\left(\int_{t_j}^{t_{j+1}} (V_s^{(\varepsilon)} - V_s)\,\mathrm{d}B_s\right)^2\right]\right)^{1/2} \left(\mathbf{E}\left[\sum_{j=0}^{n-1}\left(\int_{t_j}^{t_{j+1}} (V_s^{(\varepsilon)} + V_s)\,\mathrm{d}B_s\right)^2\right]\right)^{1/2}.$$

Itô's isometry gives that the first factor in the square root is the one in equation (7.7) so it is smaller than ε. The second equals $\mathbf{E}\left[\int_0^t (V_s^{(\varepsilon)} + V_s)^2\,\mathrm{d}s\right]$ by Itô's isometry. This is bounded uniformly for all $\varepsilon > 0$. This concludes the proof of the proposition. □

We are now ready to state Itô's formula for Itô processes. We write the result in differential form for conciseness.

Theorem 7.8 (Itô's formula for Itô processes). *Let $(B_t, t \geq 0)$ be a standard Brownian motion, and let $(X_t, t \geq 0)$ be an Itô process of the form $\mathrm{d}X_t = V_t\,\mathrm{d}B_t + D_t\,\mathrm{d}t$. Consider a function $f(t, x) \in \mathcal{C}^{1,2}([0, T] \times \mathbb{R})$. Then, we have with probability one for all $t \leq T$,*

$$\mathrm{d}f(t, X_t) = \partial_1 f(t, X_t)V_t\,\mathrm{d}B_t + \left(\partial_0 f(t, X_t) + \frac{V_t^2}{2}\partial_1^2 f(t, X_t) + D_t\,\partial_1 f(t, X_t)\right)\mathrm{d}t.$$

Again, we use for convenience the notation $\partial_0 f$ for the first derivative in time of f and $\partial_1 f$, $\partial_1^2 f$ for the first and second derivatives in space of f, respectively. There is no need to learn this formula by heart. It suffices to compute the differential of $\mathrm{d}f(t, X_t)$ classically up to the second order and apply the rules of Itô calculus as in Chapter 5, equation (5.22):

$$\begin{aligned}
(7.9) \quad \mathrm{d}f(t, X_t) &= \partial_0 f(t, X_t)\,\mathrm{d}t + \partial_1 f(t, X_t)\,\mathrm{d}X_t + \frac{1}{2}\partial_1^2 f(t, X_t)(\mathrm{d}X_t)^2 \\
&= \partial_1 f(t, X_t)\,\mathrm{d}X_t + \left(\partial_0 f(t, X_t) + \frac{1}{2}\partial_1^2 f(t, X_t)V_t^2\right)\mathrm{d}t,
\end{aligned}$$

since the rules of Itô calculus yield $(\mathrm{d}X_t)^2 = V_t^2\,\mathrm{d}t$. Because $\mathrm{d}X_t = V_t\,\mathrm{d}B_t + D_t\,\mathrm{d}t$, it is understood that the first term represents

$$\int_0^t \partial_1 f(s, X_s)\,\mathrm{d}X_s = \int_0^t \partial_1 f(s, X_s)V_s\,\mathrm{d}B_s + \int_0^t \partial_1 f(s, X_s)D_s\,\mathrm{d}s.$$

The proof of Theorem 7.8 is again a Taylor approximation together with the form of the quadratic variation of the process. We will omit it.

Example 7.9 (Ornstein-Uhlenbeck process). Consider the Ornstein-Uhlenbeck process $(Y_t, t \geq 0)$ given in Example 5.19

$$Y_t = Y_0 e^{-t} + e^{-t} \int_0^t e^s \, dB_s.$$

Note that this process is an explicit function of t and of the Itô process $X_t = Y_0 + \int_0^t e^s \, dB_s$. Indeed, we have $Y_t = f(t, X_t)$ with $f(t, x) = e^{-t} x$. This is not an explicit function of t and B_t so Itô's formula in Proposition 5.28 is not applicable. Since $\partial_1 f = e^{-t}, \partial_1^2 f = 0$, and $\partial_0 f = -f$, Theorem 7.8 or equation (7.9) implies

$$(7.10) \qquad\qquad dY_t = -Y_t \, dt + dB_t.$$

This is the SDE for the Ornstein-Uhlenbeck process.

The SDE has a very nice interpretation: The drift is positive if $Y_t < 0$ and negative if $Y_t > 0$. Moreover, the drift is proportional to the position (exactly like a spring pulling the process back to the x-axis following Hooke's law!). This is the mechanism that ensures that the process does not venture too far from 0 and is eventually stationary.

The SDE (7.10) is now easily generalized by adding two parameters for the volatility and the drift:

$$(7.11) \qquad\qquad dY_t = -kY_t \, dt + \sigma \, dB_t, \quad k \in \mathbb{R}, \sigma > 0.$$

It is not hard to check that the solution to this SDE is

$$(7.12) \qquad\qquad Y_t = Y_0 e^{-kt} + e^{-kt} \int_0^t e^{ks} \sigma \, dB_s;$$

see Exercise 7.2. Note that we include the case where $k < 0$ here so $-k > 0$. In this case, the process will not converge to a stationary distribution.

The latest version of Itô's formula is another useful tool for producing martingales from a function of an Itô process. We start with two examples generalizing martingales of Brownian motion.

Example 7.10 (A generalization of $B_t^2 - t$). Let $(V_t, t \leq T)$ be a process in $\mathcal{L}_c^2(T)$. Consider the Itô process $(X_t, t \leq T)$ given by $dX_t = V_t \, dB_t$. Note that $(X_t^2, t \leq T)$ is a submartingale by Jensen's inequality. We show that the compensated process

$$M_t = X_t^2 - \int_0^t V_s^2 \, ds, \quad t \leq T,$$

is a martingale for the Brownian filtration. (This is another instance of the Doob-Meyer decomposition; see Remark 4.31.) Itô's formula and the rules of Itô calculus give for $f(x) = x^2$

$$df(X_t) = f'(X_t) \, dX_t + \frac{1}{2} f''(X_t)(dX_t)^2$$
$$= 2X_t V_t \, dB_t + V_t^2 \, dt.$$

In integral form, this implies

$$M_t = X_0^2 + \int_0^t 2 X_s V_s \, dB_s.$$

We conclude that M_t, $t \leq T$, is a martingale, provided that $X_t V_t \in \mathcal{L}_c^2(T)$.

There is another more direct way to prove that $(M_t, t \leq T)$ is a martingale whenever $(V_t, t \leq T) \in \mathcal{L}_c^2(T)$. This is by using increments: For $t' < t \leq T$,

$$\mathbf{E}[X_{t'}^2 | \mathcal{F}_t] = \mathbf{E}[(X_t + (X_{t'} - X_t))^2 | \mathcal{F}_t] = X_t^2 + 2 X_t \mathbf{E}[X_{t'} - X_t | \mathcal{F}_t] + \mathbf{E}[(X_{t'} - X_t)^2 | \mathcal{F}_t],$$

where we use the properties of conditional expectation in Proposition 4.19. Since $(X_t, t \leq T)$ is a martingale, $\mathbf{E}[X_{t'} - X_t | \mathcal{F}_t]$ is 0. We are left with

$$\mathbf{E}[X_{t'}^2 | \mathcal{F}_t] = X_t^2 + \mathbf{E}[(X_{t'} - X_t)^2 | \mathcal{F}_t].$$

This implies $\mathbf{E}[M_{t'} | \mathcal{F}_t] = M_t$, because the following *conditional Itô isometry* holds:

$$(7.13) \qquad \mathbf{E}[(X_{t'} - X_t)^2 | \mathcal{F}_t] = \int_0^{t'} V_s^2 \, ds - \int_0^t V_s^2 \, ds = \int_t^{t'} V_s^2 \, ds.$$

If there were no conditioning, this follows from Corollary 5.15 by expanding the square. Equation (7.13) is proved in Exercise 7.13.

Numerical Project 7.6 works out an example of this type of martingale.

Example 7.11 (A generalization of geometric Brownian motion). Let $\sigma(t)$ be a continuous, deterministic function such that $|\sigma(t)| \leq 1$, $t \in [0, T]$. The process

$$M_t = \exp\left(\int_0^t \sigma(s) \, dB_s - \frac{1}{2} \int_0^t \sigma^2(s) \, ds \right), \quad t \leq T,$$

is a martingale for the Brownian filtration. To see this, note that we can write M_t as $M_t = f(t, X_t)$ where $f(t, x) = \exp(x - \frac{1}{2} \int_0^t \sigma^2(s) \, ds)$ and $X_t = \int_0^t \sigma(s) \, dB_s$. Itô's formula gives

$$(7.14) \qquad dM_t = -\frac{1}{2} \sigma^2(t) f(t, X_t) \, dt + f(t, X_t) \, dX_t + \frac{1}{2} f(t, X_t)(dX_t)^2 = \sigma(t) M_t \, dB_t .$$

Observe also that for $t \leq T$,

$$\mathbf{E}[M_t^2] = e^{-\int_0^t \sigma^2(s) \, ds} \, \mathbf{E}\left[e^{2 \int_0^t \sigma(s) \, dB_s} \right] = e^{\int_0^t \sigma^2(s) \, ds}$$

since $\int_0^t \sigma(s) \, dB_s$ is a Gaussian random variable of mean 0 and variance $\int_0^t \sigma^2(s) \, ds$. In the case $|\sigma| \leq 1$, this implies that $(\sigma(t) M_t, t \leq T) \in \mathcal{L}_c^2(T)$. We conclude from equation (7.14) that $(M_t, t \leq T)$ is a martingale. It turns out that the conclusion holds whenever $\int_0^T \sigma^2(s) \, ds < \infty$; see Exercise 7.3. This example is generalized in Example 7.11.

Theorem 7.14 provides a systematic method to construct martingales from functions of time and space if they are solutions of a certain PDE, the same way we did in Corollary 5.29. The form of the PDE will vary depending on the process. An important example is the PDE provided by geometric Brownian motion. This will be important when we discuss the Black-Scholes PDE in Chapter 10.

Example 7.12 (Martingales of geometric Brownian motion). Let

$$S_t = S_0 \exp(\sigma B_t - \sigma^2 t/2)$$

be a geometric Brownian motion. We find a PDE satisfied by $f(t, x)$ for $f(t, S_t)$ to be a martingale. It suffices to apply Itô's formula of Theorem 7.8. We get

$$df(t, S_t) = \partial_1 f(t, S_t) \, dS_t + \frac{1}{2} \partial_1 f(t, S_t)(dS_t)^2 + \partial_0 f(t, S_t) \, dt .$$

Note that we have $dS_t = \sigma S_t \, dB_t$ and $(dS_t)^2 = \sigma^2 S_t^2 \, dt$. Therefore we get

$$df(t, S_t) = \partial_1 f(t, S_t) \sigma S_t \, dB_t + \left\{ \frac{1}{2} \partial_1 f(t, S_t) \sigma^2 S_t^2 + \partial_0 f(t, S_t) \right\} dt .$$

Finally, the PDE for $f(t, x)$ is obtained by setting the factor in front of dt to 0. It is important to keep in mind that the PDE should always be written in terms of the time variable t and the space variable x. Therefore, the PDE of f as a function of time and space is

$$\frac{\sigma^2 x^2}{2} \frac{\partial^2 f}{\partial x^2}(t, x) + \frac{\partial f}{\partial t}(t, x) = 0.$$

No more randomness appears in the PDE!

Here is a specific case where we can apply Itô's formula to construct martingales of Itô processes.

Example 7.13. Consider the process given by the SDE

$$dX_t = X_t \, dB_t, \quad X_0 = 2.$$

Let's find a PDE for which $f(t, X_t)$ is a martingale for the Brownian filtration. We have by Itô's formula that

$$df(t, X_t) = \partial_0 f(t, X_t) \, dt + \partial_1 f(t, X_t) X_t \, dB_t + \frac{1}{2} \partial_1^2 f(t, X_t) X_t^2 \, dt.$$

Setting the drift term to 0 gives the PDE

$$\partial_0 f + \frac{x^2}{2} \partial_1^2 f = 0.$$

It is then easy to check that X_t is a martingale and so is $t + \log X_t^2$, since the functions $f(t, x) = x$ and $f(t, x) = t + \log x^2$ satisfy the PDE. However, the process tX_t is not, as the function $f(t, x) = xt$ is not a solution of the PDE.

Now, consider the stopping time $\tau = \min\{t \geq 0 : X_t \geq 3 \text{ or } X_t \leq 1\}$. We will show that

$$\mathbf{P}(X_\tau = 1) = 1/2.$$

By the optional stopping theorem (Corollary 4.38), assuming it is applicable here, we have

$$2 = \mathbf{E}[X_0] = \mathbf{E}[X_\tau] = 1\mathbf{P}(X_\tau = 1) + 3(1 - \mathbf{P}(X_\tau = 1)).$$

We conclude that $\mathbf{P}(X_\tau = 1) = 1/2$. We can also use another martingale to show that

$$\mathbf{E}[\tau] = \log 4 - \frac{1}{2} \log 9 = 0.287\ldots.$$

We use the process $t + \log X_t^2$. Again, by the optional stopping theorem, we have

$$\log 4 = \mathbf{E}[\tau] + \mathbf{E}[\log X_\tau^2].$$

Since $\mathbf{P}(X_\tau = 1) = 1/2$, we have $\mathbf{E}[\log X_\tau^2] = \frac{1}{2}\log 9 + \frac{1}{2}\log 1 = \frac{1}{2}\log 9$.

7.3. Multivariate Extension

Itô's formula can be generalized to several Itô processes similarly to Theorem 6.6 for Brownian motion. Let's start by stating the example of a function of two Itô processes. Such a function $f(x_1, x_2)$ will be a function of two space variables. Not surprisingly, it needs to have two derivatives in each variable and they need to be continuous function; i.e., we need $f \in \mathcal{C}^{2,2}(\mathbb{R} \times \mathbb{R})$.

Theorem 7.14 (Itô's formula for many Itô processes). *Let $(X_t, t \geq 0)$ and $(Y_t, t \geq 0)$ be two Itô processes of the form*

(7.15)
$$\begin{aligned} dX_t &= V_t\, dB_t + D_t\, dt, \\ dY_t &= U_t\, dB_t + R_t\, dt, \end{aligned}$$

where $(B_t, t \geq 0)$ is a standard Brownian motion. Then, for $f \in C^{2,2}(\mathbb{R} \times \mathbb{R})$, we have

$$\begin{aligned} df(X_t, Y_t) = {}&\partial_x f(X_t, Y_t)\, dX_t + \partial_y f(X_t, Y_t)\, dY_t \\ &+ \frac{1}{2}\partial_x^2 f(X_t, Y_t)V_t^2\, dt + \frac{1}{2}\partial_y^2 f(X_t, Y_t)U_t^2\, dt + \partial_x\partial_y f(X_t, Y_t)V_t U_t\, dt. \end{aligned}$$

The idea of the proof is the same as in Theorem 6.6: Taylor expansion and quadratic variation, together with the *cross-variation of the two processes*; i.e., for a partition $(t_j, j \leq n)$ of $[0, t]$,

$$dX_t \cdot dY_t = \lim_{n\to\infty} \sum_{j=0}^{n-1} (X_{t_{j+1}} - X_{t_j})(Y_{t_{j+1}} - Y_{t_j}) = \int_0^t U_s V_s\, ds.$$

Note that this is consistent with the rules of Itô calculus as

$$dX_t \cdot dY_t = (V_t\, dB_t + D_t\, dt) \cdot (U_t\, dB_t + R_t\, dt) = U_t V_t\, dt + 0.$$

An important example of this formula is Itô's product rule.

Example 7.15 (**Product Rule**). Let X_t and Y_t be as in equation (7.15). Then

(7.16) $$d(X_t Y_t) = Y_t\, dX_t + X_t\, dY_t + dX_t \cdot dY_t = Y_t\, dX_t + X_t\, dY_t + V_t U_t\, dt.$$

This is obtained by taking the function $f(x, y) = xy$ in Theorem 7.14.

The above can be used to determine when the product of two processes is a martingale.

Example 7.16. Let $X_t = \int_0^t B_s\, dB_s$ and $Y_t = \int_0^t B_s^2\, dB_s$. Is $(X_t Y_t, t \geq 0)$ a martingale? We have

$$d(X_t Y_t) = X_t B_t^2\, dB_t + Y_t B_t\, dB_t + B_t^3\, dt.$$

The term in dt is not 0. Therefore the product cannot be a martingale.

The following example will be important when we discuss the Girsanov theorem; see Theorem 9.11.

Example 7.17 (A generalization of geometric Brownian motion). Consider $(\int_0^t V_s \, dB_s,\ t \geq 0)$ an Itô process. Define the positive process

$$(7.17) \qquad M_t = \exp\left(\int_0^t V_s \, dB_s - \frac{1}{2}\int_0^t V_s^2 \, ds\right), \quad t \geq 0.$$

Itô's formula of Theorem 7.14 applied to the processes $X_t = \int_0^t V_s \, dB_s$ and $Y_t = \frac{1}{2}\int_0^t V_s^2 \, ds$ with the function $f(x, y) = e^{x-y}$ yields the differential form

$$dM_t = M_t V_t \, dB_t \, .$$

To see this, note that $(dX_t)^2 = V_t^2 \, dt$, $(dY_t)^2 = dX_t \cdot dY_t = 0$ by the rules of Itô calculus. Therefore, M_t is a martingale on $[0, T]$ for some $T > 0$, provided that $(M_t V_t, t \leq T)$ is in $\mathcal{L}_c^2(T)$. It is not hard to check that this is the case if V_t is deterministic or if V_t is bounded, similarly to Example 7.11 and Exercise 7.3. However, this is not true in general. A sufficient condition for $(M_t, t \leq T)$ to be a martingale is the *Novikov condition*

$$(7.18) \qquad \mathbf{E}\left[\exp\left(\frac{1}{2}\int_0^T V_s^2 \, ds\right)\right] < \infty.$$

Theorem 7.14 can be generalized to functions of time and of many Itô processes.

Theorem 7.18 (Itô's formula for many Itô processes). *Let $X_t = (X_t^{(j)}, j \leq d)$ be Itô processes constructed on $(\Omega, \mathcal{F}, \mathbf{P})$ and $f \in \mathcal{C}^{1,2}([0, T] \times \mathbb{R}^d)$. Then for $t \in [0, T]$, we have*

$$df(t, X_t) = \partial_0 f(t, X_t) \, dt + \sum_{j,k=1}^{d} \partial_j \partial_k f(t, X_t) \, dX_t^{(j)} \cdot dX_t^{(k)},$$

where the cross-variation $dX_t^{(j)} \cdot dX_t^{(k)}$ is computed using the rules of Itô calculus.

This will be used when we construct the CIR model in Example 7.24.

7.4. Numerical Simulations of SDEs

The good news is that it is not too hard to implement iterative schemes to sample paths of a diffusion. Consider $(X_t, t \leq T)$ a solution to the SDE

$$dX_t = \sigma(X_t) \, dB_t + \mu(X_t) \, dt, \quad X_0 = x.$$

To keep the notation to a minimum, we consider a time-homogeneous diffusion. For a partition $(t_j, j \leq n)$ of $[0, T]$ with $t_n = T$, consider the increment

$$(7.19) \qquad X_{t_{j+1}} - X_{t_j} = \int_{t_j}^{t_{j+1}} \sigma(X_s) \, dB_s + \int_{t_j}^{t_{j+1}} \mu(X_s) \, ds.$$

Note that if σ and μ are smooth functions, we can apply Itô's formula to $\sigma(X_s)$ and $\mu(X_s)$ for $s \in (t_j, t_{j+1}]$! We get

$$\sigma(X_s) = \sigma(X_{t_j}) + \int_{t_j}^s \sigma'(X_u)\, dX_u + \int_{t_j}^s \frac{1}{2}\sigma''(X_u)(dX_u)^2,$$

(7.20)

$$\mu(X_s) = \mu(X_{t_j}) + \int_{t_j}^s \mu'(u, X_u)\, dX_u + \frac{1}{2}\int_{t_j}^s \mu''(X_u)(dX_u)^2.$$

Now we can approximate the increment (7.19) at different levels of precision by considering different estimate for (7.20).

Example 7.19 (Euler-Maruyama scheme). This scheme consists of simply taking $\sigma(X_s)$ $\approx \sigma(X_{t_j})$ and $\mu(X_s) \approx \mu(X_{t_j})$ for $s \in [t_j, t_{j+1})$ in (7.20). Putting this back in (7.19) we get

(7.21)
$$\boxed{X_{t_{j+1}} - X_{t_j} \approx \sigma(X_{t_j})(B_{t_{j+1}} - B_{t_j}) + \mu(X_{t_j})(t_{j+1} - t_j).}$$

The process X_t can then be constructed recursively on the discrete set $(t_j, j \leq n)$ as follows:

$$X_0 = x, \qquad X_{t_1} = (X_{t_1} - X_0) + X_0, \qquad \dots, \qquad X_t = \sum_{j=0}^{n-1}(X_{t_{j+1}} - X_{t_j}) + X_0.$$

Example 7.20 (Milstein scheme). In this scheme, we go an order further for the approximation of the volatility in (7.20) and consider also the integral in dX_u. We take $dX_u = \sigma(X_{t_j})\, dB_u$. We then write

$$\sigma(X_s) \approx \sigma(X_{t_j}) + \sigma'(X_{t_j})\sigma(X_{t_j})\int_{t_j}^s dB_u = \sigma(X_{t_j}) + \sigma'(X_{t_j})\sigma(X_{t_j})(B_s - B_{t_j}).$$

If we put this back in the equation (7.19), we get

$$X_{t_{j+1}} - X_{t_j}$$
$$\approx \sigma(X_{t_j})(B_{t_{j+1}} - B_{t_j}) + \mu(X_{t_j})(t_{j+1} - t_j) + \sigma'(X_{t_j})\sigma(X_{t_j})\int_{t_j}^{t_{j+1}}(B_s - B_{t_j})\, dB_s.$$

We have by Itô's formula

$$\int_{t_j}^{t_{j+1}}(B_s - B_{t_j})\, dB_s = \frac{1}{2}\Big(B_{t_{j+1}}^2 - B_{t_j}^2 - (t_{j+1} - t_j)\Big) - B_{t_j}(B_{t_{j+1}} - B_{t_j}).$$

Putting this together, we get the Milstein approximation

(7.22)
$$\boxed{\begin{aligned} & X_{t_{j+1}} - X_{t_j} \\ & \approx \sigma(X_{t_j})(B_{t_{j+1}} - B_{t_j}) + \mu(X_{t_j})(t_{j+1} - t_j) \\ & \quad + \frac{\sigma'(X_{t_j})\sigma(X_{t_j})}{2}\Big\{(B_{t_{j+1}} - B_{t_j})^2 - (t_{j+1} - t_j)\Big\}. \end{aligned}}$$

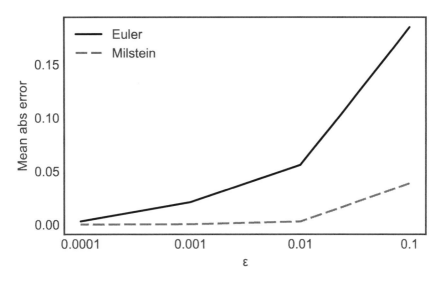

Figure 7.2. A comparison of the absolute error for the Milstein scheme and the Euler scheme for different discretization $\varepsilon = 1/n$.

The recursive nature of these two schemes makes them easy to implement numerically; see Numerical Projects 7.1 and 7.3. The Milstein scheme is not much more costly to implement as it contains only one more term than the Euler scheme.

It is of course possible to go beyond the Milstein scheme to improve the approximation. However, it turns out that the above schemes already converge quite rapidly to the process itself. To see this, consider the mean absolute error between the approximation $X_T^{(n)}$ at time T for the approximation as above and X_T. Suppose that the approximation $X^{(n)}$ is obtained for a partition with discretization $t_{j+1} - t_j = 1/n$. It is possible to show that for the Euler scheme, we have

$$\mathbf{E}[|X_T^{(n)} - X_T|] \le Cn^{-1/2},$$

whereas for the Milstein scheme

$$\mathbf{E}[|X_T^{(n)} - X_T|] \le C'n^{-1},$$

for some constant $C, C' > 0$. Note that the mean error between the two processes must be worst at the last point, since the errors add up. We refer to [**KP95**] for the proof of these statements and for more refined schemes. It is very easy to verify these statements numerically on a process where the SDE and the solution of the SDE is known, such as geometric Brownian motion. This is illustrated in Figure 7.2. See also Numerical Project 7.2.

7.5. Existence and Uniqueness of Solutions of SDEs

As for the differential equations in standard calculus, SDEs play an important role in modelling stochastic phenomena. To model a trajectory $X_t, t \le T$, it suffices to write down the variation due to the deterministic change $\mu(X_t) \, dt$ for some function $\mu(X_t)$ and the variation due to local fluctuations $\sigma(X_t) \, dB_t$ for some function $\sigma(X_t)$. Here we assume that the local drift and volatility are time-homogeneous for simplicity. This gives the SDE

(7.23) $$dX_t = \sigma(X_t)\,dB_t + \mu(X_t)\,dt,$$

for some function σ and μ. Do we get one nice Itô process for any choice of functions σ and μ? The short answer is no. Here are sufficient conditions for the existence of a unique process.

Theorem 7.21 (Existence and uniqueness of solutions to SDE). *Consider the SDE*

$$dX_t = \mu(X_t)\,dt + \sigma(X_t)\,dB_t, \quad X_0 = x, \quad t \in [0, T].$$

If the functions σ and μ grow not faster than Kx^2 for some $K > 0$ and are differentiable with bounded derivatives on $\mathbb{R}^1$, then there exists a unique solution $(X_t, t \in [0, T])$ to the SDE. In other words, there exists a continuous process $(X_t, t \leq T)$ adapted to the filtration of the Brownian motion given by

$$X_t = x + \int_0^t \mu(X_s)\,ds + \int_0^t \sigma(X_s)\,dB_s, \quad t \leq T.$$

Example 7.22. Consider the SDE

(7.24) $$dX_t = \sqrt{1 + X_t^2}\,dB_t + \sin X_t\,dt, \quad X_0 = 0.$$

There exists a unique diffusion process $(X_t, t \geq 0)$ that is a solution of this SDE. To see this, we verify the conditions of Theorem 7.21. We have

$$\sigma(x) = \sqrt{1 + x^2}, \qquad \mu(x) = \sin x.$$

Clearly, these functions satisfy the growth condition since μ is bounded and σ grows like $|x|$ for x large. As for the derivatives, we have $\sigma'(x) = \frac{1}{2\sqrt{1+x^2}}$ and $\mu'(x) = \cos x$. These two derivatives are bounded. Figure 7.3 illustrates the paths of this diffusion.

The assumptions of Theorem 7.21 are not too surprising, since similar ones are found in the classical case of ODE. For example, if we have the ODE

$$\frac{dX_t}{dt} = X_t^{1/2}, \quad X_0 = 0,$$

then clearly $X_t = 0$ for all t is a solution. But we also have by integrating that

$$X_t = \frac{t^2}{4}, \quad t \geq 0.$$

Therefore the uniqueness breaks down. Note that the function $\mu(x) = \sqrt{x}$ does not have bounded derivatives at 0. Similarly, consider the ODE

$$\frac{dX_t}{dt} = e^{X_t}, \quad X_0 = 0.$$

Here, the function $\mu(x)$ grows much faster than x^2. The solution of the ODE is by integrating

$$X_t = \log(1 - t)^{-1}.$$

---

[1] In fact, we can require that the functions σ and μ are Lipshitz; that is, $|\sigma(x) - \sigma(y)| \leq C|x - y|$ for some C independent of x, y.

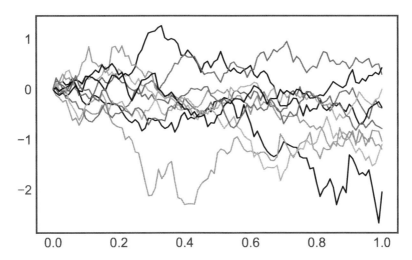

Figure 7.3. A simulations of 10 paths of the process with SDE (7.24). Note that the drift is more pronounced when X_t is far from 0, since the local drift is $\sin X_t$.

This solution explodes at $t = 1$. The same phenomenon may occur for SDE; that is, the process will almost surely go to ∞ in finite time. These times are called *explosion times*. Note that it is possible to consider the paths of the diffusion up to these explosion times. See Numerical Project 7.5.

It is important to keep in mind that the conditions of Theorem 7.21 are sufficient but not necessary. In particular, it is sometimes possible to construct explicitly a diffusion whose local volatility and local drift do not satisfy the conditions. We give two important instances of such diffusions.

Example 7.23 (The Bessel process). Let B_t be a Brownian motion in $\mathbb{R}^d$ for $d > 1$. Consider the process giving the distance at time t of B_t to the origin; i.e.,

$$R_t = \|B_t\| = \left(\sum_{j=1}^{d} B_t^{(j)^2} \right)^{1/2}, \quad t \geq 0.$$

(For $d = 1$, the distance is simply $|B_t|$. Itô's formula cannot be applied in this case, since the absolute value is not differentiable at the origin. This was the content of Tanaka's formula in Section 5.6.) In higher dimensions, the function is smooth enough as long as we stay away from the origin. This is not a problem in view of Example 6.13 as long as $R_0 > 0$. It is not hard to show, see Exercise 7.12, that the process satisfies the SDE

$$dR_t = \sum_j \frac{B_t^{(j)}}{R_t} \, dB_t^{(j)} + \frac{d-1}{2R_t} \, dt = dW_t + \frac{d-1}{2R_t} \, dt,$$

where

(7.25)
$$dW_t = \sum_j \frac{B_t^{(j)}}{R_t} \, dB_t^{(j)}.$$

It turns out that $(W_t, t \geq 0)$ is a standard Brownian motion by Lévy's characterization theorem (Exercise 7.11). This is the subject of the next section. The SDE shows that dR_t is a diffusion. The SDE makes sense for any real number $d > 1$, not only integers. Moreover, the SDE is well-defined since R_t is never equal to 0. However, the SDE does not satisfy the assumption of the existence and uniqueness of the solution of SDEs since $1/x$ diverges at 0. The solution to the SDE still exists since we just constructed it! You can sample paths of this process in Numerical Project 7.4.

Example 7.24 (The Cox-Ingersoll-Ross (CIR) model). Consider the SDE

$$(7.26) \qquad dS_t = \sigma\sqrt{S_t}\,dW_t + (a - bS_t)\,dt, \quad S_0 > 0,$$

for some parameters $a, b > 0$ where $(W_t, t \geq 0)$ is a standard Brownian motion. The local volatility $\sigma(x) = \sigma\sqrt{x}$ does not have a bounded derivative close to 0, since $\sigma'(x) = \frac{\sigma}{2\sqrt{x}}$. We will nevertheless construct a diffusion that is solution to the SDE. Consider independent Ornstein-Uhlenbeck processes $X_t^{(j)}$, $j \leq d$, with SDE

$$dX_t^{(j)} = \frac{-b}{2}X_t^{(j)}\,dt + \frac{\sigma}{2}\,dB_t^{(j)}, \quad X_0^{(j)} > 0,$$

where $B_t = (B_t^{(j)}, j \leq d)$ is a Brownian motion in $\mathbb{R}^d$. We consider the process

$$S_t = \sum_{j \leq d}(X_t^{(j)})^2\,.$$

Clearly, S_t is nonnegative for all $t \geq 0$ by design, so $\sqrt{S_t}$ is well-defined. Let's compute the SDE of the process. Itô's formula, Theorem 7.18, gives

$$dS_t = \sum_j \sigma X_t^{(j)}\,dB_t^{(j)} + \frac{d\sigma^2}{4}\,dt = \sigma\sqrt{S_t}\,dW_t + \left(\frac{d\sigma^2}{4} - bS_t\right)dt\,,$$

where we have defined

$$(7.27) \qquad dW_t = \sum_{j=1}^{d}\frac{X_t^{(j)}}{\sqrt{S_t}}\,dB_t^{(j)}.$$

It turns out that the process $(W_t, t \geq 0)$ is a standard Brownian motion by Lévy's characterization theorem (Theorem 7.26 and Exercise 7.11). If we accept this for a moment, we have the SDE

$$dS_t = \sigma\sqrt{S_t}\,dW_t + \left(\frac{d\sigma^2}{4} - bS_t\right)dt, \quad S_0 > 0.$$

This is a time-homogeneous diffusion called the Cox-Ingersoll-Ross (CIR) process. Again, notice that there are no issues with the square root, since S_t is positive by construction! The SDE also makes sense if we replace $\frac{d\sigma^2}{4}$ by a parameter a as in (7.26), as long as $a \geq \frac{d\sigma^2}{4}$. This ensures that the process remains positive. (Why?) This process is important for interest rates and stochastic volatility models. See Figure 7.4 for a simulation of its paths.

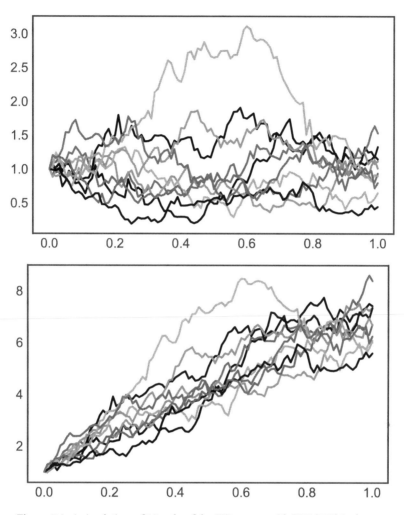

Figure 7.4. A simulations of 10 paths of the CIR process with SDE (7.30) in the case $a = 1$ and $a = 10$.

Note that the local drift is $a - bS_t$, $a \geq \frac{d\sigma^2}{4}$. This can be written as

$$a - bS_t = b\left(\frac{a}{b} - S_t\right).$$

This means that the local drift is negative if $S_t > \frac{a}{b}$ and it is positive if $S_t < \frac{a}{b}$. So the SDE exhibits the same phenomenon as for the SDE of the Ornstein-Uhlenbeck process in Example 7.9. In particular, we should expect that for t large the process should fluctuate around the mean value $\frac{a}{b}$. The CIR model is therefore an example of a *mean-reverting* process. More generally, this hints to the fact that the process is stationary in the long run, akin to the Ornstein-Uhlenbeck process. See Numerical Project 7.3.

7.6. Martingale Representation and Lévy's Characterization

We know very well by now, see Proposition 5.7, that an Itô integral is a continuous martingale with respect to the Brownian filtration, whenever the integrand is in $\mathcal{L}_c^2(T)$. What can we say about the converse? In other words, if we have a martingale with respect to the Brownian filtration, can it be expressed as an Itô integral for some integrand $(V_t, t \leq T)$? Amazingly, the answer to this question is yes.

Theorem 7.25 (Martingale representation theorem). *Let $(B_t, t \geq 0)$ be a Brownian motion with filtration $(\mathcal{F}_t, t \geq 0)$ on $(\Omega, \mathcal{F}, \mathbf{P})$. Consider a martingale $(M_t, t \leq T)$ with respect to this filtration. Then there exists an adapted process $(V_t, t \leq T)$ such that*

$$(7.28) \qquad\qquad M_t = M_0 + \int_0^t V_s \, dB_s, \quad t \leq T.$$

One striking fact of the result is that $(M_t, t \leq T)$ ought to be continuous. In other words, we cannot construct a process with jump that is a martingale adapted to a Brownian motion!

Instead of proving the theorem, we will see how the result is not too surprising with stronger assumptions. Instead of supposing that M_t is $\mathcal{F}_t$-measurable, take that M_t is $\sigma(B_t)$-measurable. In other words, $M_t = h(B_t)$ for some function h. In the case where h is smooth, then it is clear by Itô's formula that the representation (7.28) holds with $V_s = h'(B_s)$.

An important consequence of Theorem 7.25 is a third definition of standard Brownian motion. Proposition 3.1 and Definition 2.25 gave us the first two.

Theorem 7.26. *Let $(M_t, t \in [0, T])$ be a continuous martingale with respect to the filtration $(\mathcal{F}_t, t \leq T)$ with $M_0 = 0$ and with quadratic variation $\langle M \rangle_t = t$. Then $(M_t, t \leq T)$ is a standard Brownian motion.*

As an example, consider the process

$$X_t = \int_0^t \mathrm{sgn}(B_s) \, dB_s$$

that appears in Tanaka's formula, Theorem 5.32. Clearly, this is a continuous process since it is given by an Itô integral. Moreover, the quadratic variation is easily computed by Proposition 7.7. We have

$$(dX_t)^2 = (\mathrm{sgn}(B_t))^2 \, dt = dt.$$

We conclude that the process $(X_t, t \geq 0)$ is a standard Brownian motion. Other examples were given in equations (7.25) and (7.27). See Exercise 7.11.

Proof of Theorem 7.26. The martingale representation theorem implies that we can write $M_t = \int_0^t V_s \, dB_s$ for some adapted process $(V_t, t \leq T)$. For simplicity, we assume that V_s is such that the process (7.17) is a martingale. (The general case can be treated similarly by using characteristic functions; see Exercise 1.14.) We show that the increments of M are independent with the right distribution as in Proposition 3.1. In fact, we show that for $0 \leq s < t \leq T$ and $\lambda \in \mathbb{R}$, we have

(7.29) $$\mathbf{E}[\exp(\lambda(M_t - M_s))|\mathcal{F}_s] = e^{\frac{\lambda^2}{2}(t-s)}.$$

Equation (7.29) is sufficient to prove the theorem, since we can apply several conditionings to get the joint distribution of the increments.

Consider for $t \geq s$ the process

$$Y_t = \exp\left(\lambda(M_t - M_s) - \frac{\lambda^2}{2}\int_s^t V_u^2 \, du\right) = \exp\left(\lambda \int_s^t V_u \, dB_u - \frac{\lambda^2}{2}(t-s)\right).$$

Note that $Y_s = 1$. As discussed in Example 7.17, this is a martingale. This implies

$$1 = \mathbf{E}[Y_t|\mathcal{F}_s] = \mathbf{E}\left[\exp(\lambda(M_t - M_s) - \frac{\lambda^2}{2}(t-s))|\mathcal{F}_s\right],$$

thereby proving (7.29). $\qquad\square$

7.7. Numerical Projects and Exercises

7.1. **Simulating SDEs.** Simulate 100 paths for the following diffusions given by their SDEs on $[0, 1]$ using the Euler-Maruyama scheme and the Milstein scheme for a discretization of 0.01.

(a) Geometric Brownian motion:

$$dS_t = S_t \, dB_t + \left(\mu + \frac{1}{2}\right)S_t \, dt, \quad S_0 = 1,$$

for $\mu = -1/2$, $\mu = -2$, and $\mu = 0$.

(b) Ornstein-Uhlenbeck process:

$$dX_t = -X_t \, dt + dB_t, \quad X_0 = 1.$$

(c) The diffusion of Example 7.22:

$$dX_t = \sqrt{1 + X_t^2} \, dB_t + \sin X_t \, dt, \quad X_0 = 0.$$

7.2. **Euler vs Milstein.** Go back to (a) above with $\mu = 0$. Consider $S_t = \exp(B_t - t/2)$. Let $S^{(n)}$ be the approximating process defined by the Euler-Maruyama scheme in equation (7.21) for the the discretization $1/n$. Find the absolute error $\mathbf{E}[|S_1^{(n)} - S_1|]$ by averaging over 100 paths for $n = 10$, $n = 100$, $n = 1000$, $n = 10,000$. Plot the errors on a graph for the Euler-Maruyama scheme. Repeat the experiment for the Milstein scheme (7.22) and compare.

7.3. **The CIR model.** Consider the CIR process as in Example 7.24 given by the SDE

(7.30) $$dZ_t = (a - Z_t) \, dt + \sqrt{Z_t} \, dB_t, \quad Z_0 = 1.$$

(We set the parameters $\sigma = 1$, $b = 1$ here.)

(a) Plot 10 paths of this process on $[0, 1]$ with a time step of 0.001 for each of the cases $a = 0.1$, $a = 0.25$, and $a = 10$. If the path becomes negative, just set the rest of the process to 0.

(b) Sample 1,000 paths of the processes above. What is the proportion of paths that became negative for each a?

(c) Let's focus on the case $a = 10$. Plot 10 paths of this process on $[0, 10]$ with a time step of 0.001. What do you notice?

(d) Compute the average of Z_{10} on 100 paths for $a = 10$.

7.4. **Bessel process.** The Bessel process with dimension $d > 1$ introduced in Example 7.23 is the diffusion with SDE

$$dR_t = dB_t + \frac{d-1}{2R_t} \, dt, \quad R_0 > 0.$$

Use the Euler-Maruyama scheme to plot the 100 paths of Bessel process on $[0, 1]$ using a discretization of 0.001 starting at $R_0 = 1$ for $d = 2$, $d = 3$, and $d = 10$. Discard the paths that become negative. How many paths out of 100 become negative?

7.5. **Explosion time.** Consider the SDE

$$dY_t = e^{Y_t} \, dB_t, \quad Y_0 = 0.$$

Clearly, the local volatility grows much faster than the assumption of Theorem 7.21. Sample 100 paths of this process using the Euler scheme with a discretization of 0.01. What do you notice? What if you sample only one path?

7.6. **A martingale $X_t^2 - \int_0^t V_s^2 \, ds$.** Consider the Itô process

$$X_t = \int_0^t B_s \, dB_s, \quad t \in [0, 1].$$

We sample the martingale $M_t = X_t^2 - \int_0^t B_s^2 \, ds$ constructed in Example 7.10.

(a) Plot 10 paths of the quadratic variation $Q_t = \int_0^t B_s^2 \, ds$ on $[0, 1]$ with a discretization of 0.001.

(b) Plot 10 paths of the martingale $X_t^2 - \int_0^t B_s^2 \, ds$ on $[0, 1]$ with a discretization of 0.001.

Exercises

7.1. **SDE of Brownian bridge.** Consider a Brownian bridge $(Z_t, 0 \le t \le 1)$ given by $Z_0 = Z_1 = 0$ and

$$Z_t = (1 - t) \int_0^t \frac{dB_s}{1 - s}, \quad 0 \le t < 1.$$

Show that $(Z_t, 0 \le t \le 1)$ satisfies the SDE

$$dZ_t = dB_t + \frac{-Z_t}{1 - t} \, dt.$$

The Brownian bridge local volatility is the same as Brownian motion. Can you discuss the form of the local drift?

7.2. **The Ornstein-Uhlenbeck process with parameters.** Use Itô's formula to show that equation (7.12) is the solution to the SDE (7.11).

7.3. **Generalized geometric Brownian motion.** Consider the process

$$M_t = \exp\left(\int_0^t \sigma(s)\,dB_s - \frac{1}{2}\int_0^t \sigma^2(s)\,ds\right), \quad t \le T.$$

Use the properties of the increments of Brownian motion and Corollary 5.18 to directly show that the process $(M_t, t \le T)$ is a martingale for the Brownian filtration, whenever $\int_0^T \sigma^2(t)\,dt < \infty$.

7.4. **A covariance process.** Consider the processes from Exercise 5.8

$$X_t = \int_0^t (1-s)\,dB_s, \qquad Y_t = \int_0^t (1+s)\,dB_s.$$

Is the covariance process $(X_t Y_t, t \ge 0)$ a martingale?

7.5. **Practice on product rule.** We consider the Itô process

$$X_t = \int_0^t (1-2s)\,dB_s, \quad t \ge 0.$$

(a) Find the mean and the variance of X_t at all t.
(b) Use Itô's product rule to show that

$$X_t B_t = \int_0^t (1-2s)B_s\,dB_s + \int_0^t X_s\,dB_s + \int_0^t (1-2s)\,ds.$$

(c) At what times t are X_t and B_t uncorrelated?
(d) Argue that (X_1, B_1) are jointly Gaussian.
   Write B_1 as an integral and consider linear combinations of X_1 and B_1.
(e) Conclude that X_1 and B_1 are independent.

7.6. **Review of Itô integrals.** Let $(B_t, t \in [0,1])$ be a standard Brownian motion. Consider the two stochastic processes on $[0,1]$

$$X_t = \int_0^t B_s\,ds, \qquad Y_t = \int_0^t s\,dB_s, \quad t \in [0,1].$$

(a) Describe the distribution of the process $(Y_t, t \in [0,1])$.
(b) Use Itô isometry to compute the covariance $\mathbf{E}[B_t Y_t]$.
(c) Use Itô's product rule and the above to find the covariance $\mathbf{E}[X_t Y_t]$ between the two processes.
(d) Conclude that the correlation coefficient $\rho(X_1, Y_1)$ is 1/2.

7.7. **Exercise on stochastic calculus.** Consider the Itô process $(Y_t, t \ge 0)$ given by the equation

$$dY_t = e^t\,dB_t, \quad Y_0 = 0.$$

Let $\tau = \min\{t \ge 0 : Y_t = 1 \text{ or } Y_t = -1\}$.
   In this exercise, we will show that $\mathbf{E}[\tau] \le 0.347\dots$.
(a) Find a PDE for $f(t,x)$ satisfied whenever $f(t, Y_t)$ is a martingale.
(b) Verify that $f(t,x) = x^2 - \frac{e^{2t}}{2}$ satisfies the PDE and that $f(t, Y_t)$ is a martingale.

(c) Argue from the previous questions that we must have

$$\mathbf{E}[e^{2\tau}] = 3.$$

(d) Conclude from the previous question that $\mathbf{E}[\tau] \leq \frac{\log 3}{2}$.
   Hint: Jensen's inequality.

7.8. **An exercise on the CIR model.** Consider the diffusion $(X_t, t \geq 0)$ given by the SDE

$$dX_t = \frac{1}{2} dt + \sqrt{X_t}\, dB_t, \quad X_0 = 1.$$

(a) Let $f(t, x)$ be some differentiable function. Write the Itô process $f(t, X_t)$ in differential form in terms of dB_t and dt.

(b) From the previous question, find the PDE satisfied by f for which the process $f(t, X_t)$ is a martingale.

(c) Which processes below are martingales:

$$X_t, \qquad t - 2X_t, \qquad \log X_t \, ?$$

(d) Let $\tau = \min\{t \geq 0 : X_t = e \text{ or } X_t = e^{-1}\}$ be the first hitting time of the process to $e = 2.718\ldots$ or e^{-1}. Use the process $\log X_t$ to show that

$$\mathbf{P}(X_\tau = e) = \frac{1}{2}.$$

(e) Compute $\mathbf{E}[\tau]$.

7.9. **Solving an SDE.** Consider the SDE

$$dX_t = -\frac{X_t}{2 - t} dt + \sqrt{t(2 - t)}\, dB_t, \quad 0 \leq t < 1, \quad X_0 = 0.$$

Suppose the solution is of the form

$$X_t = a(t)Y_t, \qquad Y_t = \int_0^t b(s)\, dB_s,$$

for some smooth functions a, b. Apply Itô's formula to find an ODE satisfied by $a(t)$ and $b(t)$ and solve it. Use the form of the solution to show that X_t is a Gaussian process of mean zero and covariance

$$\mathbf{E}[X_t X_s] = (2 - t)(2 - s)\left\{\log \frac{4}{(2 - s)^2} - s\right\}, \quad 0 \leq s < t < 1.$$

7.10. **Practice on SDEs.**
   Let $(B_t^{(1)}, t \geq 0)$ and $(B_t^{(2)}, t \geq 0)$ be two independent standard Brownian motions. Let $(Y_t, t \geq 0)$ be a continuous, adapted process. We consider

$$W_t = \int_0^t \frac{1}{\sqrt{1 + Y_s^2}} dB_s^{(1)} + \int_0^t \frac{Y_s}{\sqrt{1 + Y_s^2}} dB_s^{(2)}.$$

(a) Argue that $(W_t, t \geq 0)$ is also a standard Brownian motion.

(b) Let $X_t = \sinh W_t = \frac{e^{W_t} - e^{-W_t}}{2}$. Find an SDE for X_t in terms of dt and dW_t.

(c) Consider now $Y_t = e^{B_t^{(2)}} \int_0^t e^{-B_s^{(2)}} dB_s^{(1)}$. Show that Y is a solution of the SDE

$$dY_t = \sqrt{1 + Y_t^2}\, dW_t + \frac{Y_t}{2}\, dt, \quad Y_0 = 0.$$

(d) Argue that $X_t = Y_t$ for all t with probability one by Theorem 7.21.

7.11. **Applications of Lévy's characterization.** Prove that the processes given in equations (7.25) and (7.27) are standard Brownian motions.

7.12. **SDE of the Bessel process.** Prove using Itô's formula that the Bessel process $R_t = \|B_t\|$ given in Example 7.23 is a solution to the SDE

$$dR_t = dW_t + \frac{d-1}{2R_t}\, dt,$$

where $(W_t, t \geq 0)$ is a standard Brownian motion.

7.13. ⋆ **Conditional Itô isometry.** We prove equation (7.13).
(a) Suppose first that the process $(V_t, s \leq T)$ is in $\mathcal{S}(T)$. Show that for $t \leq t'$

$$\mathbf{E}\left[\left(\int_0^{t'} V_s\, dB_s - \int_0^t V_s\, dB_s \right)^2 \middle| \mathcal{F}_t \right] = \int_t^{t'} V_s^2\, ds.$$

Hint: Go back to the proof of Itô's isometry in Proposition 5.7 and condition on $\mathcal{F}_t$.

(b) To prove the identity in the continuous case, mimic the proof of the martingale property in Theorem 5.12.

7.8. Historical and Bibliographical Notes

Itô processes are examples of a larger class of stochastic processes called *semimartingales*. A semimartingale $(M_t, t \geq 0)$ is a stochastic process that can be written as a sum of two processes: one martingale (or more precisely a local martingale) and a process whose paths have bounded variation. Clearly, Itô processes satisfy this definition. The Poisson process is not an Itô process. We cannot write it as an Itô integral by Theorem 7.25. However, it is a semimartingale since its paths are increasing and thus have bounded variation. Tanaka's formula in Theorem 5.32 gives a nontrivial example of a semimartingale that is not an Itô process. Note that the local time $(L_t, t \geq 0)$ has paths of finite variation since they are increasing.

As for ODEs, there are some standard techniques to integrate simple SDEs in order to find the explicit solution in terms of a function of time and Brownian motion. They essentially consist of reverse engineering Itô's formula. We refer the reader to [**Øks03**] and [**Ste01**] for more on this. Exercise 7.10 is a very nice example from [**CM15**].

The proof of Theorem 7.21 is in the spirit of the proof for the existence and uniqueness of solutions to ODEs known as the Picard-Lindelöf theorem. The idea is to approach the solution by a sequence of approximations $X^{(n)}$, $n \geq 0$, defined iteratively by

$$X_t^{(0)} = x, \quad X_t^{(n)} = x + \int_0^t \mu(X_s^{(n-1)})\, ds + \int_0^t \sigma(X_s^{(n-1)})\, dB_s.$$

Then, one would like to show that the distance between two consecutive iterations $X^{(n+1)}$ and $X^{(n)}$ *contracts.* This is a powerful idea in mathematics that is used often to prove the existence of a limit. We refer the reader to [**Øks03**] for a proof.

We suggest the nice discussion in [**Ste01**] for more on the martingale representation theorem.

Chapter 8

The Markov Property

We saw in Chapter 3 that Brownian motion has independent increments. This implies a more general property called *the Markov property*. It turns out that the diffusions we have been studying also share this property, so they may be called *Markov processes*. In general, martingales are not Markov processes, and Markov processes are not martingales. However, there are processes such as Brownian motion that enjoy both properties.

In this chapter, we start by studying the Markov property of diffusions in Section 8.1. This property is extended to random stopping times in Section 8.2. This is the so-called *strong Markov property*. Markov processes are very convenient, because many quantities can be computed by solving a PDE problem, a feature we have already exploited. This is explored further in Section 8.3 on the *Kolmogorov equations* and in Section 8.4 on the *Feynman-Kac formula*.

8.1. The Markov Property for Diffusions

Let's start by exhibiting the Markov property of Brownian motion. To see this, let $(\mathcal{F}_t, t \geq 0)$ be the natural filtration of the Brownian motion $(B_t, t \geq 0)$. Consider $g(B_t)$ for some time t and bounded function g. (For example, g could be an indicator function.) Consider also a random variable W that is $\mathcal{F}_s$-measurable for $s < t$. (For example, W could be B_s or $\mathbf{1}_{\{B_s > 0\}}$.) Let's compute $\mathbf{E}[g(B_t)W]$:

$$\mathbf{E}[g(B_t)W] = \mathbf{E}[g(B_s + B_t - B_s)W] = \int_{\mathbb{R}} \mathbf{E}[g(B_s + y)W] \frac{e^{-\frac{y^2}{2(t-s)}}}{\sqrt{2\pi(t-s)}} \, dy.$$

The second equality follows from the fact that $B_t - B_s$ is a Gaussian random variable of mean 0 and variance $t - s$ independent of $\mathcal{F}_s$. It is possible to interchange $\mathbf{E}$ with the integral by Fubini's theorem (see Remark 5.21). Since W is any $\mathcal{F}_s$-measurable random variable, this shows that the conditional expectation of $g(B_t)$ given $\mathcal{F}_s$ (Definition 4.14) is

(8.1) $$\mathbf{E}[g(B_t)|\mathcal{F}_s] = \int_{\mathbb{R}} g(B_s + y)\frac{e^{-\frac{y^2}{2(t-s)}}}{\sqrt{2\pi(t-s)}}\,\mathrm{d}y.$$

We make two important observations. First, the right-hand side is a function of s, t, and B_s only (and not of the Brownian motion before time s). In particular, we have

$$\mathbf{E}[g(B_t)|\mathcal{F}_s] = \mathbf{E}[g(B_t)|B_s].$$

This holds for any bounded function g. In particular, it holds for all indicator functions. This implies that the conditional distribution of B_t given $\mathcal{F}_s$ depends solely on B_s, and not on other values before time s. Second, the right-hand side is *time-homogeneous* in the sense that it depends on the time difference $t - s$.

   We have just shown that Brownian motion is a *time-homogenous Markov process*. The general property is defined as follows.

Definition 8.1. Consider a stochastic process $(X_t, t \geq 0)$ and its natural filtration $(\mathcal{F}_t, t \geq 0)$. It is said to be a *Markov process* if and only if for any (bounded) function $g : \mathbb{R} \to \mathbb{R}$, we have

$$\mathbf{E}[g(X_t)|\mathcal{F}_s] = \mathbf{E}[g(X_t)|X_s], \quad \text{for all } t \geq 0 \text{ and all } s \leq t.$$

This implies that $\mathbf{E}[g(X_t)|\mathcal{F}_s]$ is an explicit function of s, t, and X_s. It is said to be *time-homogeneous* if it is a function of $t - s$ and X_s. Since the above holds for all bounded g, the conditional distribution of X_t given $\mathcal{F}_s$ is the same as the conditional distribution of X_t given X_s.

   One way to compute the conditional distribution of X_t given $\mathcal{F}_s$ is to compute the conditional MGF given $\mathcal{F}_s$; i.e.,

(8.2) $$\mathbf{E}[e^{aX_t}|\mathcal{F}_s], \quad a \geq 0.$$

The process would be Markov if the conditional MGF is an explicit function of s, t, and X_s. It was shown in equation (4.13) that Brownian motion was Markov using this.

   An equivalent (but more symmetric) way to express the Markov property is to say that *the future of the process is independent of the past when conditioned on the present*; see Exercise 8.1. Concretely, this means that for any $r < s < t$, we have that X_t is independent of X_r when we condition on X_s.

   The conditional distribution of X_t given X_s is well described using *transition probabilities*. We will be mostly interested in the case where these probabilities admit a density. More precisely, for such a Markov process, we have

$$\mathbf{E}[g(X_t)|X_s = x] = \int_{\mathbb{R}} g(y)p_{s,t}(x, y)\,\mathrm{d}y.$$

Here, we explicitly write the left-hand side as a function of space, i.e., the position X_s, by fixing $X_s = x$. In words, the *transition probability density* $p_{s,t}(x, y)$ represents the probability density that starting from $X_s = x$ at time s, the process ends up at $X_t = y$ at time $t > s$. If the process is time-homogeneous, this only depends on the time difference and we write $p_{t-s}(x, y)$. From equation (8.1), we see that the transition probability density for standard Brownian motion is

$$(8.3) \qquad p_s(x, y) = \frac{e^{-\frac{(y-x)^2}{2s}}}{\sqrt{2\pi s}}, \quad s > 0, \ x, y \in \mathbb{R}.$$

This function is sometimes called the *heat kernel*, as it relates to the *heat equation*; see Example 8.10.

The Markov property is very convenient to compute quantities, as we shall see throughout the chapter. As a first example, we remark that it is easy to express joint probabilities of a Markov process $(X_t, t \geq 0)$ at different times. Consider the functions $f = \mathbf{1}_A$ and $g = \mathbf{1}_B$ from $\mathbb{R} \to \mathbb{R}$, where A, B are two intervals in $\mathbb{R}$. Let's compute $\mathbf{P}(X_{t_1} \in A, X_{t_2} \in B) = \mathbf{E}[f(X_{t_1})g(X_{t_2})]$ for $t_1 < t_2$. By the properties of conditional expectation (Proposition 4.19) and the Markov property, we have

$$\begin{aligned}
\mathbf{P}(X_{t_1} \in A, X_{t_2} \in B) &= \mathbf{E}[f(X_{t_1})g(X_{t_2})] \\
&= \mathbf{E}[f(X_{t_1})\, \mathbf{E}[g(X_{t_2})|\mathcal{F}_{t_1}]] \\
&= \mathbf{E}[f(X_{t_1})\, \mathbf{E}[g(X_{t_2})|X_{t_1}]].
\end{aligned}$$

Assuming that the process is time-homogeneous and admits a transition density $p_t(x, y)$, as for Brownian motion, this becomes

$$\begin{aligned}
\mathbf{P}(X_{t_1} \in A, X_{t_2} \in B) &= \int_{\mathbb{R}} f(x_1) \left(\int_{\mathbb{R}} g(x_2) p_{t_2 - t_1}(x_1, x_2) \, dx_2 \right) p_{t_1}(x_0, x_1) \, dx_1 \\
&= \int_A \left(\int_B p_{t_2 - t_1}(x_1, x_2) \, dx_2 \right) p_{t_1}(x_0, x_1) \, dx_1.
\end{aligned}$$

This easily generalizes to any finite-dimensional distribution of $(X_t, t \geq 0)$; see Exercises 8.2 and 8.3.

Example 8.2 (Markov vs Martingale). We already noticed in Remark 4.33 that martingales are not Markov in general and Markov processes are not martingales in general. There are processes, such as Brownian motion, that enjoy both. An example of a Markov process that is not a martingale is Brownian motion with drift $(X_t, t \geq 0)$ with $X_t = \sigma B_t + \mu t$ with $\mu \neq 0$. See Exercise 8.5. Conversely, take $Y_t = \int_0^t X_s \, dB_s$, where $X_s = \int_0^s B_u \, dB_u$ as in Example 5.17. This is a martingale for the Brownian filtration. However, it is not a Markov process.

Remark 8.3 (Functions of Markov processes). It might be tempting to think that if $(X_t, t \geq 0)$ is a Markov process, then the process defined by $Y_t = f(X_t)$ for some reasonable function f is also Markov. Indeed, one could hope to write for an arbitrary bounded function g

$$(8.4) \qquad \mathbf{E}[g(Y_t)|\mathcal{F}_s] = \mathbf{E}[g(f(X_t))|\mathcal{F}_s] = \mathbf{E}[g(f(X_t))|X_s],$$

by using the Markov property of $(X_t, t \geq 0)$. The flaw in this reasoning is that the Markov property should hold for the natural filtration $(\mathcal{F}_t^Y, t \geq 0)$ of the process $(Y_t, t \geq 0)$, and not the one of $(X_t, t \geq 0)$, $(\mathcal{F}_t^X, t \geq 0)$. It might be that the filtration of $(Y_t, t \geq 0)$ has *less information* than the one of $(X_t, t \geq 0)$, if the function f is not one-to-one. For example, if $f(x) = x^2$, then $\mathcal{F}_t^Y$ has less information than $\mathcal{F}_t^X$ as we cannot recover the sign of X_t knowing Y_t. In other words, the second equality in equation (8.4) might not hold. In some cases, a function of Brownian motion might be Markov even when f is not one-to-one; see Exercise 8.12.

It turns out that diffusions (Definition 7.2), such as the Ornstein-Uhlenbeck process (Example 7.9) and Brownian bridge (Exercise 7.1), are Markov processes.

Theorem 8.4 (Diffusions are Markov processes). *Let $(B_t, t \geq 0)$ be a standard Brownian motion. Let $\mu : \mathbb{R} \to \mathbb{R}$ and $\sigma : \mathbb{R} \to \mathbb{R}$ be differentiable functions with bounded derivatives on $[0, T]$ (as in Theorem 7.21). Then, the diffusion with SDE*

$$dX_t = \mu(X_t)\,dt + \sigma(X_t)\,dB_t, \quad X_0 = x_0,$$

defines a time-homogeneous Markov process on $[0, T]$.

An analogous statement holds for time-inhomogeneous diffusions. The proof is a generalization of the Markov property of Brownian motion: We take advantage of the independence of Brownian increments.

Proof. By Theorem 7.21, the SDE defines a unique continuous adapted process $(X_t, t \leq T)$. Let $(\mathcal{F}_t^X, t \geq 0)$ be the natural filtration of $(X_t, t \leq T)$. For a fixed $t > 0$, consider the process $W_s = B_{t+s} - B_t$, $s \geq 0$. Let $(\mathcal{F}_t, t \geq 0)$ be the filtration of $(B_t, t \geq 0)$. It turns out that the process $(W_s, s \geq 0)$ is a standard Brownian motion independent of $\mathcal{F}_t$ (Exercise 8.4). For $s \geq 0$, we consider the SDE

$$dY_s = \mu(Y_s)\,ds + \sigma(Y_s)\,dW_s, \quad Y_0 = X_t.$$

Again by Theorem 7.21, there exists a unique solution to the SDE that is adapted to the natural filtration of W. Note that $(X_{t+s}, s \geq 0)$ is *the* solution to this equation since

$$X_{t+s} = X_t + \int_t^{t+s} \mu(X_u)\,du + \int_t^{t+s} \sigma(X_u)\,dB_u$$

$$= X_t + \int_0^s \mu(X_{t+u})\,du + \int_0^s \sigma(X_{t+u})\,dW_u,$$

where the last equality follows by the change of variable $u \to u - t$. (Note that the change in the Itô integral is not completely trivial.) We conclude that for any interval A,

$$\mathbf{P}(X_{t+s} \in A | \mathcal{F}_t^X) = \mathbf{P}(Y_s \in A | \mathcal{F}_t^X).$$

But since $(Y_s, s \geq 0)$ depends on $\mathcal{F}_t^X$ only through X_t (because $(W_s, s \geq 0)$ is independent of $\mathcal{F}_t$), we conclude that $\mathbf{P}(X_{t+s} \in A | \mathcal{F}_t^X) = \mathbf{P}(X_{t+s} \in A | X_t)$, so $(X_t, t \geq 0)$ is a time-homogeneous Markov process. $\square$

8.2. The Strong Markov Property

The Doob optional stopping theorem (Corollary 4.38) extended some properties of martingales to stopping times. The Markov property can also be extended to stopping times for certain processes. These processes are called *strong Markov processes*.

First, we need the notion of sigma-field associated with a stopping τ. For τ a stopping time for the filtration $(\mathcal{F}_t, t \geq 0)$, we define the sigma-field

$$\mathcal{F}_\tau = \{A \in \mathcal{F} : A \cap \{\tau \leq t\} \in \mathcal{F}_t\}.$$

In words, $\mathcal{F}_\tau$ contains the events that pertain to the process up to time τ. For example, if $\tau < \infty$, then the event $\{B_\tau > 0\}$ is in $\mathcal{F}_\tau$. However, the event $\{B_1 > 0\}$ is not in $\mathcal{F}_\tau$ in general since $A \cap \{\tau \leq t\}$ is not in $\mathcal{F}_t$ for $t < 1$. Roughly speaking, a random variable that is $\mathcal{F}_\tau$-measurable should be thought of as an explicit function of X_τ. With this new object, we are ready to define the strong Markov property.

Definition 8.5. Let $(X_t, t \geq 0)$ be a stochastic process and let $(\mathcal{F}_t, t \geq 0)$ be its natural filtration. The process $(X_t, t \geq 0)$ is said to be *strong Markov* if for any stopping time τ for the filtration of the process and any bounded function g,

$$\mathbf{E}[g(X_{t+\tau})|\mathcal{F}_\tau] = \mathbf{E}[g(X_{t+\tau})|X_\tau], \quad t \geq 0.$$

This means that $X_{t+\tau}$ depends on $\mathcal{F}_\tau$ solely through X_τ (whenever $\tau < \infty$).

It turns out that Brownian motion is a strong Markov process. In fact, a stronger statement holds, which generalizes Exercise 8.4:

Theorem 8.6. *Let τ be a stopping time for the filtration of the Brownian motion $(B_t, t \geq 0)$ such that $\tau < \infty$. Then the process*

$$(B_{t+\tau} - B_\tau, t \geq 0)$$

is also a standard Brownian motion independent of $\mathcal{F}_\tau$.

Example 8.7 (Brownian motion is strong Markov). To see this, let's compute the conditional MGF as in equation (8.2). We have

$$\mathbf{E}[e^{aB_{t+\tau}}|\mathcal{F}_\tau] = \mathbf{E}[e^{aB_\tau}e^{a(B_{t+\tau}-B_\tau)}|\mathcal{F}_\tau] = e^{aB_\tau}\mathbf{E}[e^{a(B_{t+\tau}-B_\tau)}] = e^{aB_\tau}e^{a^2 t/2},$$

since e^{aB_τ} is $\mathcal{F}_\tau$-measurable and $B_{t+\tau} - B_\tau$ is a Gaussian of mean 0 and variance t independent of $\mathcal{F}_\tau$ by Theorem 8.6.

The reflection principle (Lemma 4.47) is also a consequence of Theorem 8.6.

Proof of Theorem 8.6. We first consider for fixed n the discrete-valued stopping time

$$\tau_n = \frac{k+1}{2^n} \qquad \text{if } \frac{k}{2^n} \leq \tau < \frac{k+1}{2^n}, \quad k \in \mathbb{N}.$$

In words, if τ occurs in the interval $[\frac{k}{2^n}, \frac{k+1}{2^n})$, we stop at the next dyadic $\frac{k+1}{2^n}$. By construction τ_n depends only on the process *in the past*. Consider the process $W_t = B_{t+\tau_n} - B_{\tau_n}$, $t \geq 0$. We show it is a standard Brownian motion independent of τ_n. This is feasible as we can decompose over the discrete values taken by τ_n. More precisely, take $E \in \mathcal{F}_{\tau_n}$ and some generic event $\{W_t \in A\}$ for the process W. Then by decomposing over the values of τ_n, we have

$$P(\{W_t \in A\} \cap E) = \sum_{k=0}^{\infty} P(\{W_t \in A\} \cap E \cap \{\tau_n = k/2^n\})$$

$$= \sum_{k=0}^{\infty} P(\{B_{t+k/2^n} - B_{k/2^n} \in A\} \cap E \cap \{\tau_n = k/2^n\})$$

$$= \sum_{k=0}^{\infty} P(B_{t+k/2^n} - B_{k/2^n} \in A) \times P(E \cap \{\tau_n = k/2^n\}),$$

since $(B_{t+k/2^n} - B_{k/2^n}, t \geq 0)$ is independent of $\mathcal{F}_{k/2^n}$ by Exercise 8.4 and since $E \cap \{\tau_n = k/2^n\} \in \mathcal{F}_{k/2^n}$ by definition of stopping time. But this process is now Brownian motion so $P(B_{t+k/2^n} - B_{k/2^n} \in A) = P(B_t \in A) = P(W_t \in A)$, dropping the dependence on k. The sum over k then yields

$$P(\{W_t \in A\} \cap E) = P(W_t \in A) \times P(E),$$

as claimed. The extension to τ is done by using continuity of the paths. We have

$$\lim_{n \to \infty} B_{t+\tau_n} - B_{\tau_n} = B_{t+\tau} - B_\tau \text{ almost surely.}$$

(Note that this only uses right-continuity!) Moreover, this implies $B_{t+\tau} - B_\tau$ is independent of $\mathcal{F}_{\tau_n}$ for all n. Again by (right)-continuity this extends to independence from $\mathcal{F}_\tau$. The limiting distribution of the process is obtained by looking at the finite-dimensional distributions of the increments of $B_{t+\tau_n} - B_{\tau_n}$ for a finite number of t's and taking the limit as above. $\qquad\square$

Most diffusions also enjoy the strong Markov property, as long as the functions σ and μ encoding the volatility and the drift are nice enough. This is the case for diffusions we have considered.

Theorem 8.8 (Most diffusions are strong Markov). *Consider a diffusion $(X_t, t \leq T)$ as in Theorem 8.4. Then the diffusion has the strong Markov property.*

The proof follows the line of the one of Theorem 8.4. Before moving on to the next section, we make a remark on $\mathcal{F}_\tau$ and the martingale property.

Remark 8.9 (An extension of optional sampling). Consider a continuous martingale $(M_t, t \leq T)$ for a filtration $(\mathcal{F}_t, t \geq 0)$ and a stopping time τ for the same filtration. Suppose we would like to compute for some T

$$E[M_T \mathbf{1}_{\{\tau \leq T\}}].$$

It would be tempting to condition on $\mathcal{F}_\tau$ and write $E[M_T | \mathcal{F}_\tau] = M_\tau$ on the event $\{\tau \leq T\}$. We would then conclude that

$$E[M_T \mathbf{1}_{\{\tau \leq T\}}] = E[M_\tau \mathbf{1}_{\{\tau \leq T\}}].$$

In some sense, we have extended the martingale property to stopping times beyond the statement of Corollary 4.38. This property can be proved under reasonable assumptions on $(M_t, t \le T)$ (e.g., it is positive). Indeed, it suffices to approximate τ by a discrete-valued stopping time τ_n as in the proof of Theorem 8.4. One can then apply the martingale property at a fixed time. Additional properties on $(M_t, t \le T)$ are needed to make sure one can take the limit $n \to \infty$ inside the expectation.

8.3. Kolmogorov's Equations

At this stage, we are well acquainted with the intimate connections between martingales and PDEs. Important examples were given in Corollary 5.29 and Example 7.12. This was also exploited in the solution of the Dirichlet problem in Chapter 6. In this section, we explain in more detail how PDEs come up when computing quantities related to Markov processes, in particular to diffusions. We start with the *heat equation*.

Example 8.10 (Heat equation and Brownian motion). Let $f(t, x)$ be a function of time and space. The heat equation in $1 + 1$-dimension (one dimension of time, one dimension of space) is the PDE

$$(8.5) \qquad \frac{\partial f}{\partial t} = \frac{1}{2} \frac{\partial^2 f}{\partial x^2}.$$

In $1 + d$ dimension (one dimension of time, d dimensions of space), the heat equation is

$$(8.6) \qquad \frac{\partial f}{\partial t} = \frac{1}{2} \Delta f,$$

where Δ is the Laplacian defined in equation (6.7). Note that this PDE differs from the martingale condition of Corollary 5.29 and of equation (6.9) by a minus sign. This is explained in more detail below. For now, we notice that solutions to this PDE can be expressed as an expectation over Brownian paths (in the same spirit as in the Dirichlet problem of Chapter 6). Indeed, the solution of the PDE can be written in the form

$$(8.7) \qquad f(t, x) = \mathbf{E}[g(B_t)|B_0 = x],$$

for the initial value

$$f(0, x) = g(x)$$

for some given function g of space. This was already noted in Exercise 6.10 using integration by parts. The form of the solution (8.7) has a natural interpretation. The function $g(x)$ can be seen as representing the temperature at position x at time 0. The value $f(t, x)$ gives the temperature at time t and position x. It can be represented as an *average of $g(B_t)$ over Brownian paths* starting at x. In other words, $f(t, x)$ is a specific type of space average. Effectively, it is a smoothing of the initial function g that models the diffusion of heat; see Figure 8.1.

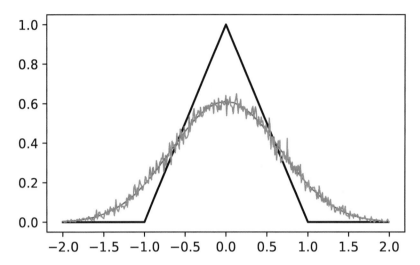

Figure 8.1. An illustration of the solution to the heat equation $f(t, x)$ with initial value $g(x) = 1 - |x|$ for $|x| \leq 1$ and 0 for $|x| > 1$ at time 0.25. The initial value function g is also depicted. The rugged function is the solution of the form (8.7) approximated with 100 paths and discretization of 0.01. There is also a solid line giving the solution (8.8). See Numerical Project 8.1.

Note that since the PDF of B_t given $B_0 = x$ is given by the *heat kernel* $p_t(x, y)$ in equation (8.3), the solution of the initial value problem of the heat equation can be written as

$$(8.8) \qquad f(t, x) = \int_{\mathbb{R}} g(y) p_t(x, y) \, dy.$$

The heat equation is a particular instance of the *Kolmogorov backward equation*, or simply *backward equation*. The solutions of these PDEs can be represented as an average over paths of diffusions, similar to the representation (8.7). This is a consequence of Itô's formula. To prove this, we will be heavily using the fact that the local volatility and drift of diffusions are functions of the diffusion's position and time only.

Theorem 8.11 (Backward equation with initial value). *Let $(X_t, t \geq 0)$ be a diffusion in $\mathbb{R}$ as in Theorem 8.4 with SDE*

$$dX_t = \sigma(X_t) \, dB_t + \mu(X_t) \, dt.$$

Let $g \in \mathcal{C}^2(\mathbb{R})$ be such that g is 0 outside an interval. Then the solution of the PDE with initial value

$$(8.9) \qquad \begin{aligned} \frac{\partial f}{\partial t}(t, x) &= \frac{\sigma(x)^2}{2} \frac{\partial^2 f}{\partial x^2}(t, x) + \mu(x) \frac{\partial f}{\partial x}(t, x), \\ f(0, x) &= g(x), \end{aligned}$$

has the representation

$$f(t, x) = \mathbf{E}[g(X_t) | X_0 = x].$$

Remark 8.12. The assumptions on g ensure that Itô's formula can be applied and that the resulting Itô integral is a bona fide martingale. They can be weakened. However, to preserve the continuity at time 0, $\lim_{t \downarrow 0} f(t, x) = g(x)$, some regularity is needed. It turns out that for the case of the Brownian motion, sufficient conditions are that g be continuous and bounded. Theorem 8.11 proves the existence of a solution to the backward equation. The solution is also unique. This can be proved by probabilistic arguments using the strong Markov property.

Proof. Let's fix t and consider the function of space $h(x) = f(t, x) = \mathbf{E}[g(X_t)|X_0 = x]$. Itô's formula (Theorem 7.8) applied to h yields

$$dh(X_s) = h'(X_s) \, dX_s + \frac{1}{2} h''(X_s)(dX_s)^2$$

$$= h'(X_s)\sigma(X_s) \, dB_s + \left\{ \frac{\sigma(X_s)^2}{2} h''(X_s) + \mu(X_s)h'(X_s) \right\} ds,$$

where we applied the rules of Itô calculus. In integral form, this is

$$h(X_s) - h(X_0) = \int_0^s h'(X_u)\sigma(X_u) \, dB_u + \int_0^s \left\{ \frac{\sigma(X_u)^2}{2} h''(X_u) + \mu(X_u)h'(X_u) \right\} du.$$

If we take the expectation on both sides, divide by s, and take the limit $s \to 0$, the right-hand side becomes

$$\lim_{s \to 0} \frac{1}{s} \int_0^s \mathbf{E}\left[\frac{\sigma(X_u)^2}{2} h''(X_u) + \mu(X_u)h'(X_u) \Big| X_0 = x \right] du = \frac{\sigma(x)^2}{2} h''(x) + \mu(x)h'(x),$$

by the fundamental theorem of calculus (and continuity of h', h'', σ, and μ). As for the left-hand side, we have

$$\lim_{s \to 0} \frac{\mathbf{E}[h(X_s)|X_0 = x] - f(t, x)}{s},$$

since $h(x) = f(t, x)$ by definition. To prove that this limit is $\frac{\partial f}{\partial t}(t, x)$, it remains to show that $\mathbf{E}[h(X_s)|X_0 = x] = \mathbf{E}[g(X_{t+s})|X_0 = x] = f(t + s, x)$. To see this, note that $h(X_s) = \mathbf{E}[g(X_{t+s})|X_s]$. And by the Markov property and the tower property, we deduce that

$$\mathbf{E}[h(X_s)|X_0 = x] = \mathbf{E}[\mathbf{E}[g(X_{t+s})|X_s] \, |X_0 = x]$$

$$= \mathbf{E}[\mathbf{E}[g(X_{t+s})|\mathcal{F}_s] \, |X_0 = x] = \mathbf{E}[g(X_{t+s})|X_0 = x],$$

as claimed. $\qquad \square$

The backward equation (8.9) can be conveniently written in terms of the *generator of the diffusion*.

Definition 8.13 (Generator of a diffusion). The generator of a diffusion with SDE $dX_t = \sigma(X_t) \, dB_t + \mu(X_t) \, dt$ is the differential operator acting on functions of space defined by

$$A = \frac{\sigma(x)^2}{2} \frac{\partial^2}{\partial x^2} + \mu(x)\frac{\partial}{\partial x}.$$

With this notation, the backward equation for the function $f(t, x)$ takes the form

(8.10)
$$\frac{\partial f}{\partial t}(t, x) = Af(t, x),$$

where it is understood that A acts only on the space variable. Theorem 8.11 gives a nice interpretation of the generator: It quantifies how much the function $f(t, x) = \mathbf{E}[g(X_t)|X_0 = x]$ changes in a small time interval.

Example 8.14 (Generator of the Ornstein-Uhlenbeck process). Example 7.9 shows that the SDE of the Ornstein-Uhlenbeck process is of the form

$$dX_t = dB_t - X_t \, dt.$$

This means that its generator is

$$A = \frac{1}{2}\frac{\partial^2}{\partial x^2} - x\frac{\partial}{\partial x}.$$

Example 8.15 (Generator of geometric Brownian motion). Recall from Example 7.4 that geometric Brownian motion of the form

$$S_t = S_0 \exp(\sigma B_t + \mu t)$$

satisfies the SDE

$$dS_t = \sigma S_t \, dB_t + \left(\mu + \frac{1}{2}\sigma^2\right) S_t \, dt.$$

In particular, the generator of geometric Brownian motion is

$$A = \frac{\sigma^2 x^2}{2}\frac{\partial^2}{\partial x^2} + \left(\mu + \frac{1}{2}\sigma^2\right) x\frac{\partial}{\partial x}.$$

For applications, in particular in mathematical finance, it is more important to solve the backward equation with terminal value instead of with initial value. The reversal of time causes the appearance of an extra minus sign in the equation.

Theorem 8.16 (Backward equation with terminal value). *Let $(X_t, t \le T)$ be a diffusion as in Theorem 8.4 with SDE*

$$dX_t = \sigma(X_t) \, dB_t + \mu(X_t) \, dt.$$

Let $g \in \mathcal{C}^2(\mathbb{R})$ be such that g is 0 outside an interval. Then the solution of the PDE with terminal value at time T

(8.11)
$$-\frac{\partial f}{\partial t}(t, x) = \frac{\sigma(x)^2}{2}\frac{\partial^2 f}{\partial x^2}(t, x) + \mu(x)\frac{\partial f}{\partial x}(t, x),$$
$$f(T, x) = g(x),$$

has the representation

$$f(t, x) = \mathbf{E}[g(X_T)|X_t = x].$$

The backward equation with terminal value is the one appearing in the martingale condition in Corollary 5.29. To see why this is the case, recall from Exercise 4.16 that one way to construct a martingale for the filtration $(\mathcal{F}_t, t \ge 0)$ is to take

$$M_t = \mathbf{E}[Y|\mathcal{F}_t],$$

where Y is some integrable random variable. The martingale property then follows from the tower property of the conditional expectation. In the setup of Theorem 8.16, the random variable Y is $g(X_T)$. By the Markov property of the diffusion, we therefore have

$$f(t, X_t) = \mathbf{E}[g(X_T)|X_t] = \mathbf{E}[g(X_T)|\mathcal{F}_t].$$

In other words, the solution to the backward equation with terminal value evaluated at $x = X_t$ yields a martingale for the natural filtration of the process. This is a different point of view on a procedure we have used many times now: To get a martingale of the form $f(t, X_t)$, apply Itô's formula to $f(t, X_t)$ and set the dt-term to 0. The PDE we obtain is the backward equation with terminal value. In fact, the proof of the theorem takes this exact route.

Proof of Theorem 8.16. Consider $f(t, X_t)$ and apply Itô's formula (Theorem 7.8):

$$df(t, X_t) = \partial_0 f(t, X_t) \, dt + \partial_1 f(t, X_t) \, dX_t + \frac{1}{2} \partial_1^2 f(t, X_t)(dX_t)^2$$

(8.12)
$$= \partial_1 f(t, X_t) \sigma(X_t) \, dB_t$$

$$+ \left\{ \partial_0 f(t, X_t) + \mu(X_t) \partial_1 f(t, X_t) + \frac{\sigma(X_t)^2}{2} \partial_1^2 f(t, X_t) \right\} dt.$$

Since $f(t, x)$ is the solution to the equation, we get that the dt-term is 0 and $f(t, X_t)$ is a martingale for the Brownian filtration (and thus also for the natural filtration of the diffusion, which contains less information). In particular, we have

$$f(t, X_t) = \mathbf{E}[f(T, X_T)|\mathcal{F}_t] = \mathbf{E}[g(X_T)|\mathcal{F}_t].$$

Since $(X_t, t \leq T)$ is a Markov process, we finally get

$$f(t, x) = \mathbf{E}[g(X_T)|X_t = x]. \qquad \square$$

Example 8.17 (Martingales of geometric Brownian motion). Let $(S_t, t \geq 0)$ be a geometric Brownian motion with SDE

$$dS_t = \sigma S_t \, dB_t + \left(\mu + \frac{\sigma^2}{2} \right) S_t \, dt.$$

As we saw in Example 8.15, its generator is

$$A = \frac{\sigma^2 x^2}{2} \frac{\partial^2}{\partial x^2} + \left(\mu + \frac{\sigma^2}{2} \right) x \frac{\partial}{\partial x}.$$

In view of Theorem 8.16, this means that if $f(t, x)$ satisfies the PDE

$$\frac{\partial f}{\partial t} + \left(\mu + \frac{\sigma^2}{2} \right) x \frac{\partial f}{\partial x} + \frac{\sigma^2 x^2}{2} \frac{\partial^2 f}{\partial x^2} = 0,$$

then processes of the form $f(t, S_t)$ would be martingales for the natural filtration. This is consistent with what we found in Example 7.12 for the case $\mu = \frac{-\sigma^2}{2}$ using Itô's formula directly.

The companion equation to the backward equation is the *Kolmogorov forward equation* or *forward equation*. It is also known as the *Fokker-Planck equation* from its physics origin. This equation is very useful as it is satisfied by the transition density function $p_t(x, y)$ of a time-homogeneous diffusion. It involves the *adjoint of the generator*.

Definition 8.18 (Adjoint of the generator). The adjoint A^* of the generator of a diffusion $(X_t, t \geq 0)$ with SDE

$$dX_t = \sigma(X_t)\,dB_t + \mu(X_t)\,dt$$

is the differential operator acting on a function of space $f(x)$ as follows:

$$(8.13) \qquad A^*f(x) = \frac{1}{2}\frac{\partial^2}{\partial x^2}\left(\sigma(x)^2 f(x)\right) - \frac{\partial}{\partial x}\left(\mu(x)f(x)\right).$$

Note the differences with the generator in Defintion 8.13: There is an extra minus sign, and the derivatives also act on the volatility and the drift.

Example 8.19 (The generator of Brownian motion is selfadjoint). In the case of standard Brownian motion, it is easy to check that

$$A^* = \frac{1}{2}\frac{\partial^2}{\partial x^2}.$$

(And $A^* = \frac{1}{2}\Delta$ in the multivariate case.) In other words, the generator and its adjoint are the same. In this case, the operator is called *selfadjoint*.

Example 8.20 (The adjoint for geometric Brownian motion). Going back to Example 8.15, we see that the adjoint of the generator acting on $f(x)$ is

$$A^*f(x) = \frac{1}{2}\frac{\partial^2}{\partial x^2}\left(\sigma^2 x^2 f(x)\right) - \frac{\partial}{\partial x}\left(\left(\mu + \frac{\sigma^2}{2}\right)xf(x)\right).$$

Using the product rule in differentiating, we get

$$A^*f(x) = \frac{\sigma^2}{2}\left(x^2 f''(x) + 4xf'(x) + 2f(x)\right) - \left(\mu + \frac{\sigma^2}{2}\right)\left(xf'(x) + f(x)\right).$$

Example 8.21 (The adjoint for the Ornstein-Uhlenbeck process). The generator of the Ornstein-Uhlenbeck process was given in Example 8.14. The adjoint acting on f is therefore

$$A^*f(x) = \frac{1}{2}f''(x) + \frac{\partial}{\partial x}\left(xf(x)\right) = \frac{1}{2}f''(x) + xf'(x) + f(x).$$

The forward equation takes the following form for a function $f(t, x)$ of time and space:

$$(8.14) \qquad \frac{\partial f}{\partial t} = A^*f.$$

For Brownian motion, since $A^* = A$, the backward and forward equations are the same. As advertised earlier, the forward equation is satisfied by the transition $p_t(x, y)$ of a diffusion. Before showing this in general, we verify it in the Brownian case.

Example 8.22 (The heat kernel as solution of the forward equation). Recall that the transition probability density $p_t(x, y)$ for Brownian motion, or heat kernel, is

$$p_t(x, y) = \frac{1}{\sqrt{2\pi t}}e^{-\frac{(y-x)^2}{2t}}.$$

Here, the space variable will be y, and x will be fixed. The relevant function is thus $f(t,y) = p_t(x,y)$. The adjoint operator acting on the space variable y is $A^* = A = \frac{1}{2}\frac{\partial^2}{\partial y^2}$. The relevant time and space derivatives are

$$\frac{\partial f}{\partial t} = -\frac{1}{2t}f + \frac{(y-x)^2}{2t^2}f, \qquad \frac{\partial^2 f}{\partial y^2} = -\frac{1}{t}f + \frac{(y-x)^2}{t^2}f.$$

We conclude that the function $f(t,y) = p_t(x,y)$ is a solution of the forward equation (8.14).

Where does the form of the adjoint operator (8.13) come from? In some sense, the adjoint operator plays a role similar to that of the transpose of a matrix in linear algebra. The adjoint acts on the function on the left. To see this, consider two functions f,g of space on which the generator A of a diffusion as in Theorem 8.4 is well-defined. In particular, let's assume that the functions are 0 outside an interval. Consider the quantity

$$\int_{\mathbb{R}} g(x)\big(Af(x)\big)\,dx = \int_{\mathbb{R}} g(x)\Big(\frac{\sigma(x)^2}{2}f''(x) + \mu(x)f'(x)\Big)\,dx.$$

This quantity can represent for example the average of $Af(x)$ over some PDF $g(x)$. In the above, A acts on the function on the right. To make the operator act on g, we integrate by parts. This gives for the second term:

$$\int_{\mathbb{R}} \mu(x)f'(x)\,dx = g(x)\mu(x)f(x)\Big|_{-\infty}^{\infty} - \int_{\mathbb{R}} \frac{d}{dx}(g(x)\mu(x))f(x)\,dx.$$

The boundary term $g(x)\mu(x)f(x)\Big|_{-\infty}^{\infty}$ is 0 by the assumptions on f,g. This explains the term in μ for the adjoint, including the extra minus sign. The term on σ is obtained by integrating by parts twice:

$$\int_{\mathbb{R}} g(x)\frac{\sigma(x)^2}{2}f''(x)\,dx = \int_{\mathbb{R}} \frac{d^2}{dx^2}\Big(g(x)\frac{\sigma(x)^2}{2}\Big)f(x)\,dx.$$

The boundary terms are 0 for the same reason as above. All in all, we have shown that

(8.15) $$\int_{\mathbb{R}} g(x)\big(Af(x)\big)\,dx = \int_{\mathbb{R}} \big(A^*g(x)\big)f(x)\,dx.$$

Example 8.23 (A strange martingale). This example explores Exercise 5.18 in more detail. There, it was shown that the process

(8.16) $$M_t = \frac{1}{\sqrt{1-t}}\exp\Big(\frac{-B_t^2}{2(1-t)}\Big), \quad \text{for } 0 \le t < 1,$$

is a martingale for the filtration of a Brownian motion $(B_t, t \le 1)$. This can be shown using Itô's formula, of course. But how can you *guess* the form of such a martingale? The simple answer is the backward-forward equations. For Brownian motion, these two equations are the same. This means that the heat kernel $p_t(0,y)$ satisfies the heat equation, as double-checked in Example 8.22. By doing the transformation $t \to 1-t$ on the time variable only, this implies that $f(t,x) = p_{1-t}(0,x)$ satisfies the backward equation with terminal value as in Theorem 8.16:

$$-\frac{\partial}{\partial t}f(t,x) = Af(t,x).$$

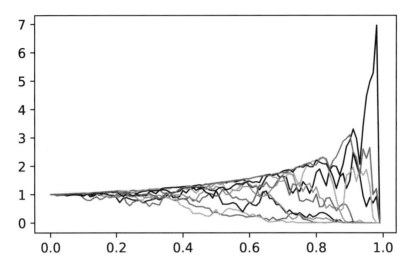

Figure 8.2. A sample of 10 paths of the martingale M_t, $t < 1$, in (8.16). See Numerical Project 8.2.

But as a consequence of Theorem 8.16, we know that $f(t, B_t) = p_{1-t}(0, B_t)$ must be a martingale for the Brownian filtration. This sheds some light on equation (8.16). It was shown in Exercise 5.18 that $\lim_{t \to 1-} M_t = 0$ almost surely, yet the expected maximum on $[0, 1)$ (the supremum on $[0, 1)$ to be precise) is $+\infty$; see Figure 8.2!

Theorem 8.24 (Forward equation and transition probability). *Let $(X_t, t \geq 0)$ be a diffusion as in Theorem 8.4 with SDE*

$$dX_t = \sigma(X_t)\, dB_t + \mu(X_t)\, dt, \quad X_0 = x_0,$$

and generator A. Let $p_t(x_0, x)$ be the transition probability density for a fixed x_0. Then the function $f(t, y) = p_t(x_0, y)$ is a solution of the PDE

$$(8.17) \qquad\qquad \frac{\partial f}{\partial t} = A^* f,$$

where A^* is the adjoint of A as in equation (8.13).

Proof. Let $h(x)$ be some arbitrary function of space that is 0 outside an interval. We compute

$$\frac{1}{\varepsilon}\Big(\mathbf{E}[h(X_{t+\varepsilon})] - \mathbf{E}[h(X_t)]\Big)$$

two different ways and take the limit $\varepsilon \to 0$. On one hand, we have by the definition of the transition density

$$\frac{1}{\varepsilon}\Big(\mathbf{E}[h(X_{t+\varepsilon})] - \mathbf{E}[h(X_t)]\Big) = \int_{\mathbb{R}} \frac{1}{\varepsilon}\Big(p_{t+\varepsilon}(x_0, x) - p_t(x_0, x)\Big) h(x)\, dx.$$

By taking the limit $\varepsilon \to 0$ inside the integral (assuming this is fine), we get

$$(8.18) \qquad\qquad \int_{\mathbb{R}} \frac{\partial}{\partial t} p_t(x_0, x)\, h(x)\, dx.$$

On the other hand, Itô's formula implies

$$\frac{1}{\varepsilon}\Big(\mathbf{E}[h(X_{t+\varepsilon})] - \mathbf{E}[h(X_t)]\Big) = \frac{1}{\varepsilon}\int_t^{t+\varepsilon} \mathbf{E}[Ah(X_s)]\,\mathrm{d}s.$$

Taking the limit $\varepsilon \to 0$ yields

$$\int_{\mathbb{R}} p_t(x_0, x) A h(x)\,\mathrm{d}x.$$

This can be written using equation (8.15) as

$$\int_{\mathbb{R}} A^* p_t(x_0, x) h(x)\,\mathrm{d}x.$$

Since h is arbitrary, we conclude that

(8.19)
$$\frac{\partial}{\partial t} p_t(x_0, x) = A^* p_t(x_0, x). \qquad \square$$

Example 8.25 (Forward equation and invariant probability). The Ornstein-Uhlenbeck process converges to a stationary distribution as noted in Examples 2.29 and 7.9. For example, for the SDE of the form

$$\mathrm{d}X_t = -X_t\,\mathrm{d}t + \mathrm{d}B_t,$$

with X_0 a Gaussian of mean 0 and variance $1/2$, the PDF of X_t is, for all t,

(8.20)
$$f(x) = \frac{1}{\sqrt{\pi}} e^{-x^2}.$$

This *invariant distribution* can be seen from the point of view of the forward equation (8.17). Indeed, since this PDF is constant in time, the forward equation simply becomes

(8.21)
$$A^* f = 0.$$

(Note that any multiple of f would actually be a solution of this equation. Why?) The reader is invited to check that f in (8.20) is a solution of (8.21), in Exercise 8.8.

Example 8.26 (Smoluchowski's equation). The SDE of the Ornstein-Uhlenbeck process can be generalized as follows. Consider $V(x)$, a smooth function of space such that $\int_{\mathbb{R}} e^{-2V(x)} < \infty$. The *Smoluchowski equation* is the SDE of the form

(8.22)
$$\mathrm{d}X_t = \mathrm{d}B_t - V'(X_t)\,\mathrm{d}t.$$

The SDE can be interpreted as follows: X_t represents the position of a particle on $\mathbb{R}$. The position varies due to the Brownian fluctuations and also due to a *force* $-V'(X_t)$ that depends on the position. The function $V(x)$ should then be thought of as the potential in which the particle moves, since the force is the (negative) derivative of the potential in Newtonian physics. The generator of this diffusion is

$$A = \frac{1}{2}\frac{\partial^2}{\partial x^2} - V'(x)\frac{\partial}{\partial x}.$$

As for the Ornstein-Uhlenbeck process, this diffusion admits an invariant distribution:

$$f(x) = C e^{-2V(x)}$$

where C is such that $\int_{\mathbb{R}} f(x)\,\mathrm{d}x = 1$. See Exercise 8.9.

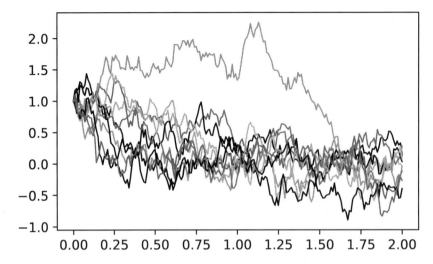

Figure 8.3. A sample of 10 paths on $[0, 2]$ of the diffusion with SDE (8.22) with $V'(x) = 2\mathrm{sgn}(x)$ and $X_0 = 1$. Note that the paths seem to converge to an invariant distribution. See Numerical Project 8.3.

8.4. The Feynman-Kac Formula

We saw in Example 8.10 that the solution of the heat equation

$$\frac{\partial f}{\partial t} = \frac{1}{2}\frac{\partial^2 f}{\partial x^2}$$

can be represented as an average over Brownian paths. This representation was extended to diffusions in Theorem 8.11, where the second derivative in the equation is replaced by the generator of the corresponding diffusion. How robust is this representation? In other words, is it possible to slightly change the PDE and still get a stochastic representation for the solution? The answer to this question is yes when a term of the form $r(x)f(t, x)$ is added to the equation, where $r(x)$ is a well-behaved function of space (for example piecewise continuous). The stochastic representation of the PDE in this case bears the name *Feynman-Kac formula*, marking a fruitful collaboration between the physicist Richard Feyman and the mathematician Mark Kac. The case when $r(x)$ is linear will be important in the applications to mathematical finance in Chapter 10, where it represents the contribution of the interest rate.

Theorem 8.27 (Initial value problem). *Let $(X_t, t \geq 0)$ be a diffusion in $\mathbb{R}$ as in Theorem 8.4 with SDE*

$$dX_t = \sigma(X_t)\,dB_t + \mu(X_t)\,dt.$$

Let $g \in \mathcal{C}^2(\mathbb{R})$ be such that g is 0 outside an interval. Then the solution of the PDE with initial value

(8.23)
$$\frac{\partial f}{\partial t}(t, x) = \frac{\sigma(x)^2}{2}\frac{\partial^2 f}{\partial x^2}(t, x) + \mu(x)\frac{\partial f}{\partial x}(t, x) - r(x)f(t, x),$$
$$f(0, x) = g(x),$$

has the representation

$$f(t, x) = \mathbf{E}\left[g(X_t) \exp\left(-\int_0^t r(X_s)\, ds \right) \middle| X_0 = x \right].$$

Proof. The proof is again based on Itô's formula. For a fixed t, we consider the process

$$M_s = f(t - s, X_s) \exp\left(-\int_0^s r(X_u)\, du \right), \quad s \leq t.$$

Write $Z_s = \exp\left(-\int_0^s r(X_u)\, du \right)$ and $V_s = f(t - s, X_s)$. A direct application of Itô's formula as in Theorem 7.8 yields

$$dZ_s = -r(X_s)Z_s\, ds,$$

$$dV_s = \partial_1 f(t - s, X_s)\sigma(X_s)\, dB_s$$
$$+ \left\{ -\partial_0 f(t - s, X_s) + \partial_1 f(t - s, X_s)\mu(X_s) + \frac{1}{2}\partial_1^2 f(t - s, X_s)\sigma^2(X_s) \right\} ds.$$

(Recall that t is fixed here and that we differentiate with respect to s in time.) Since $f(t, x)$ is a solution of the PDE, we can write the second equation as

$$dV_s = r(X_s)f(t - s, X_s)\, ds + \partial_1 f(t - s, X_s)\sigma(X_s)\, dB_s\,.$$

Now, by Itô's product rule (7.16), we finally have

$$dM_s = d(Z_s V_s) = Z_s\, dV_s + V_s\, dZ_s + 0 = \partial_1 f(t - s, X_s)\sigma(X_s)\, dB_s.$$

This proves that $(M_s, s \leq t)$ is a martingale. To be precise, this shows that it is a local martingale. It is easy to see that $(M_s, s \leq t)$ is bounded since f is, and so is the exponential term on the interval $[0, t]$. It turns out that a local martingale that is bounded is always a martingale. (This is a consequence of the definition of local martingale and Theorem 4.40.) Taking this for granted, we conclude that

$$\mathbf{E}[M_t] = \mathbf{E}[M_0]\,.$$

Using the definition of M_t, this yields

$$\mathbf{E}\left[f(0, X_t) \exp\left(-\int_0^t r(X_s)\, ds \right) \right] = \mathbf{E}\left[g(X_t) \exp\left(-\int_0^t r(X_s)\, ds \right) \right] = \mathbf{E}[M_0] = f(t, x).$$

This proves the theorem. $\qquad\qquad\qquad\qquad\qquad\qquad\qquad\qquad\qquad\qquad\quad \square$

As for the backward equation, it is natural to consider the terminal value problem for the same PDE.

Theorem 8.28 (Terminal value problem). *Let $(X_t, t \leq T)$ be a diffusion in $\mathbb{R}$ as in Theorem 8.4 with SDE*

$$dX_t = \sigma(X_t)\, dB_t + \mu(X_t)\, dt.$$

Let $g \in \mathcal{C}^2(\mathbb{R})$ be such that g is 0 outside an interval. Then the solution of the PDE with initial value

(8.24)
$$-\frac{\partial f}{\partial t}(t, x) = \frac{\sigma(x)^2}{2}\frac{\partial^2 f}{\partial x^2}(t, x) + \mu(x)\frac{\partial f}{\partial x}(t, x) - r(x)f(t, x),$$
$$f(T, x) = g(x),$$

has the representation

$$(8.25) \qquad f(t, x) = \mathbf{E} \left[g(X_T) \exp \left(- \int_t^T r(X_s) \, ds \right) \middle| X_t = x \right].$$

Proof. The proof is similar by considering instead

$$M_t = f(t, X_t) \exp \left(- \int_0^t r(X_s) \, ds \right).$$

This is left as an exercise (Exercise 8.11). □

Example 8.29 (The arcsine law). The arcsine law of Brownian motion gives the distribution of the amount of time that a Brownian path on $[0, 1]$ lies in the positive:

$$(8.26) \qquad \mathbf{P}(X \leq x) = \frac{2}{\pi} \arcsin(\sqrt{x}) = \frac{1}{\pi} \int_0^x \frac{1}{\sqrt{y(1-y)}} \, dy, \quad x \leq 1,$$

where X is the proportion of time of $[0, 1]$ spent in the positive. This was verified in Numerical Project 3.3; see also Figure 3.2. The random variable X can be written as

$$X = \int_0^1 \mathbf{1}_{[0,\infty)}(B_s) \, ds.$$

Theorem 8.27 can be applied to express the MGF of X in terms of a PDE. Indeed, we have for all $a \in \mathbb{R}$,

$$\mathbf{E}[e^{aX}] = \mathbf{E}[e^{-\int_0^1 r(B_s) \, ds}],$$

for $r(x) = -a\mathbf{1}_{[0,\infty)}(x)$. Here, the initial function is simply $f(0, x) = 1$ for all x. Thus, the MGF $\mathbf{E}[e^{aX}]$ corresponds to the solution of the initial value problem

$$\frac{\partial f}{\partial t}(t, x) = \frac{1}{2} \frac{\partial^2 f}{\partial x^2}(t, x) + a\mathbf{1}_{[0,\infty)}(x) f(t, x),$$
$$f(0, x) = 1.$$

It turns out that the solution of this problem is exactly the MGF of the arcsine distribution. The verification of this is a bit technical. We refer to [**Ste01**] for more details.

8.5. Numerical Projects and Exercises

8.1. **Temperature of a rod.** Consider the initial function $g(x) = 1 - |x|$ for $|x| \leq 1$ and 0 if $|x| > 1$. This function may represent the temperature of a rod at time 0.
  (a) Approximate the solution $f(t, x)$ to the heat equation at time $t = 0.25$ at every 0.01 in x using the representation (8.7). Use a sample of 100 paths for each x with discretization of 0.01.
  (b) Verify your approximation above by plotting the solution of the form (8.8).
  (c) Repeat the above at time $t = 1$.

8.2. **A strange martingale.** Consider the martingale $(M_t, t \leq 1)$ as in Example 8.23.
  (a) Construct a sample of 100 paths of this martingale for the discretizations 0.01, 0.001, and 0.0001.
  (b) Estimate the $\mathbf{E}[\max_{0 \leq t \leq 1} M_t]$ on each sample. What do you notice?

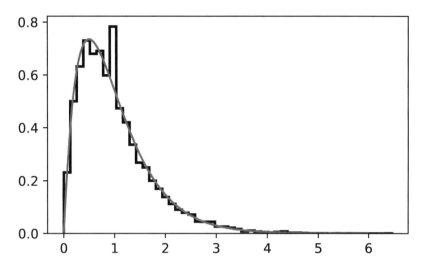

Figure 8.4. The histogram of the PDF of X_5 for the CIR diffusion (8.27) for 10,000 paths. The predicted invariant distribution $f(x) = 4xe^{-2x}$ is also plotted.

8.3. **Smoluchowski's equation.** Consider the diffusion $(X_t, t \geq 0)$ with SDE

$$dX_t = dB_t - 2\mathrm{sgn}(X_t)\,dt, \quad X_0 = 1,$$

where $\mathrm{sgn}(X_t) = 1$ if $X_t \geq 0$ and $\mathrm{sgn}(X_t) = -1$ if $X_t < 0$.
  (a) Plot 10 paths of the diffusion for $t \in [0,5]$ for a discretization of 0.01.
  (b) Build a sample of 1,000 paths on $[0,5]$ with the same discretization. Plot the histogram of the PDF of X_5 and compare to the invariant distribution

$$f(x) = 2e^{-4|x|}, \quad x \in \mathbb{R}.$$

8.4. **Invariant distribution of the CIR model.** Consider the CIR model with SDE

(8.27) $$dX_t = (1 - X_t)\,dt + \sqrt{X_t}\,dB_t, \quad X_0 = 1.$$

  (a) Plot 10 paths of the diffusion for $t \in [0,5]$ for a discretization of 0.01.
  (b) Build a sample of 1,000 paths on $[0,5]$ with the same discretization. Plot the histogram of the PDF of X_5 and compare to the invariant distribution

$$f(x) = 4xe^{-2x}, \quad x > 0.$$

See Exercise 8.10 for a theoretical justification of this.

Exercises

8.1. **Future and past are independent given the present.** Let $(X_t, t \geq 0)$ be a Markov process with its natural filtration $(\mathcal{F}_t, t \geq 0)$. Show that for any bounded functions g and h and any $0 \leq r < s < t < \infty$

$$\mathbf{E}[g(X_t)h(X_r)|X_s] = \mathbf{E}[g(X_t)|X_s] \times \mathbf{E}[h(X_r)|X_s].$$

8.2. Brownian probabilities revisited. Consider the Brownian probabilities in Exercise 3.2. Use the Markov property to write them as a double integral.

8.3. Markovariance. Let $(B_t, t \geq 0)$ be a Brownian motion.

(a) Show using the Markov property that for any intervals A_1, A_2 and $s < t$

$$\mathbf{P}(B_s \in A_1, B_t \in A_2) = \int_{A_1} \int_{A_2} \frac{e^{-\frac{(y-x)^2}{2(t-s)}}}{\sqrt{2\pi(t-s)}} \frac{e^{-\frac{x^2}{2s}}}{\sqrt{2\pi s}} \, dy \, dx \, .$$

(b) Verify that the right-hand side is equal to the integral one would get from the joint PDF of (B_s, B_t).

8.4. Shifted Brownian motion. Let $(B_t, t \geq 0)$ be a standard Brownian motion. Fix $t > 0$. Show that the process $(W_s, s \geq 0)$ with $W_s = B_{t+s} - B_s$ is a standard Brownian motion independent of $\mathcal{F}_s$.

8.5. Brownian motion with drift is a Markov process. Consider the Brownian motion with drift of the form

$$X_t = \sigma B_t + \mu t, \quad t \geq 0.$$

(a) Argue that $(X_t, t \geq 0)$ is a Markov process.
(b) Find the probability density function $p_t(x, y)$ of the process.
(c) What is its generator A and its adjoint A^*?

8.6. Adding a constant drift. Let $(X_t, t \geq 0)$ be a Markov process. Consider the process with an added drift

$$Y_t = X_t + \mu t, \quad t \geq 0,$$

for some $\mu \neq 0$. Show that $(Y_t, t \geq 0)$ is also a Markov process.
This is very different from adding a drift to a martingale!

8.7. Geometric Brownian motion as a Markov process. Let $(S_t, t \geq 0)$ be a geometric Brownian motion as in Example 8.17. Assume that $\mu = \frac{-1}{2}\sigma^2$ for simplicity.

(a) Prove that it is a Markov process using Definition 8.1.
(b) Deduce from this the transition probability density $p_t(x, y)$.
(c) Write down the backward equation for this diffusion for both initial and terminal values.
(d) Write down the forward equation with initial value for this diffusion.
(e) Verify that $f(t, y) = p_t(x, y)$ is a solution of the forward equation.

8.8. Invariant probability of the Ornstein-Uhlenbeck process. Consider the Ornstein-Uhlenbeck process with SDE

$$dX_t = dB_t - X_t \, dt, \quad t \geq 0, \quad X_0 = x,$$

as in Example 7.9.

(a) Find the transition probability density $p_t(x, y)$.
(b) What is $\lim_{t \to \infty} p_t(x, y)$?
(c) Let A^* be the adjoint of the generator of this diffusion and $f = e^{-x^2}$. Prove that $A^* f = 0$.
(d) How are the above results affected when considering the more general SDE

$$dX_t = \sigma \, dB_t - kX_t \, dt?$$

8.9. **Smoluchowski's equation.** We consider the diffusion $(X_t, t \geq 0)$ given by the SDE

$$dX_t = dB_t - V'(X_t) dt, \quad t \geq 0,$$

as in Example 8.26, where $V : \mathbb{R} \to \mathbb{R}$ is some (piecewise) smooth function such that $\int_{\mathbb{R}} e^{-2V(x)} dx < \infty$.

(a) Verify that the invariant distribution $f(x) = Ce^{-2V(x)}$ is a solution of the equation $A^* f = 0$ where A^* is the adjoint of the generator.

(b) Consider the specific example where $V(x) = |x|$. What is the SDE in this case? What is the exact invariant distribution?

8.10. **CIR model as a Markov process.** Consider the CIR model $(R_t, t \geq 0)$ as in Example 7.24 with SDE

$$dR_t = (a - bR_t) dt + \sigma \sqrt{R_t} \, dB_t$$

with $a > \sigma^2/4$. Even though the local volatility does not satisfy the assumptions of Theorem 7.21 (and thus those of Theorem 8.4), it turns out that the SDE defines a strong Markov process.

(a) Write down the generator A of this diffusion.

(b) Let f be the function

$$f(x) = x^{\frac{2a}{\sigma^2} - 1} e^{-\frac{2b}{\sigma^2} x}.$$

Show that $A^* f = 0$. (Assume $\sigma = 1$ to make the notation lighter.)

This shows that the invariant distribution of the diffusion is the Gamma distribution with PDF $f(x) = Cx^{\frac{2a}{\sigma^2} - 1} e^{-\frac{2b}{\sigma^2} x}$ *where* $C = \frac{(2b/\sigma^2)^{2a/\sigma^2}}{\Gamma(\frac{2a}{\sigma^2})}$.

8.11. **Feynman-Kac with terminal value.** Prove Theorem 8.28 using

$$M_t = f(t, X_t) \exp\left(-\int_0^t r(X_s) \, ds\right).$$

8.12. ⋆ **B_t^2 is a Markov process.** Let $(B_t, t \geq 0)$ be a standard Brownian motion.

(a) Prove that $(B_t^2, t \geq 0)$ is a Markov process.
Remember that you need to consider the natural filtration of the process in the Markov property. Writing $B_s = \text{sgn}(B_s)\sqrt{|B_s|^2}$ might be helpful.

(b) Use the above to show that the process $X_t = \int_0^t B_s \, dB_s$ is a Markov process, even though it might not be obvious from the differential

$$dX_t = B_t \, dB_t.$$

8.13. ⋆ **Generator of a Poisson process.** Let $(N_t, t \geq 0)$ be a Poisson process with rate λ as in Definition 3.19.

(a) Prove using Definition 8.1 that this is a Markov process.

(b) Show that its generator is

$$Af(x) = \lambda(f(x + 1) - f(x)).$$

8.6. Historical and Bibliographical Notes

The heat equation goes back to Joseph Fourier who was trying to model the diffusion of heat in space [**Fou09**]. His analysis led him to develop revolutionary analytical tools now known as *the Fourier series*. The connection between Brownian motion and the heat equation goes back to Einstein [**Ein05**] and Bachelier [**Bac00**]. The Smoluchowski equation was discovered around the same time in the study of particles suspended in liquid.

Interestingly, the original paper of Markov introducing processes that now bear his name was published in 1906, only one year after Einstein. The objective of Markov was to extend the law of large numbers and the central limit theorem to random variables with correlations [**Sen16**]. The CIR model turns out to be a strong Markov process, even though the SDE falls short of the assumptions of Theorem 8.4 [**DFS03**]. The transition probability density and the invariant distribution of the CIR model are derived in the original paper [**CIR85**]. The proof of Theorem 8.8 can be found in [**Øks03**]. The proof of Theorem 8.6 is based on the exposition in [**MP10**]. The Feynman-Kac formula finds several applications in probability, in PDEs, and in finance. The idea stemmed from the *Feynman's path integrals* in quantum mechanics. The idea of Feynman was to formulate quantum mechanics in terms of the likely paths taken by a particle weighted by the potential r along that path. Unfortunately, the Feynman path integrals could never have been made rigorous. This is because the exponential factor in the formulation ought to be $\exp(i \int_0^t r(X_s)\,ds)$ in quantum mechanics. Because of the $i = \sqrt{-1}$, the convergence of the integral is a problem. In the quantum mechanics framework, the heat equation is replaced by the *Schrödinger equation*.

Chapter 9

Change of Probability

We know from Chapter 1 that it is possible to define more than one probability on a given sample space Ω. The idea of modifying the probability on a sample space, and understanding how this affects the distributions of the random variables defined on that probability space, is a powerful idea in probability theory. In this section, we explore this new tool. We start by looking at the change of probability for a single random variable in Section 9.1. We then look at two important theorems: the Cameron-Martin theorem and the Girsanov theorem. The first result discussed in Section 9.2 gives an explicit way to change the deterministic drift of a Gaussian process by a change of probability. Girsanov's theorem generalizes this idea to random drift in Section 9.3. We sometimes talk about the Cameron-Martin-Girsanov theorem for the general result encompassing both cases. They play a fundamental role in the risk-neutral pricing in Chapter 10.

9.1. Change of Probability for a Random Variable

Consider a random variable X defined on $(\Omega, \mathcal{F}, \mathbf{P})$ with $\mathbf{E}[X] = 0$. We would like to change the mean of X so that $\mu \neq 0$. Of course, it is easy to change the mean of a random variable: If X has mean 0, then the random variable $X + \mu$ has mean μ. However, it might be that the variable $X + \mu$ does not share the same possible values as X. For example, take X to be a uniform random variable on $[-1, 1]$. While $X + 1$ has mean 1, the density of $X + 1$ would be nonzero on the interval $[0, 2]$ instead of $[-1, 1]$.

Our goal is to find a "good" way to change the underlying probability $\mathbf{P}$, and thus the distribution of X, so that the set of outcomes is unchanged. If X is a discrete random variable, say with $\mathbf{P}(X = -1) = \mathbf{P}(X = 1) = 1/2$, we can change the probability in order to change the mean easily. It suffices to take $\widetilde{\mathbf{P}}$ so that $\widetilde{\mathbf{P}}(X = 1) = p$ and $\widetilde{\mathbf{P}}(X = -1) = 1 - p$ for some appropriate $0 \leq p \leq 1$.

If X is a continuous random variable with a PDF f_X, the probabilities can be changed by modifying the PDF. Consider a new PDF

$$\tilde{f}_X(x) = f_X(x)g(x),$$

for some function $g(x) > 0$ such that $\int f_X(x)g(x)\,dx = 1$. Clearly, $f_X(x)g(x)$ is also a PDF and $f_X(x) > 0$ if and only if $f_X(x)g(x) > 0$, so that the possible values of X are unchanged. A convenient (and important!) choice of function g is

$$(9.1) \qquad g(x) = \frac{e^{ax}}{\int_{\mathbb{R}} e^{ax} f_X(x)\,dx} = \frac{e^{ax}}{E[e^{aX}]}, \quad a \in \mathbb{R},$$

assuming X has a well-defined MGF. Here a is a parameter that can be tuned to fit a specific mean. The normalization factor in the denominator is the MGF of X. It ensures that $f_X(x)g(x)$ is a PDF. Note that if $a > 0$, the function g gives a bigger weight to large values of X. We say that g *is biased towards the large values*.

Example 9.1 (Biasing a uniform random variable). Let X be a uniform random variable on $[0, 1]$ defined on $(\Omega, \mathcal{F}, \mathbf{P})$. Clearly, $E[X] = 1/2$. How can we change the PDF of X so that the possible values are still $[0, 1]$ but the mean is $1/4$? We have that the PDF is $f_X(x) = 1$ if $x \in [0, 1]$ and 0 elsewhere. Therefore, the mean with the new PDF with parameter a as in equation (9.1) is

$$\widetilde{E}[X] = \int_{\mathbb{R}} x\tilde{f}_X(x)\,dx = \int_0^1 x\frac{ae^{ax}}{e^a - 1}\,dx,$$

where we used the form of the MGF $E[e^{aX}] = a^{-1}(e^a - 1)$. The integral in x can be evaluated by parts and yields

$$\widetilde{E}[X] = \frac{e^a}{e^a - 1} - \frac{1}{a}.$$

For $\widetilde{E}[X]$ to be equal to $1/4$, we get numerically that $a \approx -3.6$. Note that the possible values of X remain the same under the new probability. However, the new distribution is no longer uniform! It has a bias towards values closer to 0, as it should. See Figure 9.1.

The most important example to understand the Cameron-Martin-Girsanov theorem is the following:

Example 9.2 (Biasing a Gaussian random variable). Let X be a Gaussian random variable of mean μ and variance σ^2. How can we change the PDF of X to have mean 0? Going back to (9.1), the mean under the new PDF with parameter a is

$$\widetilde{E}[X] = \int_{\mathbb{R}} x\tilde{f}_X(x)\,dx = \int_{-\infty}^{\infty} x\frac{e^{ax}}{e^{\sigma^2 a^2/2 + \mu a}}\frac{e^{-(x-\mu)^2/(2\sigma^2)}}{\sqrt{2\pi}\sigma}\,dx,$$

where we used the MGF $E[e^{aX}] = e^{\sigma^2 a^2/2 + \mu a}$. Now notice that the integral on the right reduces to

$$\widetilde{E}[X] = \int_{\mathbb{R}} x\tilde{f}_X(x)\,dx = \int_{-\infty}^{\infty} x\frac{e^{-(x-\mu-a\sigma^2)^2/(2\sigma^2)}}{\sqrt{2\pi}\sigma}\,dx.$$

Now we see that we recover the PDF of a Gaussian random variable of mean 0 for the specific choice of parameter $a = -\mu/\sigma^2$, but we can deduce more. The new PDF is also

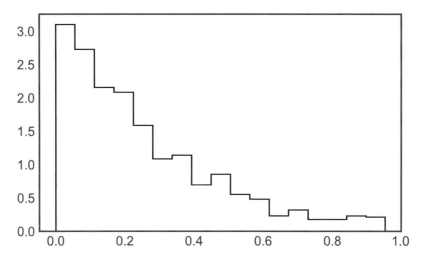

Figure 9.1. The histogram of 1,000 values of a uniform random variable X sampled proportionally to the weight $e^{-3.6X}$. The mean of the sample is $0.252\dots$. See Numerical Project 9.1.

Gaussian. This was not the case for uniform random variables. In fact, the new PDF is exactly the same as the one of $X - \mu$. In other words:

> *For Gaussians, changing the mean by recentering is equivalent to changing the probability as in* (9.1).

This is a very special property of the Gaussian distribution. The exponential and Poisson distributions have a similar property; cf. Exercises 9.2 and 9.3.

Example 9.2 is very important and we will state it as a theorem. Before doing so, we notice that the change of PDF (9.1) can be expressed more generally by changing the underlying probability on the sample space Ω on which the random variables are defined. More precisely, let $(\Omega, \mathcal{F}, \mathbf{P})$ be a probability space, and let X be a random variable defined on Ω. We define a new probability $\widetilde{\mathbf{P}}$ on Ω as follows: If $\mathcal{E}$ is an event in $\mathcal{F}$, then

$$(9.2) \qquad \widetilde{\mathbf{P}}(\mathcal{E}) = \widetilde{\mathbf{E}}[\mathbf{1}_{\mathcal{E}}] = \mathbf{E}\left[\frac{e^{aX}}{\mathbf{E}[e^{aX}]}\mathbf{1}_{\mathcal{E}}\right].$$

Intuitively, we are changing the probability of each outcome ω by the factor

$$(9.3) \qquad \frac{e^{aX(\omega)}}{\mathbf{E}[e^{aX}]}.$$

In other words, if $a > 0$, the outcomes ω for which X has large values are favored. Note that equation (9.1) for the PDF is recovered since for any function h of X we have

$$\widetilde{\mathbf{E}}[h(X)] = \mathbf{E}\left[\frac{e^{aX}}{\mathbf{E}[e^{aX}]}h(X)\right] = \int_{\mathbb{R}} h(x)\frac{e^{ax}}{\mathbf{E}[e^{aX}]}f_X(x)\,dx.$$

In this setting, the above example becomes the preliminary version of the Cameron-Martin-Girsanov theorem:

Theorem 9.3. *Let X be a Gaussian random variable with mean μ and variance σ^2 defined on $(\Omega, \mathcal{F}, \mathbf{P})$. Then under the probability $\widetilde{\mathbf{P}}$ on Ω given by*

$$(9.4) \qquad \widetilde{\mathbf{P}}(\mathcal{E}) = \mathbf{E}\left[e^{\frac{-\mu}{\sigma^2}X + \frac{1}{2}\frac{\mu^2}{\sigma^2}}\mathbf{1}_{\mathcal{E}}\right], \quad \mathcal{E} \in \mathcal{F},$$

the random variable X is Gaussian with mean 0 and variance σ^2.

Moreover, since X can be written as $X = Y + \mu$ where Y is Gaussian with mean 0 and variance σ^2, we have that $\widetilde{\mathbf{P}}$ can be written as

$$(9.5) \qquad \widetilde{\mathbf{P}}(\mathcal{E}) = \mathbf{E}\left[e^{\frac{-\mu}{\sigma^2}Y - \frac{1}{2}\frac{\mu^2}{\sigma^2}}\mathbf{1}_{\mathcal{E}}\right].$$

It is good to pause for a second and look at the signs in the exponential of equations (9.4) and (9.5). The signs in the exponential might be very confusing and is the source of many mistakes in the Cameron-Martin-Girsanov theorem. A good trick is to say that if we want to *remove* μ, then the sign in front of X or Y must be negative. Then we add to the exponential the factor needed for $\widetilde{\mathbf{P}}$ to be a probability. This is given by the MGF of X or Y depending on how we want to express it.

The probabilities $\mathbf{P}$ and $\widetilde{\mathbf{P}}$, as defined in equation (9.4), are obviously not equal since they differ by the factor in (9.3). However, they share some similarities. Most notably, if $\mathcal{E}$ is an event of positive $\mathbf{P}$-probability, $\mathbf{P}(\mathcal{E}) > 0$, then we must have that $\widetilde{\mathbf{P}}(\mathcal{E}) > 0$ as well, since the factor in (9.3) is always strictly positive. The converse is also true: If $\mathcal{E}$ is an event of positive $\widetilde{\mathbf{P}}$-probability, $\widetilde{\mathbf{P}}(\mathcal{E}) > 0$, then we must have that $\mathbf{P}(\mathcal{E}) > 0$. This is because the factor in (9.3) can be inverted, being strictly positive. More precisely, we have

$$\mathbf{P}(\mathcal{E}) = \mathbf{E}[\mathbf{1}_{\mathcal{E}}] = \mathbf{E}\left[\frac{e^{aX(\omega)}}{\mathbf{E}[e^{aX}]}\left(\frac{e^{aX(\omega)}}{\mathbf{E}[e^{aX}]}\right)^{-1}\mathbf{1}_{\mathcal{E}}\right] = \widetilde{\mathbf{E}}\left[\left(\frac{e^{aX(\omega)}}{\mathbf{E}[e^{aX}]}\right)^{-1}\mathbf{1}_{\mathcal{E}}\right].$$

The factor $\left(\frac{e^{aX(\omega)}}{\mathbf{E}[e^{aX}]}\right)^{-1}$ is also strictly positive, proving the claim. To sum it all up, the probabilities $\mathbf{P}$ and $\widetilde{\mathbf{P}}$ essentially share the same possible outcomes. Such probabilities are said to be *equivalent*.

Definition 9.4. Consider two probabilities $\mathbf{P}$ and $\widetilde{\mathbf{P}}$ on $(\Omega, \mathcal{F})$. They are said to be *equivalent* if, for any event $\mathcal{E} \in \mathcal{F}$, we have $\mathbf{P}(\mathcal{E}) > 0$ if and only if $\widetilde{\mathbf{P}}(\mathcal{E}) > 0$. Intuitively, this means that any event that is possible in $\mathbf{P}$ is also possible in $\widetilde{\mathbf{P}}$.

Keep in mind that two probabilities that are equivalent might still be very far from being equal!

9.2. The Cameron-Martin Theorem

The Cameron-Martin theorem extends Theorem 9.3 to Brownian motion on a finite interval $[0, T]$ with constant drift. The idea of the proof is very similar: It is possible to recenter a Gaussian by multiplying the density by a factor of the form in (9.3).

Theorem 9.5 (Cameron-Martin theorem for constant drift). *Let $(\widetilde{B}_t, t \in [0, T])$ be a Brownian motion with constant drift θ defined on $(\Omega, \mathcal{F}, \mathbf{P})$. Consider the probability $\widetilde{\mathbf{P}}$ on Ω given by*

$$(9.6) \qquad \widetilde{\mathbf{P}}(\mathcal{E}) = \mathbf{E}\left[e^{-\theta\widetilde{B}_T + \frac{\theta^2}{2}T}\mathbf{1}_{\mathcal{E}}\right], \quad \mathcal{E} \in \mathcal{F}.$$

Then the process $(\widetilde{B}_t, t \in [0, T])$ under $\widetilde{\mathbf{P}}$ is distributed like a standard Brownian motion.

Moreover, since we can write $\widetilde{B}_t = B_t + \theta t$ for some standard Brownian motion $(B_t, t \in [0, T])$ on $(\Omega, \mathcal{F}, \mathbf{P})$, the probability $\widetilde{\mathbf{P}}$ can also be written as

$$(9.7) \qquad \widetilde{\mathbf{P}}(\mathcal{E}) = \mathbf{E}\left[e^{-\theta B_T - \frac{\theta^2}{2}T}\mathbf{1}_{\mathcal{E}}\right], \quad \mathcal{E} \in \mathcal{F}.$$

It is a good idea to pause again and look at the signs in the exponential in equations (9.6) and (9.7). They behave the same way as in Theorem 9.3: There is a minus sign in front of B_T to *remove* the drift.

Before proving the theorem, we make some important remarks.

(i) **The endpoint.** Note that only the endpoint $\widetilde{B}_T$ of the Brownian motion is involved in the change of probability. In particular, T cannot be $+\infty$. The Cameron-Martin theorem can only be applied on a finite interval.

(ii) **A martingale.** The factor $M_T = e^{-\theta B_T - \frac{\theta^2}{2}T} = e^{-\theta\widetilde{B}_T + \frac{\theta^2}{2}T}$ involved in the change of probability is the endpoint of a $\mathbf{P}$-martingale; i.e., it is a martingale under the original probability. Indeed, we have that $M_t = e^{-\theta B_t - \frac{\theta^2}{2}t}$ is the martingale case of a geometric Brownian motion. Interestingly, the drift of $\widetilde{B}_t$ becomes the volatility factor in M_t! The fact that M_t is a martingale is very helpful in calculations. Indeed, suppose we want to compute the expectation of a function $F(\widetilde{B}_s)$ of a Brownian motion with drift at time $s < T$. Then we have by Theorem 9.5

$$\mathbf{E}[F(\widetilde{B}_s)] = \mathbf{E}[M_T M_T^{-1} F(\widetilde{B}_s)] = \widetilde{\mathbf{E}}[M_T^{-1} F(\widetilde{B}_s)] = \widetilde{\mathbf{E}}[e^{\theta\widetilde{B}_T - \frac{\theta^2}{2}T} F(\widetilde{B}_s)].$$

Now, we know from Theorem 9.5 that $(\widetilde{B}_t, t \in [0, T])$ is a standard Brownian motion under $\widetilde{\mathbf{P}}$, or a $\widetilde{\mathbf{P}}$-Brownian motion for short. Therefore, the process $e^{\theta\widetilde{B}_t - \frac{\theta^2}{2}t}$ is a martingale under the new probability $\widetilde{\mathbf{P}}$, or a $\widetilde{\mathbf{P}}$-martingale for short. By conditioning over $\mathcal{F}_s$ and applying the martingale property, we get

$$\mathbf{E}[F(\widetilde{B}_s)] = \widetilde{\mathbf{E}}[e^{\theta\widetilde{B}_s - \frac{\theta^2}{2}s} F(\widetilde{B}_s)] = \mathbf{E}[e^{\theta B_s - \frac{\theta^2}{2}s} F(B_s)].$$

The last inequality may seem wrong as we removed all the tildes. It is not! It holds because $(\widetilde{B}_t)$ under $\widetilde{\mathbf{P}}$ has the same distribution as (B_t) under $\mathbf{P}$: a standard Brownian motion. Of course, it would have been possible to directly evaluate $\mathbf{E}[F(\widetilde{B}_s)]$ here as we know the distribution of a Brownian motion with drift. However, when the function will involve more than one point (such as the maximum of the path), the Cameron-Martin theorem is a powerful tool to evaluate the expectation; see Example 9.6.

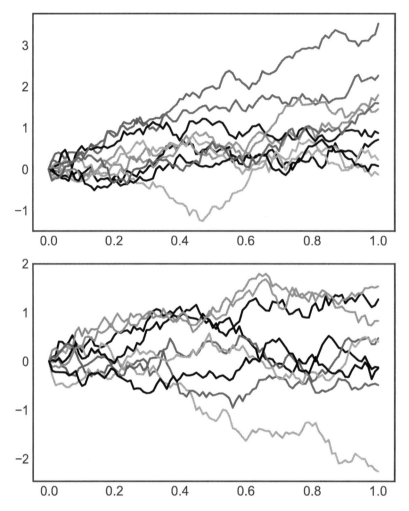

Figure 9.2. Top: 10 paths of Brownian motion with drift 1. Bottom: 10 paths of Brownian motion with drift that were sampled using the bias of the Cameron-Martin theorem. See Numerical Project 9.2.

(iii) **The paths with or without drift are the same.** Let $(B_t, t \le T)$ be a standard Brownian motion defined on $(\Omega, \mathcal{F}, \mathbf{P})$. Heuristically, it is fruitful to think of the sample space of Ω as the different continuous paths of the Brownian motion. Since the change of probability from $\mathbf{P}$ to $\widetilde{\mathbf{P}}$ simply changes the relative weights of the paths (and this change of weight is never zero, similarly to equation (9.3) for a single random variable), the theorem suggests that the paths of a standard Brownian motion and those of a Brownian with constant drift θ (with the volatility 1) *are essentially the same.* Of course, the two distributions weigh these paths differently and some paths are more likely than others in the two distributions. But, if you are given two paths generated by each of the distributions, it is impossible to distinguish between the two with 100% confidence. See Figure 9.3.

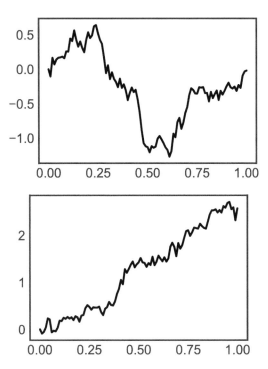

Figure 9.3. The top path was generated by a Brownian motion with drift whereas the bottom one was generated by a standard Brownian motion. This seems perhaps counterintuitive. Of course, these sorts of path are less likely for the respective distribution, but they can still occur.

The form of the factor $M_T = e^{-\theta \widetilde{B}_T + \frac{\theta^2}{2} T}$ can be easily understood at the heuristic level. For each outcome ω, it is proportional to $e^{-\theta \widetilde{B}_T(\omega)}$. (The term $e^{\frac{\theta^2}{2} T}$ is simply there to ensure that $\widetilde{\mathbf{P}}(\Omega) = 1$.) Therefore, the factor M_T penalizes the paths for which $\widetilde{B}_T(\omega)$ is large and positive (if $\theta > 0$). In particular, it is conceivable that the Brownian motion with positive drift is reduced to standard Brownian motion under the new probability. Numerical Project 9.2 details how to generate such bias sampling. See Figure 9.2 for paths of Brownian motion with drift that are sampled from the probability $\widetilde{\mathbf{P}}$.

(iv) **Changing the volatility?** What about the volatility? Is it possible to change the probability $\mathbf{P}$ to $\widetilde{\mathbf{P}}$ in such a way that a standard Brownian motion under $\mathbf{P}$ has volatility $\sigma \neq 1$ under $\widetilde{\mathbf{P}}$? The answer is no! The paths of Brownian motions with different volatilities are inherently different. Indeed, it suffices to compute the quadratic variation (Corollary 3.17). If $(B_t, t \in [0, T])$ has volatility 1 and $(\widetilde{B}_t, t \in [0, T])$ has volatility 2, then the following convergence holds for ω in a set of probability one (for a partition fine enough, say $t_{j+1} - t_j = 2^{-n}$):

$$(9.8) \quad \lim_{n \to \infty} \sum_{j=0}^{n-1} (B_{t_{j+1}}(\omega) - B_{t_j}(\omega))^2 = T, \qquad \lim_{n \to \infty} \sum_{j=0}^{n-1} (\widetilde{B}_{t_{j+1}}(\omega) - \widetilde{B}_{t_j}(\omega))^2 = 4T.$$

In other words, the distribution of the standard Brownian motion on $[0, T]$ is supported on paths whose quadratic variation is T whereas the distribution of $(\widetilde{B}_t, t \in [0, T])$ is supported on paths whose quadratic variation is $4T$. These paths are very different. We conclude that the distributions of these two processes are not equivalent. In fact, we say that they are *mutually singular*, meaning that the sets of paths on which they are supported are disjoint.

Proof of Theorem 9.5. Let $(\widetilde{B}_t, t \in [0, T])$ be a Brownian motion with constant drift θ defined on $(\Omega, \mathcal{F}, \mathbf{P})$. Let $n \in \mathbb{N}$ and $0 = t_0 < t_1 < \cdots < t_{n-1} < t_n = T$. Consider the increments $\widetilde{B}_{t_1}, \widetilde{B}_{t_2} - \widetilde{B}_{t_1}, \ldots, \widetilde{B}_{t_n} - \widetilde{B}_{t_{n-1}}$. It suffices to show that for any event of the form $\mathcal{E} = \{\widetilde{B}_{t_1} \in A_1, \ldots, \widetilde{B}_{t_n} - \widetilde{B}_{t_{n-1}} \in A_n\}$ for $A_j \subseteq \mathbb{R}$ and $n \in \mathbb{N}$, we have

$$(9.9) \qquad \widetilde{\mathbf{P}}(\mathcal{E}) = \int_{A_1} \cdots \int_{A_n} \prod_{j=0}^{n-1} \frac{e^{-\frac{y_{t_j}^2}{2(t_{j+1}-t_j)}}}{\sqrt{2\pi(t_{j+1} - t_j)}} \, dy_{t_1} \cdots dy_{t_n}.$$

This proves that increments of $(\widetilde{B}_t)$ under $\widetilde{\mathbf{P}}$ are the ones of standard Brownian motion, as we know from Proposition 3.1.

By definition of $\widetilde{\mathbf{P}}$, we have

$$(9.10) \qquad \begin{aligned} \widetilde{\mathbf{P}}(\mathcal{E}) &= \mathbf{E}\left[e^{-\theta \widetilde{B}_T + \frac{\theta^2}{2} T} \mathbf{1}_{\mathcal{E}} \right] \\ &= \mathbf{E}\left[e^{-\theta \sum_{j=0}^{n-1}(\widetilde{B}_{t_{j+1}} - \widetilde{B}_{t_j}) + \frac{\theta^2}{2}(t_{j+1}-t_j)} \mathbf{1}_{\mathcal{E}} \right], \end{aligned}$$

where we simply wrote $\widetilde{B}_T$ and T in a telescopic way. The joint distribution of the increments $(\widetilde{B}_{t_{j+1}} - \widetilde{B}_{t_j}, j \leq n)$ under $\mathbf{P}$ is known since it is a Brownian motion with drift: They are independent Gaussian random variables with mean $\theta(t_{j+1} - t_j)$ and variance $t_{j+1} - t_j$. Therefore, equation (9.10) can be written as

$$\int_{A_1} \cdots \int_{A_n} e^{-\theta \sum_{j=0}^{n-1} y_{t_j} + \frac{\theta^2}{2}(t_{j+1}-t_j)} \prod_{j=0}^{n-1} \frac{e^{-\frac{(y_{t_j} - \theta(t_{j+1}-t_j))^2}{2(t_{j+1}-t_j)}}}{\sqrt{2\pi(t_{j+1} - t_j)}} \, dy_{t_1} \cdots dy_{t_n}.$$

If we develop the square, the expression simplifies to equation (9.9). $\qquad\square$

Here are some examples of direct applications of the Cameron-Martin theorem:

Example 9.6 (Bachelier's formula for Brownian motion with drift). One of the most interesting formulas we have seen so far is Bachelier's formula for the maximum of Brownian motion in Proposition 4.45 which says that

$$\max_{t \in [0,T]} B_t \text{ has the same distribution as } |B_T|.$$

We derive the formula for the maximum of Brownian motion with drift. Let $\widetilde{B}_t = B_t + \theta t$ be a Brownian motion with constant drift θ on $(\Omega, \mathcal{F}, \mathbf{P})$. Consider the event $\mathcal{E} = \{\max_{t \in [0,T]} \widetilde{B}_t > a\}$. This event can be expressed differently using the stopping time $\tilde{\tau}_a = \min\{t \geq 0 : \widetilde{B}_t > a\}$. We then have $\mathcal{E} = \{\tilde{\tau}_a \leq T\}$. By definition of $\widetilde{\mathbf{P}}$ in Theorem 9.5, we can write

$$\mathbf{P}(\tilde{\tau}_a \leq T) = \widetilde{\mathbf{E}}\left[e^{\theta\widetilde{B}_T - \frac{\theta^2}{2}T}\mathbf{1}_{\{\tilde{\tau}_a \leq T\}}\right].$$

Theorem 9.5 now states that $(\widetilde{B}_t, t \leq T)$ is a standard Brownian motion under $\widetilde{\mathbf{P}}$. Amazingly, this means in particular that $\tilde{\tau}_a$ now has the distribution of the first-passage time of standard Brownian motion. To compute the expectation, we use the fact that $\widetilde{M}_t = e^{\theta\widetilde{B}_t - \frac{\theta^2}{2}t}$ is a $\widetilde{\mathbf{P}}$-martingale. By the optional stopping theorem (see Remark 8.9), we can condition on $\mathcal{F}_{\tilde{\tau}_a}$ and get

$$\mathbf{E}\left[\widetilde{M}_T\mathbf{1}_{\{\tilde{\tau}_a \leq T\}}\right] = \mathbf{E}\left[\widetilde{M}_{\tilde{\tau}_a}\mathbf{1}_{\{\tilde{\tau}_a \leq T\}}\right].$$

Clearly, we have $\widetilde{M}_{\tilde{\tau}_a} = e^{\theta a - \frac{\theta^2}{2}\tilde{\tau}_a}$. We have shown so far that

$$\mathbf{P}\left(\max_{t \in [0,T]} \widetilde{B}_t > a\right) = \widetilde{\mathbf{E}}\left[e^{\theta a - \frac{\theta^2}{2}\tilde{\tau}_a}\mathbf{1}_{\{\tilde{\tau}_a \leq t\}}\right].$$

The expectation is now only on the random variable $\tilde{\tau}_a$. We know its distribution by Corollary 4.46, since $(\widetilde{B}_t, t \leq T)$ is a standard Brownian motion under $\widetilde{\mathbf{P}}$. Thus we have by using the PDF of $\tilde{\tau}_a$ that

$$\mathbf{P}\left(\max_{t \in [0,T]} \widetilde{B}_t > a\right) = \int_0^T e^{\theta a - \frac{\theta^2}{2}s} \frac{a}{\sqrt{2\pi}} \frac{e^{-\frac{a^2}{2s}}}{s^{3/2}}\, ds.$$

Simplifying the argument of the exponential yields the following nice theorem.

Proposition 9.7 (Bachelier formula for Brownian motion with constant drift). *Let $\widetilde{B}_t = B_t + \theta t$ be a Brownian motion with drift θ. Let $\tilde{\tau}_a = \min\{t \geq 0 : \widetilde{B}_t > a\}$. Then we have*

$$\mathbf{P}\left(\max_{t \in [0,T]} \widetilde{B}_t > a\right) = \mathbf{P}(\tilde{\tau}_a \leq T) = \int_0^T \frac{a}{s^{3/2}} \frac{e^{-\frac{(a-\theta s)^2}{2s}}}{\sqrt{2\pi}}\, ds.$$

In particular, the PDF of $\tilde{\tau}_a$ is

(9.11)
$$f_{\tilde{\tau}_a}(t) = \frac{a}{t^{3/2}} \frac{e^{-\frac{(a-\theta t)^2}{2t}}}{\sqrt{2\pi}}.$$

Will the Brownian motion with drift always reach the level a if we take T large? If $\theta > 0$, it is not hard to check numerically that the PDF $f_{\tilde{\tau}_a}$ indeed integrates to 1. However, if $\theta < 0$, this is not the case. This reflects the fact that if the drift is negative, there are some paths that will never reach a. This was already discussed in Section 5.5. See Exercise 9.5 for more on this.

The above example can also be interpreted as giving the probability that a standard Brownian motion reaches a moving barrier. More precisely, consider the linear *moving barrier* that is a function $mt + a$. Let's look at the event that there exists a time $t \in [0, T]$ such that $B_t > mt + a$. This is the same as the event $\{\max_{t \in [0,T]} B_t - mt > a\}$. In particular, one recovers the probability of this event using Proposition 9.7 by taking $\theta = -m$.

Example 9.8 (Gambler's ruin...again). Consider $\widetilde{B}_t = B_t + \theta t$, a Brownian motion with drift on (Ω, P). We are interested in computing

$$\widetilde{\mathbf{P}}(\widetilde{B}_\tau = a),$$

where τ is the hitting time

$$\tau = \min\{t \geq 0 : \widetilde{B}_t \geq a \text{ or } \widetilde{B}_t \leq -a\}.$$

Note that $\tau < \infty$ with probability one. (Why?) We dealt with the general case in (5.27). Here we look at the symmetric case $b = a$ using the Cameron-Martin theorem. The idea is to express $\widetilde{B}_t$ as a standard Brownian motion. For this, we first need to consider a finite interval $[0, T]$. We will take $T \to \infty$ at the end. We have by definition of the probability $\widetilde{\mathbf{P}}$ in Theorem 9.5

$$\mathbf{P}(\widetilde{B}_{\tau \wedge T} = a) = \mathbf{E}\left[\mathbf{1}_{\{\widetilde{B}_{\tau \wedge T} = a\}}\right]$$
$$= \widetilde{\mathbf{E}}\left[e^{\theta \widetilde{B}_T - \frac{\theta^2}{2}T}\mathbf{1}_{\{\widetilde{B}_{\tau \wedge T} = a\}}\right].$$

Recall that $\widetilde{M}_t = e^{\theta \widetilde{B}_t - \frac{\theta^2}{2}t}$, $t \in [0, T]$, is a $\widetilde{\mathbf{P}}$-martingale, and so is the stoppped martingale $\widetilde{M}_{t \wedge \tau}$; cf. Proposition 4.37. Therefore, we have by the optional stopping theorem (Remark 8.9) applied with the stopping time $\tau \wedge T$ that

$$\mathbf{P}(\widetilde{B}_{\tau \wedge T} = a) = \widetilde{\mathbf{E}}\left[e^{\theta \widetilde{B}_{\tau \wedge T} - \frac{\theta^2}{2}T \wedge T}\mathbf{1}_{\{\widetilde{B}_{\tau \wedge T} = a\}}\right].$$

By taking $T \to \infty$ and passing it inside the probability and the expectation (which is feasible here by the dominated convergence theorem, Theorem 4.40), we have that

$$(9.12) \qquad \mathbf{P}(\widetilde{B}_\tau = a) = \widetilde{\mathbf{E}}\left[e^{\theta \widetilde{B}_\tau - \frac{\theta^2}{2}\tau}\mathbf{1}_{\{\widetilde{B}_\tau = a\}}\right] = e^{\theta a}\widetilde{\mathbf{E}}\left[e^{-\frac{\theta^2}{2}\tau}\mathbf{1}_{\{\widetilde{B}_\tau = a\}}\right].$$

What is the last expectation? Remember that $(\widetilde{B}_t)$ is a standard Brownian motion under $\widetilde{\mathbf{P}}$. Furthermore, by the optional stopping theorem,

$$\widetilde{\mathbf{E}}\left[e^{\theta \widetilde{B}_\tau - \frac{\theta^2}{2}\tau}\right] = 1.$$

Therefore, by splitting the expectation into the two events $\mathbf{1}_{\{B_\tau = a\}}$ and $\mathbf{1}_{\{B_\tau = -a\}}$, we get

$$1 = \mathbf{E}\left[e^{\theta B_\tau - \frac{\theta^2}{2}\tau}\mathbf{1}_{\{B_\tau = a\}}\right] + \mathbf{E}\left[e^{\theta B_\tau - \frac{\theta^2}{2}\tau}\mathbf{1}_{\{B_\tau = -a\}}\right]$$
$$= e^{\theta a}\mathbf{E}\left[e^{-\frac{\theta^2}{2}\tau}\mathbf{1}_{\{B_\tau = a\}}\right] + e^{-\theta a}\mathbf{E}\left[e^{-\frac{\theta^2}{2}\tau}\mathbf{1}_{\{B_\tau = -a\}}\right].$$

But $\mathbf{E}\left[e^{-\frac{\theta^2}{2}\tau}\mathbf{1}_{\{B_\tau = a\}}\right] = \mathbf{E}\left[e^{-\frac{\theta^2}{2}\tau}\mathbf{1}_{\{B_\tau = -a\}}\right]$ by the reflection of Brownian motion at time 0 (Proposition 3.3). Therefore, we conclude that $\mathbf{E}\left[e^{-\frac{\theta^2}{2}\tau}\mathbf{1}_{\{B_\tau = a\}}\right] = (e^{\theta a} + e^{-\theta a})^{-1}$. Putting this back into equation (9.12), we get

$$\mathbf{P}(\widetilde{B}_\tau = a) = \frac{e^{\theta a}}{e^{\theta a} + e^{-\theta a}}.$$

Note that this equals $1/2$ in the case $\theta = 0$, as it should.

9.3. Extensions of the Cameron-Martin Theorem

We consider now a Brownian motion with a deterministic drift θ_t, $t \in [0, T]$, that is time-dependent. More precisely, let

$$\tag{9.13} \widetilde{B}_t = B_t + \int_0^t \theta_s \, ds, \qquad t \in [0, T],$$

where $(B_t, t \in [0, T])$ is a standard Brownian motion defined on $(\Omega, \mathcal{F}, \mathbf{P})$. It is again possible to define a new probability $\widetilde{\mathbf{P}}$ on Ω such the process $(\widetilde{B}_t, t \in [0, T])$ has the distribution of a standard Brownian motion under $\widetilde{\mathbf{P}}$. However, we must suppose that $\int_0^T \theta_s^2 \, ds < \infty$ to ensure that $\widetilde{\mathbf{P}}$ is well-defined. This echoes Example 7.11 and Exercise 7.3.

Theorem 9.9 (Cameron-Martin theorem for deterministic drift). *Let $(\widetilde{B}_t, t \in [0, T])$ be the process in equation* (9.13) *on $(\Omega, \mathcal{F}, \mathbf{P})$. Suppose that $\int_0^T \theta_s^2 \, ds < \infty$. Consider the probability $\widetilde{\mathbf{P}}$ on Ω given by*

$$\tag{9.14} \widetilde{\mathbf{P}}(\mathcal{E}) = \mathbf{E}[M_T \mathbf{1}_{\mathcal{E}}], \qquad \mathcal{E} \in \mathcal{F},$$

where

$$\tag{9.15} M_t = \exp\left(-\int_0^t \theta_s \, d\widetilde{B}_s + \frac{1}{2} \int_0^t \theta_s^2 \, ds\right) = \exp\left(-\int_0^t \theta_s \, dB_s - \frac{1}{2} \int_0^t \theta_s^2 \, ds\right).$$

Then the process $(\widetilde{B}_t, t \in [0, T])$ under $\widetilde{\mathbf{P}}$ is distributed like a standard Brownian motion.

Proof. Equation (9.15) is straightforward since $d\widetilde{B}_t = dB_t + \theta_t \, dt$. We want to show that $(\widetilde{B}_t, t \in [0, T])$ is a $\widetilde{\mathbf{P}}$-Brownian motion. We cannot use the exact same approach as in the proof of Theorem 9.5, the reason being that the factor does not exactly split into increments of $\widetilde{B}_t$, as θ now depends on time. Instead, we will use the MGF of the increments. For $n \in \mathbb{N}$ and $(t_j, j \leq n)$ a partition of $[0, T]$ with $t_n = T$, we will show that

$$\tag{9.16} \widetilde{\mathbf{E}}\left[\exp\left(\sum_{j=0}^{n-1} \lambda_j (\widetilde{B}_{t_{j+1}} - \widetilde{B}_{t_j})\right)\right] = \exp\left(\frac{1}{2} \sum_{j=0}^{n-1} \lambda_j^2 (t_{j+1} - t_j)\right).$$

This proves that the increments are the ones of standard Brownian motion; cf. Proposition 3.1. Let $(\mathcal{F}_{t_j}, j \leq n)$ be the Brownian filtrations at the time of the partition. (Note that the filtrations of $(\widetilde{B}_t, t \leq T)$ and $(B_t, t \leq T)$ are the same. Why?) The proof is by successively conditioning from t_{n-1} down to t_1. As a first step we have by Proposition 4.19

$$\widetilde{\mathbf{E}}\left[e^{\sum_{j=0}^{n-1} \lambda_j (\widetilde{B}_{t_{j+1}} - \widetilde{B}_{t_j})}\right]$$

$$\tag{9.17} = \mathbf{E}\left[\mathbf{E}\left[M_{t_n} e^{\sum_{j=0}^{n-1} \lambda_j (\widetilde{B}_{t_{j+1}} - \widetilde{B}_{t_j})} \Big| \mathcal{F}_{t_{n-1}}\right]\right]$$

$$= \mathbf{E}\left[M_{t_{n-1}} e^{\sum_{j=0}^{n-2} \lambda_j (\widetilde{B}_{t_{j+1}} - \widetilde{B}_{t_j})} \mathbf{E}\left[e^{\int_{t_{n-1}}^{t_n} (-\theta_s + \lambda_{n-1}) \, d\widetilde{B}_s + \frac{1}{2} \int_{t_{n-1}}^{t_n} \theta_s^2 \, ds} \Big| \mathcal{F}_{t_{n-1}}\right]\right].$$

Here we used that by definition $M_{t_n} = M_{t_{n-1}} e^{\int_{t_{n-1}}^{t_n} (-\theta_s + \lambda_{n-1}) \, d\widetilde{B}_s + \frac{1}{2} \int_{t_{n-1}}^{t_n} \theta_s^2 \, ds}$ and that $M_{t_{n-1}}$ is $\mathcal{F}_{t_{n-1}}$-measurable. The conditional expectation is easily evaluated since the integral $\int_{t_{n-1}}^{t_n}(-\theta_s + \lambda_{n-1})\,d\widetilde{B}_s$ only depends on the increments $\widetilde{B}_{t_n} - \widetilde{B}_{t_{n-1}}$. Thus, it is independent of $\mathcal{F}_{t_{n-1}}$. We get, since $d\widetilde{B}_t = dB_t + \theta_t\,dt$,

$$\mathbf{E}\left[e^{\int_{t_{n-1}}^{t_n}(-\theta_s + \lambda_{n-1})\,d\widetilde{B}_s + \frac{1}{2}\int_{t_{n-1}}^{t_n}\theta_s^2\,ds}\Big|\mathcal{F}_{t_{n-1}}\right]$$

$$= \mathbf{E}\left[e^{\int_{t_{n-1}}^{t_n}(-\theta_s + \lambda_{n-1})\,dB_s + \int_{t_{n-1}}^{t_n}(-\frac{1}{2}\theta_s^2 + \theta_s\lambda_{n-1})\,ds}\right]$$

$$= \exp\left(\frac{1}{2}\int_{t_{n-1}}^{t_n}(-\theta_s + \lambda_{n-1})^2\,ds + \int_{t_{n-1}}^{t_n}\left(-\frac{1}{2}\theta_s^2 + \theta_s\lambda_{n-1}\right)\,ds\right)$$

$$= \exp\left(\frac{1}{2}\lambda_{n-1}^2(t_n - t_{n-1})\right).$$

The second equality comes from the fact that $\int_{t_{n-1}}^{t_n}(-\theta_s + \lambda_{n-1})\,dB_s$ is a Gaussian random variable of mean 0 and variance $\int_{t_{n-1}}^{t_n}(-\theta_s + \lambda_{n-1})^2\,ds$ by Corollary 5.18. The third equality is simply by developing the square. Putting this back into equation (9.17), we have shown that

$$\widetilde{\mathbf{E}}\left[e^{\sum_{j=0}^{n-1}\lambda_j(\widetilde{B}_{t_{j+1}} - \widetilde{B}_{t_j})}\right] = \exp\left(\frac{1}{2}\lambda_{n-1}^2(t_n - t_{n-1})\right)\cdot\mathbf{E}\left[M_{t_{n-1}}e^{\sum_{j=0}^{n-2}\lambda_j(\widetilde{B}_{t_{j+1}} - \widetilde{B}_{t_j})}\right]$$

$$= \exp\left(\frac{1}{2}\lambda_{n-1}^2(t_n - t_{n-1})\right)\cdot\widetilde{\mathbf{E}}\left[e^{\sum_{j=0}^{n-2}\lambda_j(\widetilde{B}_{t_{j+1}} - \widetilde{B}_{t_j})}\right].$$

It remains to condition on $\mathcal{F}_{t_{n-2}}$ down to $\mathcal{F}_{t_1}$ and proceed as above to obtain equation (9.16). $\square$

Example 9.10. Consider the time-inhomogeneous diffusion $(X_t, t \in [0,1])$ on $(\Omega, \mathcal{F}, \mathbf{P})$ defined by the SDE

$$dX_t = \sqrt{1 + t^2}\,dB_t + \cos t\,dt, \quad X_0 = 0.$$

Can we find a probability $\widetilde{\mathbf{P}}$ equivalent to $\mathbf{P}$ for which $(X_t, t \in [0,1])$ is a martingale? Yes. To see this, we need to rearrange the SDE as follows:

$$dX_t = \sqrt{1 + t^2}\left(dB_t + \frac{\cos t}{\sqrt{1 + t^2}}\,dt\right).$$

Note that we can divide by $\sqrt{1 + t^2}$ as it is never 0. We take

$$d\widetilde{B}_t = dB_t + \frac{\cos t}{\sqrt{1 + t^2}}\,dt.$$

Theorem 9.9 is now applicable with $\theta_t = \frac{\cos t}{\sqrt{1+t^2}}$. Note that $|\theta_t| \le 1$ so the assumption is trivially satisfied. The exponential martingale is

$$M_t = \exp\left(-\int_0^t \frac{\cos s}{\sqrt{1 + s^2}}\,d\widetilde{B}_s + \frac{1}{2}\int_0^t \frac{\cos^2 s}{1 + s^2}\,ds\right)$$

$$= \exp\left(-\int_0^t \frac{\cos s}{\sqrt{1 + s^2}}\,dB_s - \frac{1}{2}\int_0^t \frac{\cos^2 s}{1 + s^2}\,ds\right).$$

The new probability $\widetilde{\mathbf{P}}$ for which $(X_t, t \in [0,1])$ is a martingale is therefore

$$\widetilde{\mathbf{P}}(\mathcal{E}) = \mathbf{E}[M_1 \mathbf{1}_{\mathcal{E}}].$$

Under this probability, $(\widetilde{B}_t, t \in [0,1])$ is a standard Brownian motion and the SDE of $(X_t, t \in [0,1])$ is simply

$$dX_t = \sqrt{1+t^2} \; d\widetilde{B}_t.$$

The Girsanov theorem generalizes the Cameron-Martin theorem for a drift that is now an adapted process. More precisely, we consider

$$(9.18) \qquad \widetilde{B}_t = B_t + \int_0^t \Theta_s \, ds, \qquad t \in [0,T],$$

where $(B_t, t \in [0,T])$ is a standard Brownian motion defined on $(\Omega, \mathcal{F}, \mathbf{P})$ with its natural filtration $(\mathcal{F}_t, t \in [0,T])$, and $(\Theta_t, t \in [0,T])$ is an adapted process. The proof of the theorem relies heavily on the fact that

$$(9.19) \qquad M_t = \exp\left(\int_0^t \Theta_s \, dB_s - \frac{1}{2} \int_0^t \Theta_s^2 \, ds \right), \qquad t \in [0,T],$$

is a martingale with respect to the Brownian filtration. This was discussed in Example 7.17. A sufficient condition for $(M_t, t \in [0,T])$ to be a martingale is the Novikov condition (7.18).

Theorem 9.11 (Girsanov's theorem). *Let $(\widetilde{B}_t, t \in [0,T])$ be the process in equation (9.18) on $(\Omega, \mathcal{F}, \mathbf{P})$. Suppose that the adapted process $(\Theta_t, t \in [0,T])$ satisfies the Novikov condition*

$$\mathbf{E}\left[\exp\left(\frac{1}{2} \int_0^T \Theta_s^2 \, ds \right) \right] < \infty.$$

Consider the probability $\widetilde{\mathbf{P}}$ on Ω given by

$$(9.20) \qquad \widetilde{\mathbf{P}}(\mathcal{E}) = \mathbf{E}\left[M_T \mathbf{1}_{\mathcal{E}} \right], \qquad \mathcal{E} \in \mathcal{F},$$

where $(M_t, t \in [0,T])$ is the process defined in (9.19). Then the process $(\widetilde{B}_t, t \in [0,T])$ under $\widetilde{\mathbf{P}}$ is distributed like a standard Brownian motion.

Proof. The proof is very similar to the one of Theorem 9.9. We need to prove equation (9.16). Let $(\mathcal{F}_{t_j}, j \leq n)$ be the Brownian filtration at the times $t_1 < t_2 < \cdots < t_n = T$. By conditioning on $\mathcal{F}_{t_{n-1}}$ as in equation (9.17), we get

$$(9.21)$$
$$\widetilde{\mathbf{E}}\left[e^{\sum_{j=0}^{n-1} \lambda_j (\widetilde{B}_{t_{j+1}} - \widetilde{B}_{t_j})} \right]$$
$$= \mathbf{E}\left[M_{t_{n-1}} e^{\sum_{j=0}^{n-2} \lambda_j (\widetilde{B}_{t_{j+1}} - \widetilde{B}_{t_j})} \mathbf{E}\left[e^{\int_{t_{n-1}}^{t_n} (-\Theta_s + \lambda_{n-1}) \, dB_s - \frac{1}{2} \int_{t_{n-1}}^{t_n} (\Theta_s^2 - 2\lambda_{n-1}\Theta_s) \, ds} \Big| \mathcal{F}_{t_{n-1}} \right] \right].$$

The form of the last integral suggests that we complete the square. Define

$$V_s = -\Theta_s + \lambda_{n-1}.$$

(Recall that λ_{n-1} is simply a constant.) We then have that the last line of equation (9.21) can be conveniently written as

$$\widetilde{\mathbf{E}}\left[e^{\sum_{j=0}^{n-1} \lambda_j (\widetilde{B}_{t_{j+1}} - \widetilde{B}_{t_j})} \right]$$
$$= e^{\frac{1}{2}\lambda_{n-1}^2 (t_n - t_{n-1})} \cdot \mathbf{E}\left[M_{t_{n-1}} e^{\sum_{j=0}^{n-2} \lambda_j (\widetilde{B}_{t_{j+1}} - \widetilde{B}_{t_j})} \mathbf{E}\left[e^{\int_{t_{n-1}}^{t_n} V_s \, dB_s - \frac{1}{2} \int_{t_{n-1}}^{t_n} V_s^2 \, ds} \Big| \mathcal{F}_{t_{n-1}} \right] \right].$$

It turns out that the drift V_s also satisfies the Novikov condition, since Θ_s satisfies it, and λ_{n-1} is simply a constant; see Exercise 9.10. In particular, the process

$$\exp\left(\int_0^t V_s\, dB_s - \frac{1}{2}\int_0^t V_s^2\, ds\right), \quad t \in [0,T],$$

is also a martingale. This implies that the conditional expectation above is

$$\mathbf{E}\left[e^{\int_{t_{n-1}}^{t_n} V_s\, dB_s - \frac{1}{2}\int_{t_{n-1}}^{t_n} V_s^2\, ds}\Big|\mathcal{F}_{t_{n-1}}\right] = 1.$$

We have shown

$$\widetilde{\mathbf{E}}\left[e^{\sum_{j=0}^{n-1}\lambda_j(\widetilde{B}_{t_{j+1}}-\widetilde{B}_{t_j})}\right] = e^{\frac{1}{2}\lambda_{n-1}^2(t_n-t_{n-1})}\cdot \mathbf{E}\left[M_{t_{n-1}}e^{\sum_{j=0}^{n-2}\lambda_j(\widetilde{B}_{t_{j+1}}-\widetilde{B}_{t_j})}\right]$$

$$= e^{\frac{1}{2}\lambda_{n-1}^2(t_n-t_{n-1})}\cdot \widetilde{\mathbf{E}}\left[e^{\sum_{j=0}^{n-2}\lambda_j(\widetilde{B}_{t_{j+1}}-\widetilde{B}_{t_j})}\right],$$

where we used the martingale property in the last line. The proof can then be iterated by conditioning from $\mathcal{F}_{t_{n-2}}$ down to $\mathcal{F}_{t_1}$. This proves the theorem. $\qquad\square$

Note that we could prove Theorem 9.9 with the same martingale approach as the proof of Girsanov theorem using Example 7.11 and Exercise 7.3; see Exercise 9.7.

Example 9.12. Consider the time-homogeneous diffusion $(Y_t, t \in [0,1])$ on $(\Omega, \mathcal{F}, \mathbf{P})$ defined by the SDE

$$dY_t = \sqrt{1+Y_t^2}\, dB_t + dt, \quad Y_0 = 0.$$

Again, we can find a probability $\widetilde{\mathbf{P}}$ equivalent to $\mathbf{P}$ for which $(Y_t, t \in [0,1])$ is a martingale. We write the SDE as

$$dY_t = \sqrt{1+Y_t^2}\left(dB_t + \frac{1}{\sqrt{1+Y_t^2}}\, dt\right).$$

Note that we can divide by $\sqrt{1+Y_t^2}$ as it is never 0. The candidate for the Brownian motion has the SDE

$$d\widetilde{B}_t = dB_t + \frac{1}{\sqrt{1+Y_t^2}}\, dt.$$

Theorem 9.11 can be applied with the local drift $\Theta_t = \frac{1}{\sqrt{1+Y_t^2}}$. The drift is bounded by 1, so that the process $(M_t, t \in [0,1])$ is a martingale with

$$M_t = \exp\left(-\int_0^t \frac{1}{\sqrt{1+Y_s^2}}\, d\widetilde{B}_s + \frac{1}{2}\int_0^t \frac{1}{1+Y_s^2}\, ds\right)$$

$$= \exp\left(-\int_0^t \frac{1}{\sqrt{1+Y_s^2}}\, dB_s - \frac{1}{2}\int_0^t \frac{1}{1+Y_s^2}\, ds\right).$$

The new probability $\widetilde{\mathbf{P}}$ for which $(Y_t, t \in [0,1])$ is a martingale is therefore

$$\widetilde{\mathbf{P}}(\mathcal{E}) = \mathbf{E}[M_1 1_{\mathcal{E}}].$$

Under this probability, $(\widetilde{B}_t, t \in [0, 1])$ is a standard Brownian motion and the SDE of $(Y_t, t \in [0, 1])$ is simply

$$dY_t = \sqrt{1 + Y_t^2}\, d\widetilde{B}_t.$$

9.4. Numerical Projects and Exercises

9.1. **Sampling bias of a uniform variable.** Consider a uniform random variable X on $[0, 1]$ as in Example 9.1.
   (a) Generate a sample $\mathcal{S}$ of 100,000 values of X.
   (b) Using the function `numpy.random.choice`, create a subsample $\widetilde{\mathcal{S}}$ of 1,000 values among $\mathcal{S}$ proportionally to the weight

$$e^{-3.6X}.$$

   (You will need to normalize the weight so that the sum over all the weights is 1; see parameter p in random.choice.)
   (c) Draw the histogram of $\widetilde{\mathcal{S}}$ and compute the mean of $\widetilde{\mathcal{S}}$.

9.2. **Sampling bias à la Cameron-Martin.** We consider a Brownian motion with drift $\theta = 1$:

$$\widetilde{B}_t = B_t + t.$$

   (a) Generate a sample $\mathcal{S}$ of 100,000 paths for $(\widetilde{B}_t, t \in [0, 1])$ using a 0.01 discretization.
   (b) Using the function `numpy.random.choice`, sample 1,000 paths in $\mathcal{S}$ not uniformly but proportionally to their weight:

$$M(\widetilde{B}) = e^{-\widetilde{B}_1 + 1/2}.$$

   Again you will need to normalize the weights $M(\widetilde{B})$ so that the sum over the 100,000 paths is 1. Let's call this new sample $\widetilde{\mathcal{S}}$.
   (c) Draw the histogram of $\widetilde{B}_{1/2}$ on the sample $\widetilde{\mathcal{S}}$. It should look like a Gaussian PDF with mean 0 and variance 1/2.
   (d) Plot the first 10 paths from $\mathcal{S}$. Plot the first 10 paths from $\widetilde{\mathcal{S}}$.

9.3. **Sampling bias à la Girsanov.** Let $(Y_t, t \in [0, 1])$ be an Ornstein-Uhlenbeck process with SDE

$$dY_t = dB_t - Y_t\, dt, \quad Y_0 = 0.$$

   (a) Generate a sample $\mathcal{S}$ of 10,000 paths for $(Y_t, t \in [0, 1])$ using a 0.01 discretization.
   (b) Using the function `numpy.random.choice`, sample 1,000 paths in $\mathcal{S}$ not uniformly but proportionally to their weight:

$$M(Y) = e^{\int_0^1 Y_s\, dB_s - \frac{1}{2}\int_0^1 Y_s^2\, ds}.$$

   Here, you can approximate the Itô and Riemann integrals by a sum. Do not forget to normalize the weights. Let's call this new sample $\widetilde{\mathcal{S}}$.
   (c) Draw the histogram of Y_1 on the sample $\widetilde{\mathcal{S}}$. Compare it to the histogram of the whole sample $\mathcal{S}$.
   (d) Compare the variance on $\widetilde{\mathcal{S}}$ and on the whole sample $\mathcal{S}$.

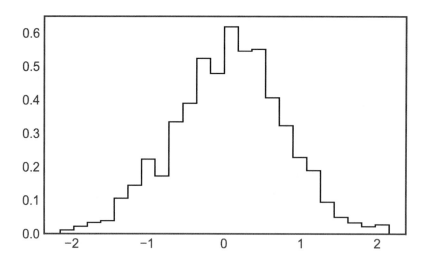

Figure 9.4. The histogram of the biased sampled of $\widetilde{B}_{1/2}$ in Numerical Project 9.2. Note that the mean is around 0 and not 1/2. In fact, it is 0.06 ....

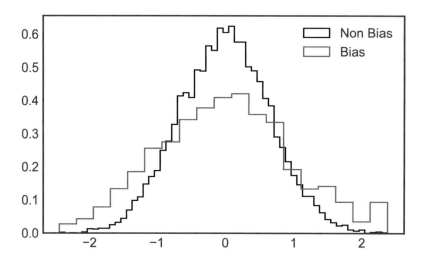

Figure 9.5. The histograms of the biased sample of Y_1 and of the whole sample in Numerical Project 9.3. The variance of the biased sample is 0.94..., consistent with a variance of 1 for standard Brownian motion at time 1, whereas the one of the whole sample is 0.44 consistent with a variance $\frac{1}{2}(1-e^{-2s}) \approx 0.43 \ldots$ of the Ornstein-Uhlenbeck process at time 1.

Exercises

9.1. **The mean of a biased distribution.** Let X be a random variable defined on $(\Omega, \mathcal{F}, \mathbf{P})$ and let $\widetilde{\mathbf{P}}$ be the probability defined by

$$\widetilde{\mathbf{P}}(\mathcal{E}) = \mathbf{E}\left[\frac{e^{aX}}{\mathbf{E}[e^{aX}]}\mathbf{1}_{\mathcal{E}}\right].$$

Show that

$$\widetilde{\mathbf{E}}[X] = \frac{\mathrm{d}}{\mathrm{d}a}(\log \mathbf{E}[e^{aX}]).$$

(You can assume that the derivative can be taken inside the expectation.)

9.2. **Biasing an exponential random variable.** Consider a random variable X with the exponential distribution with parameter 1 on $(\Omega, \mathcal{F}, \mathbf{P})$. From Exercise 1.11, we know that

$$\mathbf{E}[e^{aX}] = \frac{1}{1-a}, \quad a < 1.$$

Use this and equation (9.3) to find a probability $\widetilde{\mathbf{P}}$ under which X has mean $\frac{1}{1-a}$. What is the distribution of X under $\widetilde{\mathbf{P}}$?

9.3. **Biasing a Poisson random variable.** Let N be a Poisson random variable with parameter $\lambda > 0$.
   (a) Show that the MGF of N is

$$\mathbf{E}[e^{aN}] = \exp(\lambda(e^a - 1)).$$

   (b) Use the above to show that under the probability $\widetilde{\mathbf{P}}$ defined by $\widetilde{\mathbf{P}}(\mathcal{E}) = \mathbf{E}[M\mathbf{1}_{\mathcal{E}}]$, where $M = \frac{e^{aN}}{\mathbf{E}[e^{aN}]}$, the random variable N is also Poisson-distributed with parameter λe^a.

9.4. **Brownian motion with drift as a $\widetilde{\mathbf{P}}$-martingale.** Consider a Brownian motion with drift of the form

$$\widetilde{B}_t = \sigma B_t + \mu t, \quad t \in [0, T],$$

where $\sigma > 0$ and $\mu \in \mathbb{R}$. Find a probability $\widetilde{\mathbf{P}}$ for which $(\widetilde{B}_t, t \in [0, T])$ is a martingale. What is the distribution of the process $(\widetilde{B}_t, t \in [0, T])$ under $\widetilde{\mathbf{P}}$?

9.5. **Gambler's ruin on $[0, T]$.** Consider a Brownian motion with drift of the form

$$\widetilde{B}_t = B_t + t,$$

where $(B_t, t \geq 0)$ is a standard Brownian motion and the first passage time at $a = -1$: $\tau = \min\{t \geq 0 : \widetilde{B}_t = -1\}$.
   (a) Note that equation (5.29) gives $\mathbf{P}(\tau < \infty) = e^{-2} \approx 0.13534\dots.$
   (b) Formula (5.29) was tested in Numerical Project 5.6 on a finite interval $[0, 5]$. Use equation (9.11) to compute $\mathbf{P}(\tau < 5)$ exactly using a software.
   (c) Repeat the above for $\mathbf{P}(\tau < 1)$ and compare to the numerical results on the interval $[0, 1]$ with a discretization of 0.001 on 10,000 paths.

9.6. **Geometric Brownian motion as a $\widetilde{\mathbf{P}}$-martingale.** Consider a geometric Brownian motion

$$S_t = \exp(\sigma B_t + \mu t), \quad t \in [0, T].$$

The SDE was computed in Example 7.4. Use the SDE to find a probability $\widetilde{\mathbf{P}}$ under which $(S_t, t \in [0, T])$ is a martingale. What is the distribution of the process under this probability?

9.7. **A proof of Cameron-Martin using martingales.** Prove the Cameron-Martin theorems, Theorems 9.5 and 9.9, using Example 7.11 and the same martingale approach as in the proof of Theorem 9.11.

9.8. ★ **Ornstein-Uhlenbeck process as a $\widetilde{\mathbf{P}}$-martingale.** Consider the Ornstein-Uhlenbeck process with SDE

$$dY_t = dB_t - Y_t \, dt, \quad Y_0 = 0.$$

(a) Find a probability $\widetilde{\mathbf{P}}$ for which $(Y_t, t \in [0,1])$ is a $\widetilde{\mathbf{P}}$-martingale. (We will verify the Novikov condition below.)

(b) As a first step to verify the Novikov condition, show that

$$\exp\left(\frac{1}{2}\int_0^1 Y_s^2 \, ds\right) \leq \int_0^1 \exp\left(\frac{1}{2}Y_s^2\right) ds.$$

(c) Use the marginal distribution at Y_s to show that

$$\mathbf{E}\left[\exp\left(\frac{1}{2}Y_s^2\right)\right] = \sqrt{2(1 + e^{-2s})}.$$

(d) Conclude that $(Y_t, t \in [0,1])$ satisfies the Novikov condition.

9.9. ★ **Application of Girsanov theorem.** Let $(B_t, t \in [0,T])$ be a standard Brownian motion on $(\Omega, \mathcal{F}, \mathbf{P})$. For $b > 0$, we consider the process

$$Z_t = \exp\left(-b\int_0^t B_s \, dB_s - \frac{b^2}{2}\int_0^t B_s^2 \, ds\right), \quad t \in [0,T].$$

(a) Show that Z_t can be written as

$$Z_t = \exp\left(-\frac{b}{2}(B_t^2 - t) - \frac{b^2}{2}\int_0^t B_s^2 \, ds\right), \quad t \in [0,T].$$

(b) Deduce from this that $(Z_t, t \in [0,T])$ is a martingale for the Brownian filtration.

(c) We consider the process

$$\widetilde{B}_t = B_t + b\int_0^t B_s \, ds, \quad t \in [0,T].$$

Find $\widetilde{\mathbf{P}}$ under which $\widetilde{B}_t$ is a standard Brownian motion.
Make sure the conditions are satisfied!

(d) Argue that $(B_t, t \in [0,T])$ has the distribution of a Ornstein-Uhlenbeck process under $\widetilde{\mathbf{P}}$.

(e) Use the previous questions to show that for $a, b > 0$

$$\mathbf{E}\left[e^{-aB_t^2 - \frac{b^2}{2}\int_0^t B_s^2 \, ds}\right] = \exp\left(-\frac{bt}{2}\right)\left\{\frac{1}{2}(1 + e^{-2bt}) + \frac{a}{b}(1 - e^{-2bt})\right\}^{-1/2}.$$

Perhaps amazingly, this determines the joint distribution of $(B_t^2, \int_0^t B_s^2 \, ds)$!

9.10. ★ **Novikov condition.** Consider a process $(V_t, t \in [0,T])$ adapted to a Brownian filtration and that satisfies the Novikov condition (7.18). Prove that $V_s + 1$ also satisfies it. In other words, show that

$$\mathbf{E}\left[\exp\left(\frac{1}{2}\int_0^T V_s^2 \, ds\right)\right] < \infty \implies \mathbf{E}\left[\exp\left(\frac{1}{2}\int_0^T (V_s + 1)^2 \, ds\right)\right] < \infty.$$

Argue that this also holds for $V_s + \lambda$ for any fixed $\lambda \in \mathbb{R}$.

9.5. Historical and Bibliographical Notes

In Theorems 9.5, 9.9, and 9.11, we constructed a probability $\widetilde{\mathbf{P}}$ on Ω from $\mathbf{P}$ by adding a factor M_T to the expectation. The factor M_T is called the *Radon-Nikodym derivative* of $\widetilde{\mathbf{P}}$ with respect to $\mathbf{P}$. More generally, we say that a probability $\widetilde{\mathbf{P}}$ is absolutely continuous with respect to $\mathbf{P}$ if $\mathbf{P}(\mathcal{N}) = 0$ for some event $\mathcal{N}$ implies $\widetilde{\mathbf{P}}(\mathcal{N}) = 0$. The Radon-Nikodym theorem, see for example [**Wal12**], states that if $\widetilde{\mathbf{P}}$ is absolutely continuous with respect to $\mathbf{P}$, there must exist a random variable $M \geq 0$ such that $\widetilde{\mathbf{P}}(E) = \mathbf{E}[M\mathbf{1}_\mathcal{E}]$. The random variable is called the *Radon-Nikodym derivative* of $\widetilde{\mathbf{P}}$ with respect to $\mathbf{P}$. The factor M is also called the *likelihood ratio* in statistics and *Gibbs weight* in statistical physics. Two probabilities $\widetilde{\mathbf{P}}$ and $\mathbf{P}$ on Ω are said to be *equivalent* if $\mathbf{P}$ is absolutely continuous with respect to $\widetilde{\mathbf{P}}$ and vice versa. In that case, we must have that Radon-Nikodym derivatives must be nonzero with probability one. Why? In particular, M^{-1} is well-defined. This is the case in Theorems 9.5, 9.9, and 9.11.

The Cameron-Martin theorem was proved in 1944 in [**CM44**], not too long after the proof of the existence of Brownian motion by Wiener. The Girsanov theorem was proved later in [**Gir60**].

Chapter 10

Applications to Mathematical Finance

One of the main applications of stochastic calculus is the pricing of *derivatives* in financial engineering. Derivatives are financial assets whose value is a function of an underlying asset, for example a stock, a currency, a bond, etc. The pricing of derivatives was revolutionized by the work of Black, Scholes, and Merton, who based the pricing on the notion of *arbitrage*. In a nutshell, if there is no arbitrage in the market, two products with the same payoff should have the same price. Otherwise, it would be possible to make money without any risk of loss and with no a priori investment by buying the cheap asset and selling the expensive one. The no-arbitrage assumption enables the construction of a very rich pricing theory. In this chapter, we revisit the Black-Scholes-Merton theory of option pricing through the lens of stochastic calculus.

We first introduce simple models of a market, including the Black-Scholes model. In Section 10.2, we define the notion of derivatives and give several important examples. Such derivatives can be priced after introducing the notion of *arbitrage* and *replication* in Section 10.3. This is done for the Black-Scholes model in Section 10.4. We look at how the prices depend on the parameters of the market and of the derivatives using the *Greeks* in Section 10.5. An important quantum leap happens in Section 10.6 where we discuss how the no-arbitrage price turns out to be equivalent to averaging the payoff of derivatives under the *risk-neutral probability*. In particular, this allows us to extend the pricing theory to include exotic options whose payoff is *path-dependent*; see Section 10.7. Finally, we see how the theory can be generalized to other models including interest rate models and stochastic volatility models in Sections 10.8 and 10.9.

10.1. Market Models

The simplest *market models* are of the following form.

Definition 10.1. Let $(\Omega, \mathcal{F}, \mathbf{P})$ be a probability space. A market model is a pair of two processes $(S_t, t \geq 0)$ and $(R_t, t \geq 0)$ on this space. We write $(\mathcal{F}_t, t \geq 0)$ for the filtration defined by the two processes. In financial terms, these represent:

- **A risky asset**. This asset could represent a stock, a bond, a foreign currency, etc. The value of the risky asset is modelled by the stochastic process $(S_t, t \geq 0)$. In this chapter, the process $(S_t, t \geq 0)$ will usually be an Itô process, except in Section 10.9.

- **A risk-free interest rate**. This is the interest rate of a money market account for example. The interest rate is modelled by the stochastic process $(R_t, t \geq 0)$. We will assume that the interest rate is an Itô process in Section 10.8. Until then, we will take the interest rate to be constant; i.e., $R_t = r$ for all $t \geq 0$, where $r \geq 0$ is a fixed parameter of the model.

We note that the interest rate defines two other processes. First, the interest rate defines a *risk-free asset*. This represents cash (say in dollars \$) put in a money market account at the risk-free rate. We denote the value of the risk-free asset by $(A_t, t \geq 0)$. Assuming continuous compounding, we have that A_t satisfies the differential equation

$$\mathrm{d}A_t = R_t A_t \, \mathrm{d}t.$$

In other words, the rate of increase of the value A_t is proportional to A_t, the proportion being quantified by the rate R_t. The equation is easily solved by integrating both sides (even though R_t might be random!), and we have

$$A_t = A_0 \exp\left(\int_0^t R_s \, \mathrm{d}s\right).$$

In the simplest case where $R_t = r$ is constant, we have not surprisingly that $A_t = A_0 e^{rt}$.

Second, the interest rate defines the *discounting process*. In this chapter, it is denoted by $(D_t, t \geq 0)$ (not to be confused with the notation for the drift in Chapter 7). The discounting process is used to express the value of the different assets at different times in the same unit: dollars at time 0. Indeed, the value of \$1 changes over time because of the risk-free rate: \$1 at time 0 is worth $\$\exp\left(\int_0^t R_s \, \mathrm{d}s\right)$ at time t. Therefore, to express the value of an asset in dollars at time 0, we need to multiply the value by

$$(10.1) \qquad\qquad D_t = \exp\left(-\int_0^t R_s \, \mathrm{d}s\right).$$

In the case $R_t = r$ again, this is simply $D_t = e^{-rt}$. Note that D_t satisfies the differential

$$\mathrm{d}D_t = -R_t D_t \, \mathrm{d}t.$$

We will usually take the convention that $A_0 = 1$. With this convention, we have that

$$A_t = D_t^{-1}.$$

The most common market model is the following:

Example 10.2 (The Black-Scholes model). The value of the risky asset is modelled by a geometric Brownian motion $(S_t, t \geq 0)$ of the form

$$S_t = S_0 \exp(\sigma B_t + \mu t),$$

for some standard Brownian motion $(B_t, t \geq 0)$. It has the SDE, see Example 7.4,

$$dS_t = \sigma S_t \, dB_t + (\mu + \sigma^2/2)S_t \, dt.$$

The risk-free rate is simply constant: $R_t = r$ for all t. In particular, the discounting factor is simply $D_t = e^{-rt}$. It satisfies the differential

$$dD_t = -rD_t \, dt.$$

10.2. Derivatives

The goal of this chapter is to price *derivatives*. A *derivative* is a financial asset whose value at a given time is a function of an underlying risky asset. More precisely, in probabilistic language, we have the following:

Definition 10.3 (Derivative). A *derivative security*, or simply *derivative*, for a given market model is a financial asset whose value is given by a process $(O_t, t \in [0, T])$ on $(\Omega, \mathcal{F}, \mathbf{P})$ which is adapted to the filtration $(\mathcal{F}_t, t \in [0, T])$ of the market model. In other words, its value at time t is determined by the values of the risky asset (and possibly the risk-free rate) up to time t. The last time T is called the *expiration* or the *maturity* T of the derivative. The value of the derivative at expiration T is called the *payoff* at expiration.

The unit of the maturity T is usually *years*, since the interest rate is usually given in a yearly basis. Here are some elementary examples of derivatives.

Example 10.4 (European call option). A *European call option* with expiration T and strike price K is the derivative with payoff

$$C_T = \max\{S_T - K, 0\} = (S_T - K)^+.$$

The payoff can be interpreted as follows: The holder of the security has the option of *buying the underlying risky asset at the strike price K* at expiration date T. The value is 0 if $S_T \leq K$ since it is better to just buy the asset on the market than to exercise the option, as the market price is cheaper than the strike price. If $S_T \geq K$, then the value of the option is $S_T - K$, since we can *exercise the option*; i.e., buy the asset at the strike price K, and sell it right away on the market at the market price S_T.

Example 10.5 (European put option). A *European put option* is the derivative that gives the holder the option of *selling the underlying asset at the strike price K* at expiration date T. In other words, the payoff at expiration is

$$P_T = \max\{K - S_T, 0\} = (K - S_T)^+.$$

The value is 0 if $S_T \geq K$, since it is more advantageous in this case to sell the asset at the market price rather than exercise the option and sell at the strike price. If $S_T < K$, then it is advantageous to exercise the option, i.e., to sell the asset at the price K and buy it back on the market at the market price S_T. This gives a net value of $K - S_T$.

Example 10.6 (Forward). A *forward contract* with *delivery price K* is a derivative whose payoff at expiration T is

$$F_T = S_T - K.$$

The interpretation of the contract is as follows: The holder of the security has the *obligation* (and not the option) to buy the underlying risky asset at time T at the price K. Since the price of the asset S_T might be above or below K, the payoff at expiration of this contract may be positive or negative. This is to be contrasted with the put and call options whose payoff at expiration is always positive.

Note that all the examples above are such that the payoff at expiration is a function $g(S_T)$ of the value at time T of the underlying asset. Options of this kind can be generalized using other functions g. Other types of options have payoffs which are *path-dependent*. Such options are sometimes called *exotic*. The pricing of exotic options is studied in Section 10.7.

The starting point of financial engineering is as follows. There are two sides to an option contract: There is the *buyer or the holder* of the option and there is the *seller or writer* of the option. The payoff for the seller is simply *minus* the payoff of the holder. With this in mind, it is possible to build a *portfolio of options* by buying and selling options of different types. If these options have the same expiration T, then the payoff of the portfolio is the linear combination of the payoffs. More precisely, if the portfolio has a_1 units of options (1), a_2 units of options (2), etc., then the payoff at expiration of the portfolio is

$$a_1 O_T^{(1)} + \cdots + a_k O_T^{(k)}.$$

Now, the allocations $a_1, \ldots, a_k$ can be taken to be real numbers. If $a_j > 0$, then the investor owns a_j contracts of the option (j). This is called a *long* position in the security. If $a_j < 0$, then the investor wrote a_j contracts of the option (j). This is called a *short* position in the security. Combining the different payoffs for the different option types allows the investor to *engineer* a payoff for the portfolio. As a first example, let's look at how to build an option portfolio of put and call options that *replicates* the payoff of a forward contract.

Example 10.7 (Put-Call parity). Consider an option portfolio that consists of one *long* call option at strike price K and maturity T and one *short* put option at the same strike price and maturity. Then, the payoff of this strategy is

$$C_T - P_T = (S_T - K)^+ - (K - S_T)^+ = S_T - K.$$

This is the same payoff as a forward contract! We say that the portfolio *replicates* the forward contract. This is an important relation between calls and puts that extends at every time $t \leq T$; see Example 10.10.

Here are some other standard examples of *option portfolios* or *option strategies*.

Example 10.8 (Option strategies).

(1) *Covered call.* The strategy consists of being *short* one call option with strike price K and *long* in the underlying asset. In other words, the payoff at expiration is

$$-C_T + S_T = \begin{cases} S_T & \text{if } 0 \le S_T \le K, \\ K & \text{if } S_T > K. \end{cases}$$

The covered call allows the writer of the option to protect against a large S_T. Indeed, if the writer did not own the asset, the loss of the position would be $S_T - K$, which is unbounded in S_T.

(2) *Straddle.* The strategy consists of being *long* one call option with strike price K and *long* one put option with same strike price and same expiration. The payoff is then

$$C_T + P_T = \begin{cases} K - S_T & \text{if } 0 \le S_T \le K, \\ S_T - K & \text{if } S_T > K \end{cases} = |S_T - K|.$$

The payoff of this option gets larger as S_T deviates substantially from K.

(3) *Bull call spread.* This strategy consists of one *long* call option at strike price K (option 1) and of one *short* call options at a higher strike, say $K + a$ (option 2). Both are assumed to have the same expiration. The payoff at expiration is then

$$C_T^{(1)} - C_T^{(2)} = \begin{cases} 0 & \text{if } 0 \le S_T \le K, \\ S_T - K & \text{if } K \le S_T \le K + a, \\ a & \text{if } S_T > K + a. \end{cases}$$

(4) *Bear call spread.* This strategy consists of one *short* call option at strike price K (option 1) and of one *long* call option at a higher strike, say $K + a$ (option 2). The payoff at expiration is then

$$-C_T^{(1)} + C_T^{(2)} = \begin{cases} 0 & \text{if } 0 \le S_T \le K, \\ K - S_T & \text{if } K \le S_T \le K + a, \\ -a & \text{if } S_T > K + a. \end{cases}$$

This is the other side of a *bull call spread* contract.

(5) *Butterfly.* The strategy consists of being *long* one call option with strike price $K - a$ (option 1), *short* two call options with strike price K (option 2), and *long* one call option with strike price $K + a$ (option 3), all with the same expiration.

$$C_T^{(1)} - 2C_T^{(2)} + C_T^{(3)} = \begin{cases} 0 & \text{if } 0 \le S_T \le K - a, \\ S_T - (K - a) & \text{if } K - a \le S_T \le K, \\ K + a - S_T & \text{if } K \le S_T \le K + a, \\ 0 & \text{if } S_T > K + a. \end{cases}$$

See Numerical Project 10.2 and Exercise 10.3 for more on this. There are numerous other option strategies each with their own eccentric names: *collars, iron butterfly, condor, iron condor,* etc. Such strategies are available on most trading platforms.

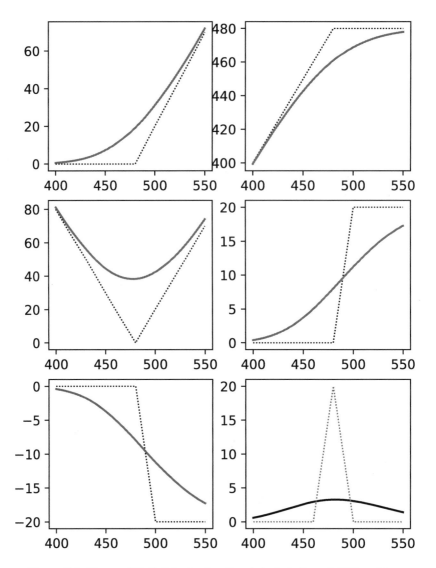

Figure 10.1. The payoff (dotted) for the six option strategies in Example 10.8 as a function of S_T with $K = 480$ and $a = 20$. The price at time 0 in the Black-Scholes model is the solid line (with parameters of Example 10.16 but with $r = 0$). See Numerical Project 10.2.

As can be seen from Figure 10.1, some strategies have a positive payoff, no matter what S_T is. Clearly, the holder of such portfolios should *pay* a price or *premium* for building such a portfolio. In the same way, the bear spread strategy has a negative payoff. Therefore, the holder of this position should *receive* a premium for building the portfolio. The *profit at expiration* is defined to be the payoff of the strategy minus the premium paid (expressed in the same dollar). A negative premium means the holder is receiving the cash. The profit diagram is therefore the same as the payoff diagram but shifted down (if a premium is paid) or up (if a premium is received).

10.3. No Arbitrage and Replication

The problem of pricing a derivative can be summarized as follows:

(10.2)   *Deduce the price O_t for all $t \leq T$ only knowing the payoff O_T of the option.*

This is an impossible task without additional assumptions on the market model. The crucial assumptions coming from mathematical finance are as follows:

(1) **The payoff can be replicated**. This means that the payoff of the derivative at time T can be written as a linear combination of the values of the risky asset and of the risk-free asset at T:

(10.3) $$O_T = a_T D_T^{-1} + \Delta_T S_T,$$

for some $a_T, \Delta_T \in \mathbb{R}$. Recall that D_T^{-1} represents the value of the risk-free asset (i.e., cash). The coefficients a_T, Δ_T represent the number of units in each asset one has to own to replicate the value of the derivative. (The notation Δ_T may appear a bit arbitrary at first, but it will make sense in the terminology of Section 10.5.) If the number of units is negative, we say the portfolio is *short* in the corresponding asset. If it is positive, we say the portfolio is *long* in the corresponding asset. If equation (10.3) holds, then we say that the payoff of the option can be replicated.

(2) **The market has no arbitrage**. An arbitrage is a portfolio, i.e., a linear combination of assets in the market model, whose value $(X_t, t \geq 0)$ is such that for some time τ (possibly random)

$$X_0 = 0, \quad \mathbf{P}(X_\tau \geq 0) = 1, \quad \mathbf{P}(X_\tau > 0) > 0.$$

In other words, an arbitrage is an asset that costs zero but will have positive value with some probability at time T without any downside of being negative. This is a nice asset to own!

It is easy to see that if there is no arbitrage, then we must have that

(10.4) $$O_t = a_t D_t^{-1} + \Delta_t S_t, \quad t \leq T,$$

for some $a_t, \Delta_t \in \mathbb{R}$. Indeed, if we observe that $O_\tau > a_\tau A_\tau + \Delta_\tau S_\tau$ for some time τ, then we can sell the option (the expensive asset) and buy the portfolio $a_\tau A_\tau + \Delta_\tau S_\tau$ and receive a strictly positive amount of cash. At expiration, the position is worth 0 by equation (10.3), but we end up with a positive amount of cash.

The portfolio in equation (10.3) allocating a_t units of the risk-free asset and Δ_t units of the risky asset is called *the replicating portfolio* of the option. Of course, these units must be $(\mathcal{F}_t)$-adapted and also may change with time. The action of changing the allocation of the portfolio composed of the risk-free asset and the risky asset is called *dynamic hedging*. We can sum up the no-arbitrage pricing of options with the following *law of one-price*:

If $(O_t, t \leq T)$ and $(O'_t, t \leq T)$ have the same payoff at expiration T, i.e., $O_T = O'_T$, then

$$O_t = O'_t \quad \text{for all } t \leq T.$$

Example 10.9 (The price of a forward). In Example 10.6, we saw that a forward $(F_t, t \leq T)$ has payoff $S_T - K$ at expiration. We suppose for simplicity that the interest rate $R_t = r$ is constant. The general case is handled in Section 10.8. A replicating portfolio for this payoff consists simply of holding one unit of the underlying asset and borrowing (i.e., holding a negative amount of cash) K dollars suitably discounted. In other words, the replicating portfolio at time $t \leq T$ has value

$$(10.5) \qquad\qquad\qquad S_t - e^{-r(T-t)}K.$$

The replicating portfolio at time t allocates $\Delta_t = 1$ unit of the risky asset and $a_t = -e^{-r(T-t)}K$ dollars in risk-free asset. Using the no-arbitrage pricing, we conclude that the value of the forward contract F_t at time T is

$$F_t = S_t - e^{-r(T-t)}K.$$

At time 0, this is $F_0 = S_0 - e^{-rT}K$. In particular, if we pick the particular delivery price $K = e^{rT}S_0$, we have $F_0 = 0$! In this case, no money is exchanged at time 0 when entering the contract. This particular delivery price is called *the forward price*.

Example 10.10 (Put-Call parity). Recall the put-call parity relation at expiration in Example 10.7. The no-arbitrage assumption implies that this relation must hold at any time $t \leq T$. More precisely, we have for a constant risk-free rate r that

$$C_t - P_t = F_t = S_t - e^{-r(T-t)}K, \quad t \leq T.$$

This relation must be satisfied by the prices of calls, puts, and the asset at all times.

The no-arbitrage assumption also implies that P_t and C_t must be both greater than or equal to 0 at all times, since their payoff is positive. Together with put-call parity, this implies the following bounds for the prices of puts and calls:

$$(10.6) \qquad \begin{aligned} S_t - e^{-r(T-t)}K &\leq C_t < S_t, \\ e^{-r(T-t)}K - S_t &\leq P_t < e^{-r(T-t)}K. \end{aligned}$$

Example 10.11 (Price of option strategies). We now look at the consequences of the no-arbitrage assumptions on the option strategies of Example 10.8.

- *Bull call spread.* This strategy is such that $C_T^{(1)} - C_T^{(2)} \geq 0$. By the no-arbitrage assumption, it must be that

$$C_t^{(1)} - C_t^{(2)} \geq 0 \text{ for all } t \leq T.$$

  Recall that the option (2) had a greater strike price. We conclude that the *value of a call is a decreasing function of the strike K.*

- *Butterfly.* This strategy is such that $C_T^{(1)} - 2C_T^{(2)} + C_T^{(3)} \geq 0$. In other words, we must have that

$$C_t^{(3)} - C_t^{(2)} \geq C_t^{(2)} - C_t^{(1)}.$$

  Option (3) had strike price $K + a$, option (2) had strike price K, and option (1) had strike price $K - a$. This is evidence that C_t is a convex function of K.

We will go back to these properties in Section 10.6.

The assumption of replication as expressed in equation (10.4) is often paired with the condition of *self-financing*. In words, this means that the variation of O_t in a small interval of time can only be caused by a variation of S_t or D_t^{-1}, and not from the variation of a_t and Δ_t. In differential notation, this becomes

(10.7) $$\mathrm{d}O_t = a_t\,\mathrm{d}(D_t^{-1}) + \Delta_t\,\mathrm{d}S_t = a_t R_t\,\mathrm{d}t + \Delta_t\,\mathrm{d}S_t.$$

It is now possible to eliminate a_t from the equations by putting equations (10.4) and (10.7) together to get

(10.8) $$\boxed{\mathrm{d}O_t = \Delta_t\,\mathrm{d}S_t + (O_t - \Delta_t S_t)R_t\,\mathrm{d}t.}$$

We call this the *valuation equation*. It has a nice interpretation: The replication portfolio is made of Δ_t units of the risky asset. The rest of the value, $O_t - \Delta_t S_t$, is put in cash in the risk-free asset. Therefore the variation of O_t is due to the variation of the risky asset (multiplied by Δ_t) and to the variation of the cash due to the interest rate.

Equipped with the above assumptions (and in particular the valuation equation (10.8)), we can solve the problem (10.2) of finding O_t knowing O_T. We shall see two main approaches to solve this problem. One is based on obtaining a PDE for f whenever O_t can be expressed as a function of t and S_t. The other method is based on *risk-neutral pricing*.

10.4. The Black-Scholes Model

We begin by solving problem (10.2) for the Black-Scholes model of Example 10.2 in terms of a PDE. We shall need the extra assumption that the value of the option at time t is a function of t and of the value S_t of the underlying asset at time t (sometimes called the *spot price*). This assumption will be motivated in Section 10.6, Example 10.31.

Theorem 10.12. *Consider a derivative* $(O_t, t \in [0, T])$ *in the Black-Scholes model. Suppose that there is a function* $f \in \mathcal{C}^{1,2}([0, T] \times \mathbb{R})$ *such that* $O_t = f(t, S_t)$. *Then we must have that* $f(t, x)$ *satisfies the Black-Scholes PDE:*

(10.9) $$\frac{\partial f}{\partial t} + \frac{1}{2}\sigma^2 x^2 \frac{\partial^2 f}{\partial x^2} + rx\frac{\partial f}{\partial x} - rf = 0.$$

Proof. We may apply Itô's formula of Theorem 7.8 as in Example 7.12 with $\mathrm{d}S_t = \sigma S_t\,\mathrm{d}B_t + (\mu + \frac{1}{2}\sigma^2)S_t\,\mathrm{d}t$ (Example 7.4). This gives on one hand

$$\mathrm{d}O_t = \mathrm{d}f(t, S_t) = \partial_0 f(t, S_t)\,\mathrm{d}t + \partial_1 f(t, S_t)\,\mathrm{d}S_t + \frac{\sigma^2 S_t^2}{2}\partial_1^2 f(t, S_t)\,\mathrm{d}t.$$

On the other hand, this equation must correspond to equation (10.8). By equating the $\mathrm{d}S_t$-terms and the $\mathrm{d}t$-terms, we find the two equations
(10.10)
$$\Delta_t = \partial_1 f(t, S_t),$$

$$r(f(t, S_t) - \Delta_t S_t) = r(f(t, S_t) - \partial_1 f(t, S_t)S_t) = \partial_0 f(t, S_t) + \frac{\sigma^2 S_t^2}{2}\partial_1^2 f(t, S_t).$$

The PDE is obtained by replacing S_t by the space variable x in the second equation. Note that it is possible to get the same PDE using Proposition 5.28 instead of Theorem

7.8. To see this, it suffices to write $f(t, S_t)$ as $f(t, S_0 e^{\sigma B_t + \mu t})$. In particular, we get an explicit function of time and Brownian motion: $f(t, S_t) = g(t, B_t)$ where $g(t, x) = f(t, S_0 e^{\sigma x + \mu t})$. Proposition 5.28 is applied to $g(t, B_t)$ to recover the PDE.   □

We make some important remarks.

(i) The PDE (10.9) is linear. In other words, if f_1 and f_2 are solutions of the PDE, then $cf_1 + df_2$ for $c, d \in \mathbb{R}$ is also a solution. This makes sense if we think of $cf_1 + df_2$ as the value of a portfolio of options.

(ii) The drift μ has disappeared. There are only two parameters of the model left: r and the volatility σ. In practice, r is known but σ is not. However, since many types of options are traded, the Black-Scholes formula is often used in reverse. Knowing the price, we can extract the volatility σ using the formula. This is called the *implied volatility*. We will say more on this in Section 10.5.

Of course, as is often the case for differential equations, there are many solutions to the PDE unless we specify some initial or terminal conditions. In our case, these conditions are determined by the payoff at expiration which depends on the option. If the payoff of the option is an explicit function of S_T, say $O_T = g(S_T)$, then solving problem (10.2) is equivalent to *solving the following PDE problem with terminal value*: we have $O_t = f(t, S_t)$ where f is the solution of

$$(10.11) \quad \frac{\partial f}{\partial t}(t, x) + \frac{1}{2}\sigma^2 x^2 \frac{\partial^2 f}{\partial x^2}(t, x) + rx\frac{\partial f}{\partial x}(t, x) - rf(t, x) = 0, \qquad f(T, x) = g(x).$$

This PDE has the following explicit solution:

Proposition 10.13. *The solution of the Black-Scholes PDE with terminal condition* $f(T, x) = g(x)$ *as in equation* (10.11) *is given by*

$$(10.12) \qquad f(t, x) = e^{-r(T-t)} \int_{-\infty}^{\infty} g\Big(x e^{\sigma\sqrt{T-t}z + (r - \frac{\sigma^2}{2})(T-t)}\Big) \frac{e^{-z^2/2}}{\sqrt{2\pi}}\, dz.$$

In other words, if Z is a standard Gaussian random variable, then

$$(10.13) \qquad\qquad f(t, x) = e^{-r(T-t)} \mathbf{E}\Big[g\big(x e^{\sigma\sqrt{T-t}Z + (r - \frac{\sigma^2}{2})(T-t)}\big)\Big].$$

Proof. This is a bit involved, but it is greatly facilitated by using *Gaussian integration by parts*; see Exercise 10.4.   □

Note that the solution of the PDE is linear in g. This reflects the linearity of the Black-Scholes PDE. Indeed, if we consider a portfolio of options with terminal condition $cg_1(S_T) + dg_2(S_T)$, then the price of the portfolio at time t is $cf_1(t, S_t) + df_2(t, S_t)$ where f_1 and f_2 are the expectations for g_1 and g_2, respectively. Let's apply the solution to some specific examples.

Example 10.14 (Price of a forward). In this case, the terminal condition is $g(x) = x - K$. In this case, it is easy to check that the solution reduces to

$$f(t, x) = e^{-r(T-t)} \mathbf{E} \left[x e^{\sigma\sqrt{T-t}Z + (r - \frac{\sigma^2}{2})(T-t)} - K \right]$$

$$= e^{-r(T-t)} \left(x e^{\frac{\sigma^2}{2}(T-t) + (r - \frac{\sigma^2}{2})(T-t)} - K \right)$$

$$= x - K e^{-r(T-t)},$$

where we used the MGF of a standard Gaussian in the second equality. We recover the price $f(t, S_t) = S_t - K e^{-r(T-t)}$ computed in Example 10.10.

Example 10.15 (Price of a European call). In this case, the payoff function is $g(x) = (x - K)^+$. From equation (10.13), we get that

$$(10.14) \qquad f(t, x) = e^{-r(T-t)} \mathbf{E} \left[\left(x e^{\sigma\sqrt{T-t}Z + (r - \frac{\sigma^2}{2}(T-t))} - K \right)^+ \right].$$

Of course, this can be numerically evaluated. However, it is good to rearrange the formula a bit more to obtain the explicit replicating portfolio. The idea is to get rid of the positive part $(\cdot)^+$ by introducing the event where the quantity inside is positive:

$$\left\{ x e^{\sigma\sqrt{T-t}Z + (r - \frac{\sigma^2}{2}(T-t))} > K \right\} = \left\{ -Z < d_- \right\},$$

where we solved for $-Z$ with

$$(10.15) \qquad d_- = \frac{1}{\sigma\sqrt{T-t}} \left(\log \frac{x}{K} + \left(r - \frac{\sigma^2}{2} \right)(T - t) \right).$$

The expectation in (10.14) can be restricted on this event, since the integrand is 0 on the complement. Therefore, we have that the expectation is

$$(10.16) \qquad \mathbf{E} \left[\left(x e^{\sigma\sqrt{T-t}Z + (r - \frac{\sigma^2}{2})(T-t)} - K \right) \mathbf{1}_{\{-Z < d_-\}} \right]$$

$$= x e^{(r - \frac{\sigma^2}{2})(T-t)} \mathbf{E} \left[e^{\sigma\sqrt{T-t}Z} \mathbf{1}_{\{-Z < d_-\}} \right] - K \mathbf{P}(-Z < d_-).$$

The second term is $-K N(d_-)$ where

$$(10.17) \qquad N(x) = \int_{-\infty}^{x} \frac{e^{-z^2}}{\sqrt{2\pi}} \, dz$$

is the CDF of a standard Gaussian. Here we used the fact that $-Z$ has the same distribution as Z. It remains to evaluate

$$\mathbf{E} \left[e^{\sigma\sqrt{T-t}Z} \mathbf{1}_{\{-Z < d_-\}} \right] = \int_{-\infty}^{d_-} \frac{e^{-\sigma\sqrt{T-t}z - \frac{z^2}{2}}}{\sqrt{2\pi}} \, dz = e^{\frac{\sigma^2}{2}(T-t)} \int_{-\infty}^{d_-} \frac{e^{-\frac{1}{2}(z + \sigma\sqrt{T-t})^2}}{\sqrt{2\pi}} \, dz,$$

by completing the square. The integral is $N(d_+)$ where

$$(10.18) \qquad d_+ = d_- + \sigma\sqrt{T - t} = \frac{1}{\sigma\sqrt{T-t}} \left(\log \frac{x}{K} + \left(r + \frac{\sigma^2}{2} \right)(T - t) \right).$$

If we put all the above together, we obtain the solution to the PDE problem:

$$f(t, x) = x N(d_+) - K e^{-r(T-t)} N(d_-).$$

This is a very compelling expression, since by replacing x by S_t, we get the price of the call in the simple form:

(10.19)
$$\boxed{f(t, S_t) = C_t = S_t N(d_+) - Ke^{-r(T-t)}N(d_-).}$$

Here, $d_\pm$ are evaluated at $x = S_t$ also. It is important to keep in mind that S_t (the current market price or *spot price*) is known at time t. Therefore, it is a deterministic parameter that can be used to compute the price of the option at time t. Thus, the replicating portfolio (10.4) of a European call in the Black-Scholes model has the allocation

$$\Delta_t = N(d_+), \quad a_t = -Ke^{-r(T-t)}N(d_-).$$

In other words, it consists of being long $N(d_+)$ shares of the risky asset at time t and short $KN(d_-)e^{-r(T-t)}$ dollars at time t.

It is not hard to check that the price of a put with strike price K is

(10.20)
$$P_t = -S_t N(-d_+) + Ke^{-r(T-t)}N(-d_-).$$

See Exercise 10.6. Thus, the replicating portfolio (10.4) of a European put in the Black-Scholes model has the allocation

$$\Delta_t = -N(-d_+), \quad a_t = Ke^{-r(T-t)}N(-d_-).$$

The reader is invited to write a calculator for the price of a call and a put using these formulas in Numerical Project 10.1.

Example 10.16 (A numerical example). Let's get some numbers in to make the formula more tangible. Consider a Black-Scholes model where the risk-free rate is $r = 0.05$, the volatility is $\sigma = 0.1$, and $S_0 = \$500$. Let's find the price of a European call with strike price $K = \$480$ and expiration $T = 1$. A quick computation shows that $d_+ = 0.96\ldots, d_- = 0.86\ldots$. Therefore, equation (10.19) becomes [1]

$$C_0 = 500\, N(0.96) - 480e^{-0.05}N(0.86) \approx 48.13.$$

The price of a put with same strike price is by put-call parity

$$P_0 = C_0 - S_0 + 480e^{-0.05} \approx 48.13 - 500 + 480e^{-0.05} = 4.72\ldots.$$

Example 10.17 (Demystifying the d's). Let's take a closer look at $d_\pm$ entering into the Black-Scholes formula:

(10.21)
$$d_\pm = d_\pm(t, S_t, r, \sigma, T, K) = \frac{1}{\sigma\sqrt{T-t}}\left(\log\frac{S_t}{K} + \left(r \pm \frac{\sigma^2}{2}\right)(T-t)\right).$$

The d's are explicit functions of t, S_t, r, σ, T, K. It is instructive to look at their value when t is close to expiration; i.e., $t \approx T$. In this case, the last term is negligible. In particular, d_+ and d_- are almost equal and we have

$$d_\pm \approx \frac{1}{\sigma\sqrt{T-t}}\log\frac{S_t}{K}.$$

If the *spot price* is smaller than the strike price, that is, $S_t < K$, we say that the call option is *out of the money* (OTM). In that case, the logarithm is negative. Since $\sqrt{T-t}$ is close to 0, we must have that d_+ and d_- are both very large in the negative. Going

---

[1] To evaluate the CDF of a standard normal random variable you can import the command norm from scipy.stats in Python. Then, $N(0.96)$ is obtained by norm.cdf(0.96). The PDF N' at the same point is norm.pdf(0.96).

back to the Black-Scholes formula for a call (10.19), remembering that $N(-\infty)$ is 0, we conclude that OTM calls close to expiration are such that $C_t \approx 0$. This is not at all surprising since it is very likely that these calls expire worthless.

If at time t we have $S_t > K$, the call option is said to be *in the money* (ITM). In that case, the logarithm is positive. Therefore, $d_\pm$ are large positive numbers. We conclude from equation (10.19) that $C_t \approx S_t - K$. Again this is not at all surprising since this is close to the value of the payoff at expiration. Finally, a call option at time t with $S_t \approx K$ is said to be *at the money* (ATM). This is an interesting case: The logarithm is approximately 0, and since $N(0) = 1/2$, we have that

$$C_t \approx \frac{1}{2}(S_t - K).$$

In particular, the delta of the option is 1/2. We will discuss delta and the Greeks in Section 10.5. The reader is strongly encouraged to check that fact in real options data:

ATM call options close to expiration have a delta of 1/2.

Example 10.18 (Asset-or-nothing options). In the Black-Scholes model, consider an option with payoff at expiration

$$O_T = \begin{cases} S_T & \text{if } S_T > K, \\ 0 & \text{if } S_T \leq K, \end{cases}$$

where $(S_t, t \leq T)$ is the price of the underlying asset. In words, the holder of this option receives the asset if S_T is larger than the strike price and nothing if not. Let's find the price for this option. From equation (10.13), we get

$$O_t = e^{-r(T-t)}\mathbf{E}[g(S_t e^{\sigma\sqrt{T-t}Z+(r-\sigma^2/2)(T-t)})],$$

where $g(x) = x$ if $x > K$ and 0 otherwise. (Here the expectation is only on Z. The spot price S_t is known.) But this can be written using the same event $\{-Z < d_-\}$ as in equation (10.16). This gives

$$O_t = S_t e^{-r(T-t)}E[e^{\sigma\sqrt{T-t}Z+(r-\sigma^2/2)(T-t)}\mathbf{1}_{\{-Z<d_-\}}].$$

Again, this is exactly as the first term of (10.16). This leads not surprisingly to the simple answer

$$O_t = S_t N(d_+).$$

Note that we could define the complementary option

$$U_T = \begin{cases} 0 & \text{if } S_T > K, \\ S_T & \text{if } S_T \leq K. \end{cases}$$

We then have the relation $U_T + O_T = S_T$ between the payoffs of these two options. This leads to another *put-call parity* for this type of option:

$$U_t + O_t = S_t, \quad t \leq T.$$

Let's run some numbers again to get a feel for the price. Using the same setup as in Example 10.16 with $S_0 = 500$, $K = 480$ yields

$$O_0 = 500N(0.96) \approx 500 \cdot 0.83 = 415,$$

so that $U_0 \approx 85$.

10.5. The Greeks

In view of equations (10.19) and (10.20), it is clear that the price of a derivative depends explicitly on the parameters t, S_t, r, σ, as well as the defining parameters T, K. It is important to understand how the price of the options varies as these parameters vary. This is the role of the *Greeks*. They are the partial derivatives of the price of the options at a given time t. If we include the dependence on these parameters in the notation, we have that the price O_t of an option at time t and expiration T is

$$O_t = O_t(S_t, r, \sigma, K, T).$$

(The parameter t already appears in the notation as an index.) The name *Greeks* usually refers to the partial derivatives with respect to S_t, t, r, and σ. These are the standard ones:

- *Delta*. This is the derivative of O_t with respect to the price of the underlying asset at time t:

(10.22) $$\Delta_t = \frac{\partial O_t}{\partial S_t}.$$

  This explains the notation Δ_t in the replication portfolio (10.4). This derivative is exactly the number of units of risky assets needed to replicate the derivative.

- *Gamma*. This is the second-order derivative with respect to the spot price:

(10.23) $$\Gamma_t = \frac{\partial^2 O_t}{\partial S_t^2}.$$

  Delta and Gamma can be used together to build a portfolio that is *delta-neutral* and *gamma-neutral*. Such a portfolio will be less sensitive to the variation of the price of the underlying asset; see Example 10.21.

- *Theta*. This is the variation with respect to time:

(10.24) $$\Theta_t = \frac{\partial O_t}{\partial t}.$$

- *Rho*. This is the variation with respect to the interest rate:

(10.25) $$\rho_t = \frac{\partial O_t}{\partial r}.$$

- *Vega* (not a Greek letter!). This is the variation with respect to the volatility:

(10.26) $$\text{vega}_t = \frac{\partial O_t}{\partial \sigma}.$$

Keep in mind that the Greeks are themselves functions of the underlying parameters. There are other partial derivatives with fancy names including: *Vanna, Charm, Speed, Vomma, Veta, Zomma, Ultima, Charm, Vera , Color.*

Example 10.19 (Greeks in Black-Scholes). Let's compute some Greeks for a European call in the Black-Scholes model. Equation (10.19) gives that the price at time t of such an option is

$$C_t = S_t N(d_+) - K e^{-r(T-t)} N(d_-).$$

Now it is very tempting to conclude right away that $\frac{\partial C_t}{\partial S_t} = N(d_+)$! We know that this is the number of units needed in the risky asset to replicate the call. The answer is correct, but the reasoning is wrong as d_+ and d_- depend on the spot price S_t; see equations (10.18) and (10.15) for $x = S_t$. There is a bit of magic happening here, since we have the identity:

$$(10.27) \qquad S_t N'(d_+) = K e^{-r(T-t)} N'(d_-).$$

To see this, note that, since the derivative of the CDF of N is the PDF of a standard Gaussian, we have

$$N'(d_+) = \frac{e^{-(d_+)^2/2}}{\sqrt{2\pi}} = \frac{e^{-(d_- + \sigma\sqrt{T-t})^2/2}}{\sqrt{2\pi}} = \frac{K}{S_t} e^{-r(T-t)} \frac{e^{-d_-^2/2}}{\sqrt{2\pi}},$$

by simply developing the square. Therefore, the delta of the option is, by differentiating all terms and using the identity (10.27),

$$\frac{\partial C_t}{\partial S_t} = N(d_+) + S_t N'(d_+) \cdot \frac{\partial d_+}{\partial S_t} - K e^{-r(T-t)} N'(d_-) \cdot \frac{\partial d_-}{\partial S_t} = N(d_+),$$

where we also use $\frac{\partial d_+}{\partial S_t} = \frac{\partial d_-}{\partial S_t}$.

Applying the derivative again yields Γ. Similar computations give the other Greeks for the call:

$$\Delta_t = N(d_+),$$

$$\Gamma_t = \frac{1}{S_t \sigma \sqrt{T-t}} \frac{1}{\sqrt{2\pi}} e^{-d_+^2/2},$$

$$(10.28) \qquad \Theta_t = \frac{S_t \sigma}{2\sqrt{2\pi(T-t)}} e^{-d_+^2/2} - r K e^{-r(T-t)} N(d_-),$$

$$\rho_t = K(T-t) e^{-r(T-t)} N(d_-),$$

$$\text{vega}_t = S_t \sqrt{T-t} \frac{1}{\sqrt{2\pi}} e^{-d_+^2/2}.$$

The reader is invited to verify these in Exercise 10.5 and to extend them to puts in Exercise 10.6.

Example 10.20 (Numerical examples). Let's go back to Examples 10.16 and 10.18. The delta and gamma are easily evaluated:

$$\Delta_0 = N(0.96) \approx 0.83, \qquad \Gamma_0 = \frac{1}{500 \cdot 0.1} \frac{1}{\sqrt{2\pi}} e^{-0.96^2/2} \approx 0.005.$$

For the delta and gamma of the asset-or-nothing option, denoted by $\widetilde{\Delta}_0$ and $\widetilde{\Gamma}_0$, it suffices to take the derivative with respect to S_0 of $O_0 = S_0 N(d_+)$:

$$\widetilde{\Delta}_0 = N(d_+) + \frac{e^{-d_+^2/2}}{\sigma\sqrt{2\pi T}} = 3.35\ldots,$$

$$\widetilde{\Gamma}_0 = \frac{e^{-d_+^2/2}}{S_0 \sigma\sqrt{2\pi T}} - \frac{e^{-d_+^2/2}}{\sigma\sqrt{2\pi T}} \cdot d_+ \cdot \frac{\partial d_+}{\partial S_0} = -\frac{e^{-d_+^2/2}}{S_0^2 T \sqrt{2\pi}} \cdot d_- = -0.043\ldots.$$

What are the Greeks good for? It turns out that most trading platforms offer the values of these parameters in real time. They are valuable instruments to build portfolios that are more robust to fluctuations of the underlying parameters. The goal is to *hedge* against a sudden variation of these parameters. Let's work out an example.

Example 10.21 (Delta hedging). Consider the setup of Example 10.16. We are given the following table for the values of the delta and gamma of the call at a given time:

Options	Price	Δ	Γ
Call	48.13	0.831	0.005

Recall that the spot price is $S_0 = \$500$. Suppose we currently hold a portfolio $V^{(1)}$ consisting of 100 short calls and the cash from the sale of these calls. This means that the value of the portfolio $V_0^{(1)} = 0$. It is not hard to find the amount of cash in the risk-free asset. Since the allocation in the call is -100, the value of the short calls is $\$(-4{,}813)$. Therefore we have $\$4{,}813$ in cash.

First, we build a portfolio $V^{(2)}$ that has the same number of calls but that is *delta-neutral*. This means it will be less sensitive to the variation of the price of the underlying asset. For this, we need to buy the underlying asset, but how much of it? We suppose that $V_0^{(2)} = 0$, so that the portfolio is self-financing. This gives us a first equation for the portfolio allocation:

$$0 = V_0^{(2)} = xS_0 + y + (-100)C_0 = 500x + y - 4{,}813$$

It remains to find the allocation x, y in the asset and in cash. The second equation is given by the delta neutrality condition $\frac{\partial V_0^{(2)}}{\partial S_0} = 0$. By taking the derivative with respect to S_0 of the portfolio value we get

$$x + (-100)\Delta = 0.$$

Therefore, $x = 83.1$. Since $S_0 = 500$, the equation $V_0^{(2)} = 0$ finally gives $y = -36{,}737$.

Example 10.22 (Gamma hedging). Let's now build a portfolio that is gamma neutral, still contains 100 short calls, and for which $V_0^{(3)} = 0$. The gamma neutrality condition gives a third equation: $\frac{\partial V_0^{(3)}}{\partial S_0} = 0$. It is even less sensitive to variation of the spot price than the delta-hedged portfolio. We now have three equations; therefore we need a third unknown to pin down a solution. In finance terms, this means we need a third asset in our portfolio. Let's use the asset-or-nothing call as given in Example 10.18. The parameters were already computed:

Options	Price	$\widetilde{\Delta}$	$\widetilde{\Gamma}$
Asset-or-Nothing	415.51	3.35	-0.043

We need to determine the allocation (x, y, w) in the asset, in cash, and in the asset-or-nothing call. The three equations of gamma hedging are now

$$xS_0 + y + (-100)C_0 + wO_0 = 0,$$
$$x + (-100)\Delta + w\widetilde{\Delta} = 0,$$
$$(-100)\Gamma + w\widetilde{\Gamma} = 0.$$

We conclude that the allocations are respectively

$$x = 122, \quad y = -51,382, \quad \text{and} \quad w = 11.6.$$

Although they are not Greeks per se, the derivatives with respect to the strike price also play an important role in the theory.

Example 10.23 (Derivatives with respect to strike price). In the Black-Scholes model framework, we can compute $\frac{\partial C_t}{\partial K}$, $\frac{\partial^2 C_t}{\partial K^2}$, and the analogues for puts. We get from equation (10.19)

$$\frac{\partial C_t}{\partial K} = S_t N'(d_+) - Ke^{-r(T-t)}N'(d_+) - e^{-r(T-t)}N(d_+) = -e^{-r(T-t)}N(d_+),$$

where we used the identity (10.27) again. In particular, we have $\frac{\partial C_t}{\partial K} < 0$. In other words, C_t is a *decreasing function* of the strike price. This was already observed in Example 10.11. If we take one more derivative, we have

$$\frac{\partial^2 C_t}{\partial K^2} = -e^{-r(T-t)}N'(d_+) \cdot \frac{\partial d_+}{\partial S_0} = e^{-r(T-t)}d_+ \frac{1}{\sigma\sqrt{2\pi(T-t)}} > 0.$$

This implies that C_t is a *convex function of the strike price*. These facts are true not only in the Black-Scholes model, but anytime the no-arbitrage assumption holds.

K	330	335	340	345	350	355	360
Calls	32.80	30.50	27.60	25.30	23.10	21.10	19.30
Puts	21.00	23.30	25.70	28.20	31.20	34.40	37.10

Figure 10.2. The option chain for Call/Puts of Zoom on March 4, 2021, with $S_0 = 342.67$ with expiration on April 9, 2021. Note that the price of calls is decreasing in K and the price of puts is increasing in K. Are the prices convex in K?

Source: etrade.com.

Example 10.24 (Vega and the implied volatility). We saw in equation (10.28) that the vega of a call option in the Black-Scholes model is

(10.29) $$\text{vega}_t = \frac{\partial C_t}{\partial \sigma} = S_t\sqrt{T-t}\frac{1}{\sqrt{2\pi}}e^{-d_+^2/2} > 0.$$

This means that C_t as a function of σ is a strictly increasing function of σ. This fact has a very important application. In practice, the options are traded so the price of a call with given expiration T and strike price K is set by the market. Let's denote it by C_{market}. Now, we can compute the *implied volatility*, that is, the value σ that yields C_{market} when plugged into the Black-Scholes expression $C_0(S_0, r, \sigma, K, T)$. In other words, we can solve for σ in the equation

(10.30) $$\boxed{C_0(S_0, r, \sigma, K, T) = C_{\text{market}}.}$$

There is a single solution σ_{imp} to this equation by equation (10.29). Interestingly, if the Black-Scholes model was entirely correct, we should get the *same* implied volatility σ for all the options derived from a risky asset, irrespective of expiration and strike price.

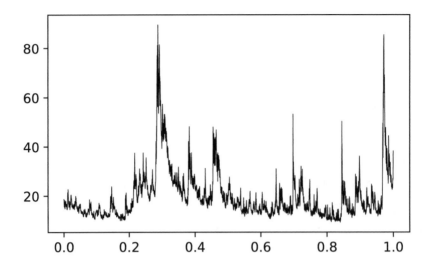

Figure 10.3. The daily high of the VIX volatility index from January 2004 to September 2020 normalized on $[0, 1]$. Can you guess where 2008 and March 2020 are located? Source: `https://datahub.io/core/finance-vix#readme`

In reality, this is not what is observed. In fact, the implied volatility σ_{imp} is a function of both T and K. The function $\sigma_{\text{imp}}(T, K)$ is called *the volatility surface.*

The implied volatilities of the different options on the market can give us a snapshot of the market as a whole. This is the idea behind the *VIX volatility index* that tracts the implied volatilities of many options in real time. See Figure 10.3.

10.6. Risk-Neutral Pricing

The no-arbitrage and replication assumptions of Section 10.3 can be conveniently recast in a probabilistic framework known as the *risk-neutral probability* [2]. The idea is in the spirit of Chapter 9. As a motivation, consider a simple one-step model for the pricing of an option.

Example 10.25 (Pricing in a one-step model). Suppose that $r = 0$ and the price of the asset at time 0 is $S_0 = 15$. At time 1, it is either $S_1 = 20$ with probability p or $S_1 = 10$ with probability $1 - p$. A call option with strike price $K = 12$ in this model has value $C_1 = 8$ or $C_1 = 0$ at expiration, depending on the outcome. The replicating portfolio of the option is $C_1 = \Delta S_1 + a$. The two outcomes yield two equations:

$$8 = 20\Delta + a,$$
$$0 = 10\Delta + a.$$

A quick calculation gives the allocation $\Delta = 0.8$ and $a = -8$. In particular, we have that the price of the call is $C_0 = \Delta S_0 + a = 4$.

An equivalent way to price this option is to use the *risk-neutral probability*. This is the probability for which *the price of the asset is constant on average*. Write $\tilde{p}$ for the

---

[2]A *risk-neutral probability* is also called *equivalent martingale measure*.

risk-neutral probability. We must have

$$15 = S_0 = \widetilde{\mathbf{E}}[S_1] = 20\tilde{p} + 10(1 - \tilde{p}) \Longrightarrow \tilde{p} = 1/2.$$

The expected value of the call turns out to be the average of the payoff over this probability

$$\widetilde{\mathbf{E}}[C_1] = 8\tilde{p} + 0(1 - \tilde{p}) = 4 = C_0,$$

the same as the price obtained by replication. In other words, *pricing using replication is equivalent to pricing under the risk-neutral probability.*

The idea of risk-neutral valuation remains the same for more complicated models of the asset price $(S_t, t \geq 0)$:

> *The risk-neutral probability is the probability for which the process $(S_t, t \geq 0)$ (after proper discounting) remains constant on average, that is, for which the process is a* **martingale***!*

In Example 10.25, we changed the probability p on the outcomes of S_1, but we kept the same possible outcomes for S_1. In other words, the probability was *equivalent* to the original one in the sense of Definition 9.4. For general market models as in Definition 10.1, the risk-neutral probability plays the same role.

Definition 10.26. A risk-neutral probability $\widetilde{\mathbf{P}}$ for the market model is a probability on $(\Omega, \mathcal{F})$ such that the following hold:

- $\widetilde{\mathbf{P}}$ is equivalent to $\mathbf{P}$ in the sense of Definition 9.4. Intuitively, this means that the paths of the processes under $\widetilde{\mathbf{P}}$ are the same paths as the ones under $\mathbf{P}$. In other words, $\widetilde{\mathbf{P}}$ simply changes the weight or the probability of each path.
- For D_t as in equation (10.1), the discounted process $(\tilde{S}_t, t \leq T)$, where

$$\tilde{S}_t = D_t S_t,$$

is a martingale for $\widetilde{\mathbf{P}}$ with respect to the filtration $(\mathcal{F}_t, t \leq T)$. We will often say for short that $(\tilde{S}_t, t \leq T)$ is a $\widetilde{\mathbf{P}}$-martingale. In particular, for $t \leq T$,

$$\tilde{S}_t = \widetilde{\mathbf{E}}[\tilde{S}_T | \mathcal{F}_t],$$

where $\widetilde{\mathbf{E}}$ stands for the expectation under $\widetilde{\mathbf{P}}$.

When do we know if a risk-neutral probability exists for a given market model? This is not a simple question in general. When the risky asset is modelled by an Itô process, this can be handled using the different version of the Cameron-Martin-Girsanov theorem given in Chapter 9. As we shall see in examples, the good news is that once the existence of a risk-neutral probability is assumed or known, then the pricing of options is fairly straightforward. As a start, we show that the risk-neutral probability exists in the Black-Scholes model.

Example 10.27 (Risk-neutral probability for the Black-Scholes model)**.** For the Black-Scholes model in Example 10.2, we have that the discounted process is $(\tilde{S}_t, t \geq 0)$ where

$$\tilde{S}_t = e^{-rt} S_t = S_0 \exp(\sigma B_t + (\mu - r)t).$$

Itô's formula in Proposition 5.28 applied to the function $f(t, x) = S_0 \exp(\sigma x + (\mu - r)t)$ gives

$$d\tilde{S}_t = \sigma \tilde{S}_t \, dB_t + (\mu - r + \sigma^2/2) \tilde{S}_t \, dt.$$

Here is a simple but important step. Since $\sigma > 0$, we can write

$$d\tilde{S}_t = \sigma \tilde{S}_t \Big(dB_t + \frac{(\mu - r + \sigma^2/2)}{\sigma} \, dt \Big).$$

Now, if we define the Brownian motion with drift $\widetilde{B}_t = B_t + \theta t$ where $\theta = \frac{1}{\sigma}(\mu - r + \sigma^2/2)$, we have the SDE

$$(10.31) \qquad\qquad d\tilde{S}_t = \sigma \tilde{S}_t \, d\widetilde{B}_t, \quad \tilde{S}_0 = S_0.$$

Therefore, $(\tilde{S}_t, t \geq 0)$ would be a $\widetilde{\mathbf{P}}$-martingale if $\widetilde{\mathbf{P}}$ turns the Brownian motion with drift $(\widetilde{B}_t, t \geq 0)$ into a standard Brownian motion. The Cameron-Martin theorem, Theorem 9.5, does exactly that on $[0, T]$. Therefore, the risk-neutral probability for the Black-Scholes model on $[0, T]$ is the one obtained from $\widetilde{\mathbf{P}}$ with the factor

$$(10.32) \qquad\qquad e^{-\theta B_T - \theta^2 T/2}, \quad \theta = \frac{1}{\sigma}(\mu - r + \sigma^2/2).$$

Interestingly, in the case $\mu - r + \sigma^2/2 = 0$, the risk-neutral probability is the same as the original probability on the model.

Under $\widetilde{\mathbf{P}}$, the discounted process $(\tilde{S}_t, t \leq T)$ satisfies the SDE (10.31). In other words, it is a martingale geometric Brownian motion

$$(10.33) \qquad\qquad \tilde{S}_t = S_0 \exp(\sigma \widetilde{B}_t - \sigma^2 t/2), \quad t \leq T.$$

Note that the drift μ has disappeared from the equations, as it did in deriving the Black-Scholes PDE in Theorem 10.12. The form of the process $(S_t, t \leq T)$ under $\widetilde{\mathbf{P}}$ is simply

$$(10.34) \qquad S_t = e^{rt} \tilde{S}_t = S_0 \exp\Big(\sigma \widetilde{B}_t + (r - \frac{\sigma^2}{2})t\Big), \quad t \leq T,$$

and in differential form

$$(10.35) \qquad\qquad dS_t = \sigma S_t \, d\widetilde{B}_t + r S_t \, dt.$$

Suppose a market model of the form of Definition 10.1 admits a risk-neutral probability $\widetilde{\mathbf{P}}$. How can this be used in general to price a derivative $(O_t, t \leq T)$? First, by Itô's product formula in equation (7.16), the discounted process $\tilde{S}_t = D_t S_t$ satisfies the SDE

$$d\tilde{S}_t = D_t \, dS_t - R_t \tilde{S}_t \, dt.$$

Because of the valuation equation (10.8), we also have

$$dO_t = \Delta_t \, dS_t + (O_t - \Delta_t S_t)R_t \, dt,$$

and by the product rule (7.16),

$$d\widetilde{O}_t = D_t \, dO_t + O_t \, dD_t = D_t \, dO_t - D_t O_t R_t \, dt.$$

The three above equations lead us to the elegant relation

$$(10.36) \qquad\qquad d\widetilde{O}_t = \Delta_t \, d\tilde{S}_t.$$

In other words, we showed that any discounted derivative process $(\widetilde{O}_t, t \leq T)$ satisfying the valuation equation is a $\widetilde{\mathbf{P}}$-martingale (assuming Δ_t is well-behaved) under the market filtration $(\mathcal{F}_t, t \le T)$, much like the discounted process of the risky asset. This has an important implication: the martingale property of $(\widetilde{O}_t, t \le T)$ yields the *pricing formula*

(10.37)
$$\boxed{\widetilde{O}_t = \widetilde{\mathbf{E}}[\widetilde{O}_T | \mathcal{F}_t].}$$

In words, the pricing formula asserts the following:

> *The price O_t of a derivative at time t is simply the expected payoff (after proper discounting) under the risk-neutral probability given the information up to time t.*

This is a natural valuation equation. Conversely, if instead of assuming first equation (10.8) and then deriving (10.37) we start by assuming (10.37), then the process $(\widetilde{O}_t, t \le T)$ would automatically be a martingale by Exercise 4.16.

The pricing formula seems harmless, but its power is illustrated by the examples below. As a first example, we observe that the put-call parity in Example 10.10 is a simple consequence of the linearity of expectation.

Example 10.28 (Put-call parity and risk-neutral pricing). Recall from Example 10.7 that the payoff functions of a European call and of a European put satisfy

$$C_T - P_T = S_T - K.$$

We suppose for simplicity that the interest rate $R_t = r$ is constant. By the pricing formula (10.37), we have by linearity of conditional expectation and the fact that $(\tilde{S}_t, t \le T)$ is $\widetilde{\mathbf{P}}$-martingale for the $(\mathcal{F}_t, t \le T)$-filtration

$$C_t - P_t = e^{rt}\widetilde{\mathbf{E}}[D_T(C_T - P_T)|\mathcal{F}_t] = e^{rt}\widetilde{\mathbf{E}}[\tilde{S}_T - e^{-rT}K|\mathcal{F}_t] = S_t - e^{-r(T-t)}K.$$

This is the same relation obtained in Example 10.10.

Example 10.29 (Convexity of calls and puts with respect to strike price). In Example 10.23, we showed that C_t and P_t are convex functions of the strike price K. They are also decreasing and increasing functions of K, respectively. This was done by computing the first and second derivatives of the prices in the Black-Scholes model. This is a general property that holds for all models, as we first observed in Example 10.11. It is easy to see why this holds using the pricing formula (10.37). Indeed, for a call option, we have

$$C_t = D_t^{-1}\widetilde{\mathbf{E}}[D_T(S_T - K)^+|\mathcal{F}_t].$$

The payoff $(S_T - K)^+$ is obviously a decreasing convex function of K (draw the graph!). But C_t is simply an averaging of that payoff. Since a sum of decreasing convex functions must also be decreasing and convex, we conclude that C_t inherits these properties. The same analysis holds for put options: P_t is an increasing convex function of K.

The prices of derivatives in the Black-Scholes model were first computed using the solution to the Black-Scholes PDE in Proposition 10.13. We show how these results can be recovered directly using the pricing formula. We know from Example 10.27 that the Black-Scholes model admits a risk-neutral probability whenever $\sigma > 0$, so equation (10.37) can be used.

Example 10.30 (Pricing formula in the Black-Scholes model). Recall that the $\widetilde{\mathbf{P}}$-distribution of $(\tilde{S}_t, t \leq T)$ in (10.31) is the one of a martingale geometric Brownian motion. The price of a forward was already computed in Example 10.28. More generally, we can treat the case where the payoff is $O_T = g(S_T)$ for some function g. We will recover equation (10.12) of Proposition 10.13. The pricing formula gives

$$O_t = e^{-r(T-t)}\widetilde{\mathbf{E}}[g(S_T)|\mathcal{F}_t].$$

By definition of the martingale geometric Brownian motion, we have

$$\tilde{S}_T = \tilde{S}_t \exp(\sigma(\widetilde{B}_T - \widetilde{B}_t) - \sigma^2(T-t)/2).$$

This implies

$$O_t = e^{-r(T-t)}\widetilde{\mathbf{E}}\left[g\Big(S_t e^{\sigma(\widetilde{B}_T - \widetilde{B}_t) + (r-\sigma^2/2)(T-t)}\Big)\Big|\mathcal{F}_t\right].$$

Now S_t is $\mathcal{F}_t$-measurable. Moreover, the increment $\widetilde{B}_T - \widetilde{B}_t$ is independent of $\mathcal{F}_t$ and has the distribution of a Gaussian with mean 0 and variance $T - t$. Therefore, the conditional expectation is reduced to the expectation and we get

(10.38) $$O_t = e^{-r(T-t)}\mathbf{E}\left[g\Big(S_t e^{\sigma\sqrt{T-t}Z + (r-\sigma^2/2)(T-t)}\Big)\right],$$

where the expectation is only on Z, a standard Gaussian, and S_t is known, as in equation (10.13).

We should be gradually convinced at this point that the framework of risk-neutral pricing is equivalent to the no-arbitrage pricing. As an additional example, we show that the Black-Scholes PDE can be derived from the pricing formula.

Example 10.31 (Black-Scholes PDE from risk-neutral pricing). By Itô's formula and equation (10.31), we have as usual that

$$dS_t = d(e^{rt}\tilde{S}_t) = \sigma S_t\, d\widetilde{B}_t + rS_t\, dt.$$

Consider an option with payoff $O_T = g(S_T)$. The pricing formula asserts that

(10.39) $$O_t = e^{-r(T-t)}\widetilde{\mathbf{E}}[g(S_T)|\mathcal{F}_t].$$

We know the distribution of $(S_t, t \leq T)$ under $\widetilde{\mathbf{P}}$ in equation (10.34). In particular, it is a Markov process! The Markov property of geometric Brownian motion implies that there is a function $f(t, x)$ of time and space such that

$$f(t, S_t) = e^{-r(T-t)}\widetilde{\mathbf{E}}[g(S_T)|\mathcal{F}_t].$$

We also have by the pricing formula that $(e^{-rt}O_t, t \leq T)$ must be a $\widetilde{\mathbf{P}}$-martingale for the filtration $(\mathcal{F}_t, t \leq T)$. Applying Itô's formula for a function $f(t, x)$ with the Itô process S_t as in Theorem 7.8, we get

$$d(e^{-rt}f(t, S_t)) = \{-rf + \partial_0 f\}e^{-rt}\, dt + e^{-rt}\partial_1 f\, dS_t + \frac{e^{-rt}}{2}\partial_1^2 f \cdot (dS_t)^2$$

$$= (\partial_1 f)e^{-rt}\sigma S_t\, d\widetilde{B}_t + \left\{\partial_0 f + \frac{\sigma^2 S_t^2}{2}\partial_1^2 f + rS_t\partial_1 f - rf\right\}e^{-rt}\, dt.$$

Therefore, since $(\widetilde{B}_t, t \leq T)$ is a $\widetilde{\mathbf{P}}$-Brownian motion, we would have a $\widetilde{\mathbf{P}}$-martingale if the dt-term is set to 0. This gives that $f(t, x)$ must satisfy the PDE

$$\frac{\partial f}{\partial t} + \frac{\sigma^2 x^2}{2}\frac{\partial^2 f}{\partial x^2} + rx\frac{\partial f}{\partial x} - rf = 0,$$

as in Theorem 10.12.

Example 10.32 (Monte-Carlo pricing). The pricing formula (10.37) is also useful to derive prices using numerics. This method is implemented in Numerical Projects 10.3 and 10.4. Indeed, suppose that we want to price an option $(O_t, t \leq T)$. For simplicity, consider the case where $R_t = r$ is constant. Then the pricing formula becomes

$$O_t = e^{-r(T-t)}\widetilde{\mathbf{E}}[O_T|\mathcal{F}_t].$$

In particular, at $t = 0$, we have

$$O_0 = e^{-rT}\widetilde{\mathbf{E}}[O_T].$$

To obtain a numerical approximation, it suffices to sample N paths of the risky asset $(S_t, t \leq T)$ according to the $\widetilde{\mathbf{P}}$-distribution of this process. The price O_0 is then the payoff O_T averaged over these N paths and multiplied by e^{-rT}.

Remark 10.33 (Black-Scholes PDE and the Feynman-Kac formula). It is a beautiful fact that the Black-Scholes PDE and its solution (Theorem 10.12) is a particular instance of the *Feynman-Kac formula* as explained in Theorem 8.28. To see this, recall that from the pricing formula (10.37), we must have

(10.40) $$f(t, S_t) = \widetilde{\mathbf{E}}[e^{-r(T-t)}g(S_T)|\mathcal{F}_t] = \widetilde{\mathbf{E}}[e^{-r(T-t)}g(S_T)|S_t],$$

where the second equality follows from the fact that $(S_t, t \leq T)$ is a Markov process, as can be seen from its SDE (10.35). Looking at the statement of Theorem 8.28, we conclude that $f(t, x)$ must then be the solution to the terminal value problem

$$\frac{\partial f}{\partial t} = Af(t, x) - rxf(t, x).$$

Here, the function $r(x)$ of Theorem 10.12 is simply rx, for a constant r, and the generator A is the one of the diffusion with SDE (10.35); i.e.,

$$A = \frac{\sigma^2 x^2}{2}\frac{\partial^2}{\partial x^2} + rx\frac{\partial}{\partial x},$$

as seen from Example 8.15. In fact, the derivation in Example 10.31 is the proof of Theorem 8.28 given in Exercise 8.11 applied to the particular case of geometric Brownian motion. Note that we worked in reverse here: we use Theorem 8.28 to infer that equation (10.40) must be a solution of the PDE problem.

How does the risk-neutral pricing framework apply to general market models as in Definition 10.1? We now discuss the concepts of risk-neutral probability, no arbitrage, and replication in a general setting.

Let's first look at the existence of a risk-neutral probability. This was done in Example 10.27 for the Black-Scholes model. The risky asset in Definition 10.1 is usually modelled by an Itô process of the form

(10.41) $$dS_t = \sigma_t S_t \, dB_t + \mu_t S_t \, dt, \quad S_0,$$

so that local volatility and the local drift are proportional to S_t. Here $(\sigma_t, t \geq 0)$ and $(\mu_t, t \geq 0)$ are some adapted processes. Mimicking the example for the Black-Scholes model, we use Itô's product formula (7.16) to get

$$d\tilde{S}_t = d(D_t S_t) = -R_t \tilde{S}_t \, dt + \sigma_t \tilde{S}_t \, dB_t + \mu_t \tilde{S}_t \, dt = \tilde{S}_t\Big(\sigma_t \, dB_t + (\mu_t - R_t) \, dt\Big).$$

One way for $(\tilde{S}_t, t \leq T)$ to be a $\tilde{\mathbf{P}}$-martingale would be to construct a probability $\tilde{\mathbf{P}}$ for which the process

$$d\tilde{B}_t = dB_t + \Theta_t \, dt, \quad \Theta_t = \frac{\mu_t - R_t}{\sigma_t},$$

is a standard Brownian motion. This is feasible in general by Girsanov's theorem (Theorem 9.11) whenever $(\Theta_t, t \leq T)$ satisfies the Novikov condition (7.18). This might depend on the specific market model that we are working with. Of course, many things could prevent this. For one, it could be that $\sigma_t = 0$ for some t in which case Θ is not even defined.

Example 10.34 (Bachelier model). The Bachelier model is a market model where the risky asset $(S_t, t \leq T)$ has the SDE

$$(10.42) \qquad\qquad dS_t = rS_t \, dt + \sigma \, d\tilde{B}_t,$$

where $(\tilde{B}_t, t \leq T)$ is a standard Brownian motion under the risk-neutral probability. The risk-free rate here is constant. The above SDE is to be compared with the SDE of the Black-Scholes model in equation (10.35). The major difference is that $(S_t, t \leq T)$ allows for negative values of the risky asset. This is not an outlandish assumption considering that securities and interest rates have been known to take negative values from time to time.

Let's price a call option in this model with equation (10.42) as a starting point. Observe that the SDE (10.42) has the form of an Ornstein-Uhlenbeck SDE (7.11) with solution (7.12). Therefore we have

$$(10.43) \qquad\qquad S_t = e^{rt} \left(S_0 + \int_0^t e^{-rs} \sigma \, d\tilde{B}_s \right).$$

In particular, the distribution of S_t is Gaussian with mean $S_0 e^{rt}$ and variance

$$\frac{\sigma^2}{2r}(e^{2rt} - 1).$$

The price of a European call is given by the pricing formula (10.37)

$$C_t = e^{-r(T-t)} \tilde{\mathbf{E}}[(S_T - K)^+ | \mathcal{F}_t].$$

Let's pin down the conditional distribution of S_T given $\mathcal{F}_t$. Going back to equation (10.43), we can write

$$S_T = e^{r(T-t)} \left(S_0 e^{rt} + e^{rt} \int_0^t e^{-rs} \sigma \, d\tilde{B}_s + e^{rt} \int_t^T e^{-rs} \sigma \, d\tilde{B}_s \right)$$

$$= e^{r(T-t)} \left(S_t + \int_0^{T-t} e^{-ru} \sigma \, d\tilde{W}_u \right),$$

where we set $\tilde{W}_u = \tilde{B}_{u+t} - \tilde{B}_t$. Note that $(\tilde{W}_u, u \leq T - t)$ is a Brownian motion independent of $\mathcal{F}_t$; see Exercise 8.4. We conclude that S_T given $\mathcal{F}_t$ has the same distribution as an Ornstein-Uhlenbeck process at time $T - t$ starting at S_t. In particular, it is Gaussian with mean $\mu = S_t e^{r(T-t)}$ and variance $v^2 = \frac{\sigma^2}{2r}(e^{2r(T-t)} - 1)$. Thus, we have

$$C_t = e^{-r(T-t)} \tilde{\mathbf{E}}[(S_T - K)^+ | \mathcal{F}_t] = e^{-r(T-t)} \int_{\mathbb{R}} (vz + \mu - K)^+ \frac{e^{-z^2/2}}{\sqrt{2\pi}} \, dz.$$

Similarly to Example 10.15, to get rid of the positive part, we consider the subset $\{z > -b\}$ where

$$b = \frac{\mu - K}{\upsilon} = \frac{S_t e^{r(T-t)} - K}{\upsilon} = \frac{S_t - Ke^{-r(T-t)}}{\sigma\sqrt{\frac{1-e^{-2r(T-t)}}{2r}}}.$$

With this notation, the above becomes (using the symmetry of the Gaussian distribution),

$$C_t = e^{-r(T-t)} \int_{-b}^{\infty} (\upsilon z + \mu) \frac{e^{-z^2/2}}{\sqrt{2\pi}} \, dz - Ke^{-r(T-t)} N(b)$$

$$= (S_t - Ke^{-r(T-t)})N(b) + e^{-r(T-t)}\upsilon \int_{-b}^{\infty} z \frac{e^{-z^2/2}}{\sqrt{2\pi}} \, dz$$

$$= (S_t - Ke^{-r(T-t)})N(b) + \sigma\sqrt{\frac{1 - e^{-2r(T-t)}}{2r}} N'(b),$$

where the last line follows by direct integration. The above reasoning holds for $r \neq 0$. The case $r = 0$ is easier and is left to the reader in Exercise 10.10. Exercise 10.11 looks at the risk-neutral probability in the Bachelier model, and Exercise 10.12 details the PDE side of the model.

How does the existence of a risk-neutral probability relate to the no-arbitrage assumption? We have the following elegant result:

Theorem 10.35 (Fundamental theorem of asset pricing. I). *If there exists a risk-neutral probability for a market model, then the market does not admit an arbitrage; that is, there is no portfolio $(X_t, t \geq 0)$ and a stopping time τ for which $X_0 = 0$ and*

$$\mathbf{P}(X_\tau \geq 0) = 1, \qquad \mathbf{P}(X_\tau > 0) > 0.$$

Proof. We prove the case when the stopping time τ is not random: $\tau = T$. From the pricing formula (10.37) we have that the price at 0 is

$$(10.44) \qquad\qquad \widetilde{\mathbf{E}}[D_t X_t] = X_0 = 0 \quad \forall t.$$

Now if $\mathbf{P}(X_T \geq 0) = 1$ for some time T, then we must also have that $\widetilde{\mathbf{P}}(X_T \geq 0) = 1$ since $\widetilde{\mathbf{P}}$ and $\mathbf{P}$ are equivalent. Thus we have $\widetilde{\mathbf{P}}(X_T < 0) = 0$. But by equation (10.44), this implies $\widetilde{\mathbf{P}}(X_T > 0) = 0$, since $D_t X_t \geq 0$. By equivalence of the probability, we conclude that $\mathbf{P}(X_T > 0) = 0$. Therefore, $(X_t, t \geq 0)$ cannot be an arbitrage. The same proof holds for a stopping time if it can be established that $\widetilde{\mathbf{E}}[D_\tau X_\tau] = 0$, which would follow from Corollary 4.38. □

One might ask if the converse of the theorem holds: does the absence of arbitrage imply the existence of a risk-neutral probability? The answer is no in general (see the bibliographical notes at the end of the chapter).

It turns out that the uniqueness of the risk-neutral probability is related to the replication of the derivatives. The following general result holds:

Theorem 10.36 (Fundamental theorem of asset pricing. II). *Suppose that a market model admits at least one risk-neutral probability. The risk-neutral probability is unique if and only if the market is complete; that is, every $(\mathcal{F}_t, t \geq 0)$-adapted security can be replicated.*

Proof. We will not prove the result in detail, but rather we will motivate why one should expect these surprising facts to hold. For simplicity, we also suppose that the interest rate $(R_t, t \geq 0)$ is constant, so that the only source of the randomness is on the Brownian motion underlying $(S_t, t \geq 0)$: $dS_t = \sigma_t S_t \, dB_t + \mu_t S_t \, dt$. In particular, the filtration $(\mathcal{F}_t, t \geq 0)$ is simply the one of the risky asset.

Suppose first that the market is complete and that it admits two risk-neutral probabilities $\widetilde{\mathbf{P}}_1$ and $\widetilde{\mathbf{P}}_2$. We will derive a contradiction. Consider $\mathcal{E}$ any event in $\mathcal{F}_T$. Then consider the derivative with payoff $O_T = \mathbf{1}_{\mathcal{E}} D_T^{-1}$. By assumption, this security can be replicated. In particular, it must be a martingale for both probabilities, by equation (10.36). Therefore we have

$$\widetilde{\mathbf{E}}_1[D_T O_T] = O_0 = \widetilde{\mathbf{E}}_2[D_T O_T].$$

This implies $\widetilde{\mathbf{P}}_1(\mathcal{E}) = \widetilde{\mathbf{P}}_2(\mathcal{E})$ for any event $\mathcal{E} \in \mathcal{F}_T$, so $\widetilde{\mathbf{P}}_1 = \widetilde{\mathbf{P}}_2$ on $\mathcal{F}_T$ for any T.

For the converse, suppose that a risk-neutral probability exists and is unique. Then, by the pricing formula (10.37), $(\widetilde{O}_t, t \leq T)$ is a $\widetilde{\mathbf{P}}$-martingale for the filtration $(\mathcal{F}_t, t \leq T)$ for any derivative. The key now is the *martingale representation theorem* (Theorem 7.25). It asserts that there exists an $(\mathcal{F}_t)$-adapted process $(\Gamma_t, t \leq T)$ such that

$$\widetilde{O}_t = \Gamma_t \, d\widetilde{B}_t, \quad t \leq T.$$

Here, $(\widetilde{B}_t, t \leq T)$ is the Brownian motion underlying the Itô process of the market model in the risk-neutral probability $d\widetilde{S}_t = \sigma_t \, d\widetilde{B}_t$. (This is where we use uniqueness; there is only one choice for this Brownian motion.) We conclude that the derivative can be replicated and its delta is

$$\Delta_t = \frac{\Gamma_t}{\sigma_t}, \quad t \leq T. \qquad \square$$

Suppose we now have a market model as in Definition 10.1 that admits a risk-neutral probability. For simplicity, let's take a constant interest rate $R_t = r$. We consider a derivative $(O_t, t \leq T)$ that can be replicated. Therefore, we know that the pricing formula (10.37) applies. Is it true in general that the price of the derivative is given by a PDE as in the Black-Scholes model? The answer is no. However, if the two following conditions are satisfied, then we have a PDE:

(i) The risky asset $(S_t, t \leq T)$ is a *diffusion* (see Definition 7.2) under the $\widetilde{\mathbf{P}}$-probability. In particular, the processes are Markov as seen in Section 8.1.

(ii) The payoff of the option is a function of the price at expiration O_T only; i.e., $O_T = g(S_T)$.

If the two conditions are satisfied, then the pricing formula implies

$$O_t = e^{-r(T-t)}\widetilde{\mathbb{E}}[O_T|\mathcal{F}_t] = e^{-r(T-t)}\widetilde{\mathbb{E}}[g(S_T)|\mathcal{F}_t].$$

Now, by the Markov property, we must have that the right-hand side is an explicit function of t and S_t. In other words, we have

$$O_t = f(t, S_t), \quad t \leq T.$$

How is the PDE derived? By Itô's formula, of course! Indeed, exactly as in Example 10.31, we ask that

$$d(e^{-rt}O_t) = d(e^{-rt}f(t, S_t))$$

has no dt-term when expressed in terms of $d\widetilde{B}_t$. The specific PDE will depend on the Itô process $(S_t, t \geq 0)$ considered.

The next two sections look at the cases where the payoff is *path-dependent* and when the interest rate is no longer constant.

10.7. Exotic Options

The pricing formula (10.37) coming from the risk-neutral framework is more robust than the PDE setting as it can handle *exotic options* with no sweat. In this section, we focus on the following two types:

(i) *Asian options.* Asian options are options whose payoff is an explicit function of the *average of the price of the underlying asset* up to expiration; that is,

$$O_T = g\left(\frac{1}{T}\int_0^T S_t \, dt\right).$$

(ii) *Lookback options.* Lookback options are options whose payoff is an explicit function of the *maximum of the price of the underlying asset* up to expiration; that is,

$$O_T = g\left(\max_{t \leq T} S_t\right).$$

Thanks to the Bachelier formula in Proposition 4.45, we can often get an explicit formula for the price of such options.

Instead of developing a general theory for those, we work through examples.

Example 10.37 (An Asian option). Consider the simple case

$$O_T = \frac{1}{T}\int_0^T S_t \, dt,$$

where the payoff of the option is the average of the value of the risky asset on $[0, T]$. Let's price this in the Black-Scholes model of Example 10.2. To find O_0, we apply the pricing formula

$$O_0 = e^{-rT}\widetilde{\mathbb{E}}\left[\frac{1}{T}\int_0^T S_t\right].$$

Assuming we can exchange the expectation and the integral, we get

$$O_0 = e^{-rT} \frac{1}{T} \int_0^T e^{rt} \widetilde{\mathbf{E}}[\tilde{S}_t] \, dt.$$

Since $\tilde{S}_t$ is a martingale we have $\widetilde{\mathbf{E}}[\tilde{S}_t] = S_0$. Therefore, we are left with

$$O_0 = \frac{S_0 e^{-rT}}{T} \int_0^T e^{rt} \, dt = \frac{S_0}{rT}(1 - e^{-rT}).$$

The replicating portfolio can be read off this price at time 0. Indeed, we have $\Delta_0 = \frac{1}{rT}$ and $a_0 = \frac{-S_0}{rT} e^{-rT}$. The general case O_t is given in Exercise 10.13.

Example 10.38 (A lookback option). Consider the following binary option for some strike price K:

$$O_T = \mathbf{1}_{\{\max_{t \leq T} S_t > K\}} = \begin{cases} 1 & \text{if } \max_{t \leq T} S_t > K, \\ 0 & \text{if } \max_{t \leq T} S_t \leq K. \end{cases}$$

Let's find the price at $t = 0$ of this option in the Black-Scholes model. The pricing formula yields

$$O_0 = e^{-rT} \widetilde{\mathbf{E}}[\mathbf{1}_{\{\max_{t \leq T} S_t > K\}}] = e^{-rT} \widetilde{\mathbf{P}}\left(\max_{t \leq T} S_t > K\right) = e^{-rT} \widetilde{\mathbf{P}}\left(\max_{t \leq T} e^{rt} \tilde{S}_t > K\right).$$

Recall that $\tilde{S}_t = S_0 \exp(\sigma \widetilde{B}_t - \sigma^2 t / 2)$ where $(\widetilde{B}_t, t \leq T)$ is a $\widetilde{\mathbf{P}}$-Brownian motion. This gives

$$O_0 = e^{-rT} \widetilde{\mathbf{P}}\left(\max_{t \leq T}\{\widetilde{B}_t + \theta t\} > a\right)$$

for $\theta = \frac{1}{\sigma}(r - \frac{\sigma^2}{2})t$ and $a = \frac{1}{\sigma} \log \frac{K}{S_0}$. Here we used the fact that the function exp is increasing so that $\max_{t \leq T} \exp(\dots) = \exp(\max_{t \leq T}(\dots))$. But we know the distribution of the maximum of Brownian motion with drift, thanks to Proposition 9.7. This gives (amazingly) the explicit formula

$$O_0 = e^{-rT} \int_0^T \frac{a}{\sqrt{2\pi} s^{3/2}} e^{-(a-\theta s)^2/(2s)} \, ds.$$

Exercises 10.14, 10.15, and 10.16 give more examples of pricing of exotic options.

10.8. Interest Rate Models

The framework of risk-neutral pricing we have developed so far applies to market models for which the risk-free rate is not constant. In this section, we look at concrete example of interest rates beyond a constant one.

We consider a market model as in Definition 10.1. We suppose that this market admits a risk-neutral probability $\widetilde{\mathbf{P}}$. It is customary to express the interest rate model $(R_t, t \geq 0)$ directly in terms of $\widetilde{\mathbf{P}}$; i.e., the SDE for $(R_t, t \geq 0)$ is usually given in the form

$$(10.45) \qquad\qquad dR_t = v(t, R_t) \, d\widetilde{W}_t + u(t, R_t) \, dt,$$

where $(\widetilde{W}_t, t \geq 0)$ is a standard Brownian motion under $\widetilde{\mathbf{P}}$. This is the SDE of a diffusion with local drift $u(t, R_t)$ and local volatility $v(t, R_t)$ for some smooth functions u, v. The randomness of $\widetilde{W}$ might be different from the randomness of the risky assets in the model. This is why we change the notation to $(\widetilde{W}_t, t \geq 0)$. equation (10.45) might look odd, since

the interest rate process is not a $\widetilde{\mathbf{P}}$-martingale in general!

This seems to violate the very definition of risk-neutral probability in Definition 10.26. It does not. The definition formally stated that the risky assets must be $\widetilde{\mathbf{P}}$-martingales. The reason why the interest rate is exempt from this is that it is not a tradable asset per se. However, we shall see that we can construct tradable assets on the interest rate, such as bonds and derivatives on bonds.

First, let's list some common examples of interest rate models that fit the mold of equation (10.45). Note that all these examples are *diffusions*. In particular, they are Markov processes as seen in Section 8.1.

Example 10.39 (Vasicek model). For this model, the SDE of $(R_t, t \geq 0)$ is of the form

$$dR_t = (a - bR_t)\,dt + \sigma\,d\widetilde{W}_t, \quad R_0,$$

for some $a, b \in \mathbb{R}$ and $\sigma > 0$. This is the Ornstein-Uhlenbeck process of equation (7.11) in disguise! Indeed, only the drift term $a\,dt$ has been added. As in the CIR model in Example 7.24, the drift has a *mean-reverting* effect: the process will fluctuate around b/a.

This SDE can be integrated to get

$$R_t = R_0 e^{-bt} + \int_0^t e^{-b(t-s)} a\,ds + \int_0^t e^{-b(t-s)} \sigma\,d\widetilde{W}_s.$$

This is a Gaussian process. In particular, R_t could be negative.

Example 10.40 (Hull-White model). This is a slight generalization of the above where a, b, and σ are now taken to be deterministic functions of t denoted by $a(t)$, $b(t)$, and $\sigma(t)$. The SDE is then

$$dR_t = (a(t) - b(t)R_t)\,dt + \sigma(t)\,d\widetilde{W}_t, \quad R_0.$$

Again, this SDE can be integrated, though it might not be obvious. We get

$$R_t = R_0 e^{-\int_0^t b(s)\,ds} + \int_0^t \left\{ e^{-\int_u^t b(v)\,dv} \right\} a(u)\,du + \int_0^t \left\{ e^{-\int_u^t b(v)\,dv} \right\} \sigma(u)\,d\widetilde{W}_u.$$

Exercise 10.17 is about double-checking this. This process is also Gaussian.

Example 10.41 (Cox-Ingersoll-Ross (CIR) model). This process was introduced in Example 7.24. The SDE of the process $(R_t, t \geq 0)$ is of the form

$$dR_t = (a - bR_t)\,dt + \sigma\sqrt{R_t}\,d\widetilde{W}_t.$$

There it was shown that $R_t > 0$ if $a > \sigma^2/2$. The CIR process is not Gaussian.

The most basic tradable security involving the interest rate is a *zero-coupon bond* with maturity T.

Example 10.42 (Price of a zero-coupon bond). We write $(\mathcal{B}_t(T), t \leq T)$ for the price of this security, making explicit the dependence on the maturity. The payoff of a *zero-coupon bond* at maturity is simply a fixed amount of cash, say \$1:

$$\mathcal{B}_T(T) = 1.$$

The framework of risk-neutral pricing still applies. In particular, the pricing formula (10.37) implies that

$$(10.46) \qquad D_t \mathcal{B}_t(T) = \widetilde{\mathbf{E}}[D_T 1 | \mathcal{F}_t] = \widetilde{\mathbf{E}}[e^{-\int_0^T R_s \, ds} | \mathcal{F}_t],$$

since $D_t = e^{-\int_0^t R_s \, ds}$. Here, $(\mathcal{F}_t, t \leq T)$ is the filtration of the whole market model. In particular, it includes the information of $(R_t, t \leq T)$. Therefore, the discounting factor D_t is $\mathcal{F}_t$-measurable and can be slipped inside the conditional expectation. We conclude that the price $\mathcal{B}_t(T)$ is

$$(10.47) \qquad \mathcal{B}_t(T) = \widetilde{\mathbf{E}}[e^{-\int_t^T R_s \, ds} | \mathcal{F}_t], \quad t \leq T.$$

In the simplest case where $R_t = r$ is constant, the bond price is simply $\mathcal{B}_t(T) = e^{-r(T-t)}$ as it should be. The price of the bond can in general be approximated with Monte-Carlo pricing (Example 10.32); see Numerical Projects 10.6 and 10.7.

Another way to price the bond is to take advantage of the fact that $(R_t, t \geq 0)$, at least in the above models, are Markov processes. We assume from now on that the SDE in equation (10.45) is the one of a diffusion, so that $U_t = u(t, R_t)$ and $V_t = v(t, R_t)$. This means that there must be a function of time and space, $f(t, x)$, such that

$$\mathcal{B}_t(T) = f(t, R_t).$$

equation (10.47) also implies that $(\mathcal{B}_t(T), t \leq T)$ is a $\widetilde{\mathbf{P}}$-martingale. (Again, this is by Exercise 4.16.) The PDE is obtained by computing $d(D_t f(t, R_t))$ using Itô's formula and writing the condition for the process to be a $\widetilde{\mathbf{P}}$-martingale. We have by the product rule (7.16)

$$(10.48) \qquad d(D_t f(t, R_t)) = D_t \, df(t, R_t) + f(t, R_t) \, dD_t + dD_t \cdot df(t, R_t).$$

Note that the term $dD_t \cdot df(t, R_t) = 0$ since $dD_t = -R_t D_t \, dt$. We now use Itô's formula of Theorem 7.8 to compute $df(t, R_t)$:

$df(t, R_t)$

$$= \partial_0 f(t, R_t) \, dt + \partial_1 f(t, R_t) \, dR_t + \frac{1}{2} \partial_1^2 f(t, R_t)(dR_t)^2$$

$$= \partial_0 f(t, R_t) \, dt + \partial_1 f(t, R_t)\{v(t, R_t) \, d\widetilde{W}_t + u(t, R_t) \, dt\} + \frac{v^2(t, R_t)}{2} \partial_1^2 f(t, R_t) \, dt$$

$$= \left\{ \partial_0 f(t, R_t) + u(t, R_t)\partial_1 f(t, R_t) + \frac{v^2(t, R_t)}{2} \partial_1^2 f(t, R_t) \right\} dt + v(t, R_t)\partial_1 f(t, R_t) \, d\widetilde{W}_t.$$

Putting this back in equation (10.48) we obtain

$d(D_t f(t, R_t))$

$$= D_t \left\{ \partial_0 f(t, R_t) + \frac{v^2(t, R_t)}{2} \partial_1^2 f(t, R_t) + u(t, R_t)\partial_1 f(t, R_t) - R_t f(t, R_t) \right\} dt$$

$$+ D_t \partial_1 f(t, R_t) v(t, R_t) \, d\widetilde{W}_t.$$

The term on the last line is a $\widetilde{\mathbf{P}}$-martingale (assuming the process in front is in $\mathcal{L}^2_c(T)$). Therefore, for $(D_t f(t, R_t), t \le T)$ to be a $\widetilde{\mathbf{P}}$-martingale, the dt-term must be 0. By replacing R_t by a generic space variable x, we get that the function $f(t, x)$ as a function of time and space must satisfy the PDE

$$(10.49) \qquad \frac{\partial f}{\partial t}(t, x) + \frac{v^2(t, x)}{2} \frac{\partial^2 f}{\partial x^2}(t, x) + u(t, x) \frac{\partial f}{\partial x}(t, x) - x f(t, x) = 0.$$

By the above reasoning, we have that $\mathcal{B}_t(T) = f(t, R_t)$, where f is the solution to the PDE (10.49) with terminal condition given by the payoff (10.47); i.e.,

$$f(T, R_T) = 1.$$

After all these computations, it is good to do a sanity check. In the simplest case where $R_t = r$ is constant, the PDE reduces to

$$\frac{\partial f}{\partial t}(t, x) - x f(t, x) = 0,$$

since $u = v = 0$. The solution is $f(t, x) = C e^{xt}$. The terminal condition then implies $f(t, r) = e^{-r(T-t)}$ as expected.

Example 10.43 (PDE for the Vasicek model). Looking at the SDE of the Vasicek model in Example 10.39, we see that the PDE (10.49) for the price of derivatives on the rate $(R_t, t \ge 0)$ is of the form

$$\frac{\partial f}{\partial t} + \frac{\sigma^2}{2} \frac{\partial^2 f}{\partial x^2} + (a - bx) \frac{\partial f}{\partial x} - x f = 0.$$

The PDE for the Hull-White model is very similar. It is possible to find the explicit solution of the PDE in the case of the zero-coupon bond. See Exercise 10.18.

Example 10.44 (PDE for the CIR model). For the CIR model in Example 10.41, the PDE (10.49) for the price of derivatives on the rate $(R_t, t \ge 0)$ is of the form

$$\frac{\partial f}{\partial t} + \frac{\sigma^2 x}{2} \frac{\partial^2 f}{\partial x^2} + (a - bx) \frac{\partial f}{\partial x} - x f = 0.$$

Remark 10.45. Like options, bonds are traded. Therefore, the prices obtained from the market models are not used to get correct prices. Rather, the models are *calibrated* with the market data. In the above models, there are three parameters to be chosen: a, b, and σ. For a given model, it is possible to obtain the *yield curve* defined by

$$Y(t, T) = -\frac{1}{T - t} \log \mathcal{B}_t(T).$$

This is the effective rate at time t of the bond if the rate were constant from time t to T. This curve can be compared to the yield curve of the market.

Example 10.46 (Forward and put-call parity). It is now possible to revisit Examples 10.9 and 10.28 in the framework of random interest rates. We suppose that we are in the setup of Definition 10.1 with a risky asset $(S_t, t \ge 0)$, say a stock, and an interest rate $(R_t, t \ge 0)$ which is a diffusion. We saw that a forward contract on the risky asset $(F_t, t \le T)$ has payoff $S_T - K$ at expiration. The pricing formula (10.37) implies that

$$F_t = D_t^{-1} \widetilde{\mathbf{E}}[D_T(S_T - K) | \mathcal{F}_t], \quad t \le T.$$

By the linearity of the conditional expectation, this becomes

$$F_t = D_t^{-1}\widetilde{\mathbf{E}}[\tilde{S}_T|\mathcal{F}_t] - K\widetilde{\mathbf{E}}[D_t^{-1}D_T|\mathcal{F}_t] = S_t - K\mathcal{B}_t(T),$$

since $(\tilde{S}_t, t \le T)$ is a $\widetilde{\mathbf{P}}$-martingale and since $\widetilde{\mathbf{E}}[D_t^{-1}D_T|\mathcal{F}_t]$ is the price $\mathcal{B}_t(T)$ of a bond at time t with expiration exactly as in Example 10.42.

We can also extend the put-call parity in this setting. Indeed, since we always have the relation $C_T - P_T = S_T - K$, linearity of conditional expectation in (10.37) yields the *put-call parity*

(10.50) $$C_t - P_t = S_t - K\mathcal{B}_t(T).$$

We now consider two examples of derivatives where the underlying asset is a zero-coupon bond with maturity T_2, $(\mathcal{B}_t, t \le T_2)$. Again, we suppose that the interest rate process is Markov, so that the pricing problem can be restated as a PDE problem with terminal condition. In both examples, we suppose that the maturity T_2 of the bond is greater than the expiration of the derivative T_1.

Example 10.47 (Forward contract on a zero-coupon bond). We consider a forward contract on a zero-coupon bond $(\mathcal{B}_t, t \le T_2)$ with maturity T_2; i.e., we have the obligation of buying the bond at a delivery price K at the time $T_1 < T_2$. The payoff of the forward at expiration T_1 is

$$O_{T_1} = \mathcal{B}_{T_1}(T_2) - K.$$

The pricing formula (10.37) then gives that the value $(O_t, t \le T_1)$ is

$$O_t = \widetilde{\mathbf{E}}[D_t^{-1}D_{T_1}(\mathcal{B}_{T_1}(T_2) - K)|\mathcal{F}_t], \quad t \le T_1.$$

By linearity of the conditional expectation and the fact that $(D_t\mathcal{B}_t, t \le T_1)$ is a $\widetilde{\mathbf{P}}$-martingale, this reduces to

$$O_t = \mathcal{B}_t(T_2) - K\mathcal{B}_t(T_1).$$

Here, we used again the fact from Example 10.42 that $\mathcal{B}_t(T_1) = \widetilde{\mathbf{E}}[D_t^{-1}D_{T_1}|\mathcal{F}_t]$.

The PDE framework is very similar to the one for the zero-coupon bond. Indeed, we must have that $O_t = g(t, R_t)$ by the Markov property for some function $g(t, x)$ of time and space. In particular, since D_tO_t, $t \le T_1$, must be a $\widetilde{\mathbf{P}}$-martingale, the computation of $d(D_tg(t, R_t))$ is exactly the same as for the bond. However, the terminal condition differs. It is

$$g(T_1, R_{T_1}) = \mathcal{B}_{T_1}(T_2) - K = f(T_1, R_{T_1}) - K,$$

where $f(t, x)$ is the solution to the PDE (10.49) with terminal condition $f(T_2, R_{T_2}) = 1$. We conclude that $O_t = g(t, R_t)$, $t \le T_1$, where g is the solution to the PDE (10.49) with terminal condition $f(T_1, R_{T_1}) - K$. Since the PDE is linear, the solution to this problem is simply the sum of the solutions for the terminal condition $f(T_1, R_{T_1})$ and for the terminal condition $-K$, respectively. Not surprisingly, this gives $\mathcal{B}_t(T_2) - K\mathcal{B}_t(T_1)$.

Example 10.48 (European call on a zero-coupon bond). We now consider a European call on a zero-coupon bond with maturity T_2; i.e., we have the option of buying the bond at a strike price K at the time $T_1 < T_2$. The payoff of this option at expiration T_1 is

$$C_{T_1}(\mathcal{B}_{T_1}(T_2) - K)^+.$$

The pricing formula (10.37) then gives that the values $(C_t, t \le T_1)$ are given by

$$C_t = \widetilde{\mathbf{E}}[e^{-\int_t^T R_s \, ds}(\mathcal{B}_{T_1}(T_2) - K)^+ | \mathcal{F}_t], \quad t \le T_1.$$

We must have that $C_t = h(t, R_t)$ by the Markov property for some function $h(t, x)$ of time and space. In particular, since $D_t C_t, t \le T_1$, must be a $\widetilde{\mathbf{P}}$-martingale, the computation of $d(D_t h(t, R_t))$ is exactly the same as for the bond. The terminal condition is now

$$h(T_1, R_{T_1}) = (\mathcal{B}_{T_1}(T_2) - K)^+ = (f(T_1, R_{T_1}) - K)^+,$$

where $f(t, x)$ is the solution to the same PDE (10.49) with terminal condition $f(T_2, R_{T_2})$ $= 1$. We conclude that $C_t = h(t, R_t), t \le T_1$, where h is the solution to the PDE (10.49) with terminal condition $(f(T_1, R_{T_1}) - K)^+$.

Remark 10.49 (Relation to the Feynman-Kac formula). The pricing in Examples 10.42, 10.47, and 10.48 all involved the same PDE (10.49). In hindsight, this is not all too surprising since all prices were of the form

$$f(t, R_t) = \widetilde{\mathbf{E}}[e^{-\int_t^T R_s \, ds}g(R_T) | \mathcal{F}_t].$$

We now make the same observation as in Remark 10.33: if $(R_t, t \ge 0)$ is a Markov process under $\widetilde{\mathbf{P}}$ (which is the case in Examples 10.39, 10.40, and 10.41), then the above is the solution of a terminal value problem as in Theorem 8.28. Here, the function $r(x)$ is simply x. Therefore, the PDE should be of the form

$$\frac{\partial f}{\partial t} = Af(t, x) - xf(t, x),$$

where A is the generator of $(R_t, t \ge 0)$. This is exactly the PDE (10.49) with the generator A of the diffusion with SDE $dR_t = v(t, R_t) \, d\widetilde{W}_t + u(t, R_t) \, dt$.

10.9. Stochastic Volatility Models

Example 10.24 introduced the notion of implied volatility. There, we observed that the volatility of European options of a given underlying asset is constant no matter the strike K and the expiration T. This is not what is observed in real markets where the implied volatility σ_{imp}, i.e., the solution of equation (10.30), is a function of T, K: $\sigma_{\text{imp}} = \sigma_{\text{imp}}(T, K)$. For a given expiration T, it is observed that $\sigma_{\text{imp}}(K)$ has higher values for small and large K's. This is the famous *volatility smile*. In this section, we investigate the Heston model, which is a market model whose option prices exhibit a smile, unlike the Black-Scholes model.

We suppose that the risk-free rate $R_t = r$ is constant. The Heston model is an example of a *two-factor* model, that is, a market model that has two sources of randomness. More precisely, consider two independent standard Brownian motions $(\widetilde{B}_t^{(1)}, \widetilde{B}_t^{(2)})$, $t \ge 0$, as in Section 6.1. The process $(S_t, t \ge 0)$ of the risky asset of the model is given by the system of SDEs

(10.51)
$$dS_t = rS_t \, dt + \sqrt{V_t}S_t \, d\widetilde{B}_t^{(1)}, \qquad S_0 > 0,$$
$$dV_t = (a - bV_t) \, dt + \eta\sqrt{V_t} \, d\widetilde{B}_t^{(2)}, \qquad V_0 > 0.$$

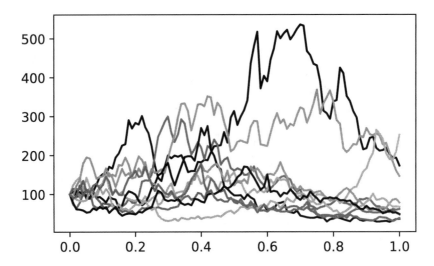

Figure 10.4. A sample of 10 paths of the Heston model with parameter $\eta = 0.1$, $S_0 = 100$, $V_0 = 0.5$, $a = 0.5$, $r = 0$, and $b = 0.1$. See Numerical Project 10.8.

Note that volatility $\sqrt{V_t}$ of S_t is itself a stochastic process. In fact, the process $(V_t, t \geq 0)$ is a CIR model as in Example 7.24 with volatility $\eta > 0$. It is also possible to define the model where the Brownian motions are correlated (see Example 6.2) without complicating the pricing problem much. Why is the CIR model a good choice for modelling volatility? As for interest rates, it is reasonable to assume that the volatility of an asset is a *stationary process*. This can be observed, at least qualitatively, in the market. See Figure 10.3.

How can we price derivatives in the Heston model? We follow the same procedure as in Section 10.6 to derive a PDE for the price of an option. Equation (10.51) is already written in terms of a risk-neutral probability $\widetilde{\mathbf{P}}$ since $d(e^{-rt}S_t) = \sqrt{V_t}S_t\,d\widetilde{B}_t^{(1)}$. Note that $(V_t, t \leq T)$ is not a $\widetilde{\mathbf{P}}$-martingale very much like the interest rate process $(R_t, t \leq T)$ in equation (10.45). Again, this is because the volatility is not itself a tradable asset. However, one might design a tradable asset whose value depends on the volatility.

Following the same principle as in Section 10.6, we have that the price of a European option $(O_t, t \leq T)$ with payoff $g(S_T)$ in the model is given by

$$e^{-rt}O_t = e^{-rT}\widetilde{\mathbf{E}}[g(S_T)|\mathcal{F}_t].$$

Here, the filtration $(\mathcal{F}_t, t \leq T)$ contains the information of both processes $(S_t, t \leq T)$ and $(V_t, t \leq T)$. From the look of the SDE of S_t, we infer that $(S_t, t \leq T)$ is a Markov process since the drift and the volatility of S_t in the SDE is a function of the position at time t. To be precise, the pair (V_t, S_t), $t \leq T$, forms a Markov process. This implies that $\widetilde{\mathbf{E}}[g(S_T)|\mathcal{F}_t]$ must be a function of t, S_t, and V_t. In particular, we can write

$$O_t = f(t, S_t, V_t), \quad \text{for some smooth function } f : [0, T] \times \mathbb{R} \times \mathbb{R}.$$

It is then possible to derive the PDE for $f(t, x_1, x_2)$ by computing $d\widetilde{O}_t$ as was done in Example 10.31 for the Black-Scholes model. We use Itô's formula in Theorem 7.18.

More precisely, by doing a Taylor's expansion up to second order, we get

$$d(e^{-rt}f(t, S_t, V_t))$$

$$= e^{-rt}\Big\{ -rf\, dt + \partial_0 f\, dt + \partial_1 f\, dS_t + \frac{1}{2}\partial_1^2 f\ dS_t \cdot dS_t$$

$$+ \partial_2 f\, dV_t + \frac{1}{2}\partial_2^2 f\ dV_t \cdot dV_t + \partial_1 \partial_2 f\ dS_t \cdot dV_t \Big\},$$

where it is understood that f above is evaluated at (t, S_t, V_t). The rules of Itô calculus and equation (10.51) give

$$dS_t \cdot dS_t = V_t S_t^2\, dt, \quad dV_t \cdot dV_t = \eta^2 V_t\, dt, \quad dV_t \cdot dS_t = 0.$$

The last equality is from independence of the Brownian motions. (If they were correlated as in Example 6.2, we would get $\rho\, dt$. This would give an extra term in the PDE below.) Putting this back in the above, we get

$$d(e^{-rt}f(t, S_t, V_t))$$

$$= e^{-rt}\Big\{ -rf + \partial_0 f + rS_t\partial_1 f + \frac{1}{2}V_t S_t^2 \partial_1^2 f + (a - bV_t)\partial_2 f + \frac{\eta^2}{2}V_t\partial_2^2 f\Big\} dt$$

$$+ e^{-rt}\Big\{ \sqrt{V_t}S_t\partial_1 f\, d\widetilde{B}_t^{(1)} + \sqrt{V_t}\partial_2 f\, d\widetilde{B}_t^{(2)}\Big\}.$$

The last term is the sum of two $\widetilde{\mathbf{P}}$-martingales, so it is a martingale. The PDE for $f = f(t, x_1, x_2)$ is obtained by setting the first parenthesis to 0 and replacing the space variables S_t by x_1 and V_t by x_2. We finally get

$$(10.52) \qquad \frac{\partial f}{\partial t} + rx_1\frac{\partial f}{\partial x_1} + \frac{1}{2}x_1^2 x_2 \frac{\partial^2 f}{\partial x_1^2} + (a - bx_2)\frac{\partial f}{\partial x_2} + \frac{1}{2}\eta^2 x_2 \frac{\partial f^2}{\partial x_2^2} = rf.$$

Therefore, the process $(O_t, t \le T)$ is given by

$$O_t = f(t, S_t, V_t),$$

for f the solution of the PDE (10.52) with terminal condition $f(T, S_T, V_T) = g(S_T)$.

What is the replicating portfolio of the option in this model? We follow the idea of Section 10.3. There is a substantial difference, as it is necessary to include in the replicating portfolio an additional asset that depends on the volatility. (This is always the case for multifactor models with more than one source of randomness.) Let's write $(U_t, t \le T)$ for the second asset. For example, this can be another option, in the same spirit as Example 10.22. If so, then we can take $U_t = h(t, S_t, V_t)$ where h is a solution of the PDE (10.52) with some terminal condition that might depend on S_T and V_T. With this in mind, it is reasonable to assume that the equivalent of the valuation equation (10.8) is of the form

$$(10.53) \qquad dO_t = \Delta_t\, dS_t + \Delta_t'\, dU_t + (O_t - \Delta_t S_t - \Delta_t' U_t)r\, dt.$$

This equation is interpreted as follows: the option is replicated by allocating Δ_t units in the first risky asset, Δ_t' units in the second risky asset, and saving the rest of the value in cash. Applying Itô's formula the same way as above, we also have

$$(10.54) \quad dU_t = dh(t, S_t, V_t) = \partial_1 h\, dS_t + \partial_2 h\, dV_t + \Big\{ \partial_0 h + \frac{1}{2}S_t^2 V_t\, \partial_1^2 h + \frac{1}{2}\eta^2 V_t\, \partial_2^2 h \Big\} dt.$$

Putting this back into (10.53) yields

(10.55)
$$dO_t = \left\{\Delta_t + \Delta_t' \partial_1 h\right\} dS_t + \left\{\Delta_t' \partial_2 h\right\} dV_t$$
$$+ \left\{\Delta_t'\left(\partial_0 h + \frac{1}{2}S_t^2 V_t\, \partial_1^2 h + \frac{1}{2}\eta^2 V_t\, \partial_2^2 h - rh\right) + rf - rS_t\Delta_t\right\} dt.$$

This is to be compared with the equation

(10.56) $\quad dO_t = df(t, S_t, V_t) = \partial_1 f\, dS_t + \partial_2 f\, dV_t + \left\{\partial_0 f + \frac{1}{2}S_t^2 V_t\, \partial_1^2 f + \frac{1}{2}\eta^2 V_t\, \partial_2^2 f\right\} dt.$

Identifying the coefficients of dS_t and of dV_t gives the allocations Δ_t and Δ_t':

(10.57) $\qquad \Delta_t' = \dfrac{\partial_2 f}{\partial_2 h}, \qquad \Delta_t = \partial_1 f - \Delta_t' \partial_1 h = \partial_1 f - \dfrac{\partial_2 f}{\partial_2 h}\partial_1 h.$

The coefficient of dt reduces to the PDE (10.52). To see this, note that because h also satisfies the PDE, the dt-term in equation (10.55) is

$$\Delta_t'\left(\partial_0 h + \frac{1}{2}S_t^2 V_t\, \partial_1^2 h + \frac{1}{2}\eta^2 V_t\, \partial_2^2 h - rh\right) + rf - rS_t(\partial_1 f - \Delta_t'\partial_1 h)$$
$$= -(a - bV_t)\partial_2 f + rf - rS_t\partial_1 f.$$

This should be equal to the dt-term of equation (10.56). This boils down to the PDE (10.52) for f.

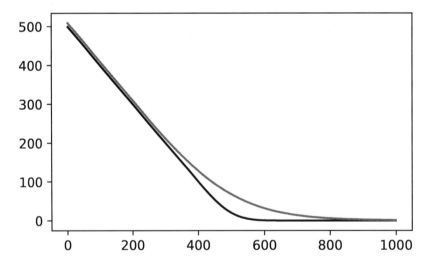

Figure 10.5. The price of a European call as a function of K on $[1, 1{,}000]$ for the Heston model and the Black-Scholes model. The curve of the Heston model lies above. See Numerical Project 10.9.

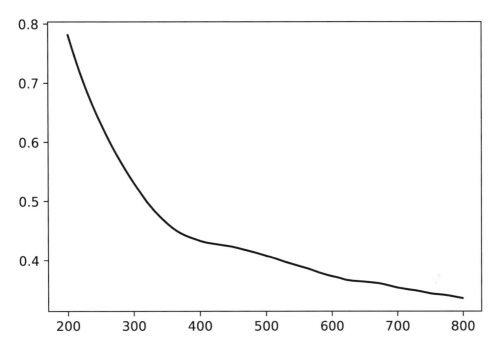

Figure 10.6. The implied volatility (solution of equation (10.30)) as a function of K when C_{market} is generated using the Heston model. Note that it is not constant. This is the *volatility smile* mentioned in Example 10.24. See Numerical Project 10.10.

10.10. Numerical Projects and Exercises

10.1. **Black-Scholes calculator.** Define a function in Python using def that takes as inputs the parameters T, K, S_t, t, r, σ and returns the prices of a European call and a European put in the Black-Scholes model as in equation (10.19).
To evaluate the CDF of a standard normal random variable you can import the command norm *from* scipy.stats *in Python.*

10.2. **Option strategies.** Consider the option strategies in Example 10.8 including the European call option. Consider the parameters $\sigma = 0.1$, $T = 1$, and $K = 480$ as in Example 10.16.
   (a) Draw the graph of the payoff of the six options as a function of S_T, the price of the underlying asset at expiration for $S_T = [400, 550]$. Use $a = 20$.
   (b) Use the Black-Scholes formula of a call in equation (10.19) to generate the value of each of the strategies at time 0 at every 0.1, from $S_0 = 400$ to $S_0 = 600$. Plot it for $r = 0$ and $r = 0.05$. Notice the difference.

10.3. **Monte-Carlo vs Black-Scholes formula.** Consider the Black-Scholes model with parameters

$$\sigma = 0.1, \quad S_0 = 500, \quad r = 0.05.$$

Use Monte-Carlo pricing (Example 10.32) to price a European call with $T = 1$ and $K = 480$ using samples of $N = 100, 1,000, 10,000, 100,000$ paths with time step 0.001. Compare your answer to the price derived using the Black-Scholes formula in Example 10.16.

10.4. **Monte-Carlo exotic option pricing.** Consider the Black-Scholes model with parameters

$$\sigma = 0.1, \quad S_0 = 500, \quad r = 0.05.$$

Use Monte-Carlo pricing in Example 10.32 to price the following options with expiration $T = 1$ using the average over 1,000 paths with a discretization of 0.001:
(a) $O_1 = \max_{t \leq 1} S_t$.
(b) $O_1 = \exp\left(\int_0^1 \log S_t \, dt\right)$.

10.5. **Implied volatility.** Let's build a code to evaluate the implied volatility as explained in Example 10.24. This can be done with the numerical solver `fsolve` available in the Python package `scipy.optimize`.

Use `fsolve` to solve the equation of the implied volatility (10.30) when $C_{\text{market}} = 0.4625$ and $T = 1, r = 0.05, S_0 = 100, t = 0, K = 120$.
You need to enter an a priori guess for the solution when using `fsolve`.

10.6. **Vasicek bond pricing.** Consider the following Vasicek model for the interest rate $(R_t, t \geq 0)$ under the risk-neutral probability:

$$dR_t = (0.1 - R_t)\,dt + 0.05\,d\widetilde{W}_t, \quad R_0 = 0.1.$$

(a) Plot 100 paths of the above process on $[0, 1]$ with a 0.001 discretization.
(b) Use Monte-Carlo pricing (Example 10.32) and the pricing formula (10.47) to find the price $\mathcal{B}_0(1)$ at time 0 of a bond with maturity $T = 1$ and $\mathcal{B}_1(1) = 100$ by averaging over 10,000 paths.
(c) Compare the price obtained in (b) to the price of the same bond when the interest rate $R_t = r = 0.1$ is constant.

10.7. **CIR bond pricing.** Consider the following CIR model for the interest rate $(R_t, t \geq 0)$ under the risk-neutral probability:

$$dR_t = (0.1 - R_t)\,dt + 0.5\sqrt{R_T}\,d\widetilde{W}_t, \quad R_0 = 0.1.$$

(a) Plot 100 paths of the above process on $[0, 1]$ with a 0.001 discretization.
(b) Use Monte-Carlo pricing (Example 10.32) and the pricing formula (10.47) to find the price $\mathcal{B}_0(1)$ at time 0 of a bond with maturity $T = 1$ and $\mathcal{B}_1(1) = 100$ by averaging over 10,000 paths.
(c) Compare the price obtained in (b) to the price of the same bond when the interest rate $R_t = r = 0.1$ is constant.

10.8. **The Heston model.** Consider the Heston model as in equation (10.51) with parameters

$$\eta = 0.1, \quad a = 0.5, \quad b = 1, \quad S_0 = 100, \quad V_0 = 0.5, \quad r = 0.$$

Plot 10 paths of this model.
You might want to define a function returning CIR paths based on Numerical Project 7.3.

10.9. **Black-Scholes vs Heston.** Consider the Heston model as in equation (10.51) with parameters

$$\eta = 0.5, \quad a = 0.1, \quad b = 1, \quad S_0 = 500, \quad V_0 = 0.1, \quad r = 0$$

and the Black-Scholes model with parameters

$$\sigma = 0.1, \quad S_0 = 500, \quad r = 0.$$

Plot the price of a European call for K between 1 and 1,000 for each of the models. For the Heston model, use Monte-Carlo pricing on 1,000 paths with time step 0.01. For the Black-Scholes model, you can either use Monte-Carlo pricing or the Black-Scholes formula. See Figure 10.5.

10.10. **Volatility smile.** Consider the Heston model as in equation (10.51) with parameters

$$\eta = 0.5, \quad a = 0.1, \quad b = 1, \quad S_0 = 500, \quad V_0 = 0.1, \quad r = 0.$$

Plot the implied volatility computed from equation (10.30) of a European call for $T = 1$ and for K between 200 and 800. For the price C_{market}, use Monte-Carlo pricing on 1,000 paths with time step 0.01.
Numerical Project 10.5 is useful.

Exercises

10.1. **Option portfolio.** Find a portfolio with payoff at time T equal to

$$V_T = \begin{cases} 2S_T + 30, & \text{if } 0 \leq S(T) \leq 10, \\ -3S_T + 80, & \text{if } 10 \leq S(T) \leq 30, \\ S_T - 40, & \text{if } 30 \leq S(T). \end{cases}$$

Here, S_T denotes the price of the underlying asset at time T. You can hold cash, the asset, calls, and puts.

10.2. **Option portfolio.** We consider the following portfolio V of options for an underlying asset: 1 long put with $K = 100$, 2 short puts with $K = 150$, 1 long put with $K = 200$. All options have the same expiration T.
  (a) Find the payoff V_T of the option at expiration T as a function of S_T.
  (b) Draw the graph of the payoff.
  (c) This strategy is called a *skip strike butterfly with puts*. Give it a catchier name.
  (d) Draw the typical graph of the price of a put as a function of the strike price. Conclude from this that the value of the portfolio at time 0 is positive.

10.3. **Iron condor.** For a given underlying asset $(S_t, 0 \leq t \leq T)$, we consider the following portfolio: long on 1 put at price $K_1 = 100$, short on 1 put at price $K_2 = 110$, short on 1 call at price $K_3 = 120$, and long on 1 call at price $K_4 = 130$.
  (a) Write $P_0^{(1)}, P_0^{(2)}$ for the price of the puts and $C_0^{(3)}, C_0^{(4)}$ for the price of the calls at time 0. Is the cost of the portfolio positive or negative?
  (b) Write an expression for the value of the portfolio at time T in terms of S_T and the strike price.

(c) Using the previous question, plot the value of the portfolio as a function of S_T.

(d) Suppose $P_0^{(1)} = 2.50$, $P_0^{(2)} = 5$, $C_0^{(3)} = 5$, $C_0^{(4)} = 2.50$ and that the risk-free rate is 0. Plot the value of the portfolio including the cost of the portfolio (in dollars at time T). For what value of S_T do we have a positive payoff?

10.4. **Solution of the Black-Scholes PDE.** Use the Gaussian integration by parts formula

$$\mathbf{E}[F'(Z)] = \mathbf{E}[ZF(Z)],$$

where Z is a standard Gaussian variable, to prove that the function $f(t, x)$ in equation (10.13) is the solution of the Black-Scholes PDE (10.11).

10.5. **Greeks of Black-Scholes.** Verify equation (10.28) for the Greeks of a European call in the Black-Scholes model.

10.6. **European puts in Black-Scholes model.** Use put-call parity of Example 10.10 to answer the following questions in the Black-Scholes model:

(a) Prove that a European put $(P_t, t \leq T)$ with strike price K has the price

$$P_t = -S_t N(-d_+) + K e^{-r(T-t)} N(-d_-),$$

where $d_{\pm}$ are evaluated at $x = S_t$.

(b) Derive formulas for the Greeks of the put based on equation (10.28).

(c) Compute $\frac{\partial P_t}{\partial K}$ and $\frac{\partial^2 P_t}{\partial K^2}$. Conclude that P_t is a convex increasing function of K.

10.7. **Delta hedging.** Consider a portfolio $V^{(1)}$ which is short 100 calls with maturity $T = 60/365$ and $K = 100$. The price of the underlying asset is $S_0 = 100$, the risk-free rate is $r = 0.05$, and the volatility is $\sigma = 0.1$.

(a) If $V_0^{(1)} = 0$, determine how much money is to be put in risk-free assets to construct $V^{(1)}$.

(b) Find a delta-neutral portfolio $V^{(2)}$ with 100 long calls and $V^{(2)}(0) = 0$.

(c) Compare the values of the portfolio $V^{(1)}$ and $V^{(2)}$ after one day when $S_{1/365} = 100, 99$, and 101.

(d) Plot the graph of $V_{1/365}^{(1)}$ and $V_{1/365}^{(2)}$ as a function of $S_{1/365}$ for the interval $[98, 102]$.

10.8. **Cash-or-nothing option.** We consider a *cash-or-nothing* call option with strike price K and value at expiry T

$$O_T = \begin{cases} 1 & S_T > K, \\ 0 & S_T \leq K. \end{cases}$$

(a) Use the pricing formula (10.37) to show that in the Black-Scholes model.

$$O_t = e^{-r(T-t)} N(d_-), \qquad t \in [0, T].$$

(b) Consider a *cash-or-nothing* put with strike price K and value at expiry T

$$U_T = \begin{cases} 0 & S_T > K, \\ 1 & S_T \leq K. \end{cases}$$

Find a relation between the put and the call that holds at T for any outcome (a sort of put-call parity). Use this to find the price of the put at any time.

(c) Verify that the previous price of the put is the same as the one given by the pricing formula.

(d) Use a cash-or-nothing call and an asset-or-nothing call (Example 10.18) to replicate a European call.

10.9. **Easy pricing in Black-Scholes.** Consider the Black-Scholes model under its risk-neutral probability $\widetilde{P}$

$$d\widetilde{S}_t = \sigma \widetilde{S}_t \, d\widetilde{B}_t, \qquad S_0 > 0, \qquad D_t = e^{-rt},$$

where $(\widetilde{B}_t, t \geq 0)$ is a standard $\widetilde{P}$-Brownian motion. Using risk-neutral pricing, we would like to price the option $(O_t, t \leq T)$ with payoff at expiration T

$$O_T = \log S_T.$$

(a) Show that

$$O_0 = e^{-rT}(\log S_0 + (r - \sigma^2/2)T).$$

(b) Show that

$$O_t = e^{-r(T-t)}(\log S_t + (r - \sigma^2/2)(T - t)).$$

(c) Suppose $r = 0$. Is O_t smaller or greater than $\log S_t$. Is this consistent with what you expect directly from the pricing formula?
   Hint: Jensen's inequality.

10.10. **Bachelier at $r = 0$.** Consider the Bachelier model with SDE in the risk-neutral probability for $(S_t, t \leq T)$ given by

$$dS_t = \sigma \, d\widetilde{B}_t.$$

Show that the price of a European call in this model is

$$C_t = (S_t - K)N(b) + \sigma\sqrt{T - t}N'(b), \quad b = \frac{S_t - K}{\sigma\sqrt{T - t}}.$$

10.11. **Risk-neutral probability for Bachelier.** Suppose that the risky asset in the Bachelier model follows the SDE

$$dS_t = \mu \, dt + \sigma \, dB_t.$$

(a) Find the risk-neutral probability $\widetilde{P}$ for which the process $\widetilde{S}_t = e^{-rt}S_t, t \leq T$, is a $\widetilde{P}$-martingale.
   Hint: Theorem 9.11 is useful here.

(b) Show that the discounted process $(\widetilde{S}_t, t \leq T)$ can be written in the form

$$\widetilde{S}_t = \int_0^t e^{-rs}\sigma \widetilde{B}_s,$$

where $(\widetilde{B}_t, t \leq T)$ is a $\widetilde{P}$-Brownian motion.

(c) Conclude as in Example 10.34 that $(S_t, t \leq T)$ is an Ornstein-Uhlenbeck process in the $\widetilde{P}$-probability.

10.12. **PDE of Bachelier.** Use Theorem 8.28 to show that the price of a European call in the Bachelier model is given by $C_t = f(t, S_t)$, where $f(t, x)$ satisfies the PDE problem with terminal value

$$\frac{\partial f}{\partial t} + \frac{\sigma^2}{2}\frac{\partial^2 f}{\partial x^2} + rx\frac{\partial f}{\partial x} - rf = 0,$$

$$f(T, x) = (x - K)^+.$$

10.13. **Asian option.** Consider Example 10.37. Show that

$$O_t = e^{-r(T-t)}\frac{1}{T}\int_0^t S_u\,du + \frac{S_t}{rT}(1 - e^{-r(T-t)}), \quad t \le T.$$

What is the delta of this option? What happens if $r = 0$?

10.14. **Lookback option.** We consider the following Bachelier model for the price of an asset:

$$dS_t = 2\,dt + 2\,dB_t, \qquad S_0 = 0,$$
$$D_t = 1, \qquad\qquad\quad r = 0.$$

Using risk-neutral valuation, we are interested in pricing lookback options with expiration $T = 1$.

(a) What is the risk-neutral probability $\widetilde{\mathbf{P}}$ for this model on $[0,1]$? What is the distribution of $(S_t, t \in [0,1])$ under $\widetilde{\mathbf{P}}$?

(b) Using the risk-neutral probability, find O_0 for the option with value at expiration given by

$$O_1 = \max_{t \le 1} S_t.$$

(c) Using a similar method, show that the price at time 0 of an *asset-or-nothing lookback call* with value at expiration and strike price $K > 0$

$$C_1 = \begin{cases} \max_{t \le 1} S_t & \text{if } \max_{t \le 1} S_t > K, \\ 0 & \text{if } \max_{t \le 1} S_t \le K \end{cases}$$

is given by

$$C_0 = \frac{4}{\sqrt{2\pi}}e^{-K^2/8}.$$

(d) Find an adequate put-call parity relation. Use the two previous questions to price a lookback put at time 0 with value at expiration

$$P_1 = \begin{cases} 0 & \text{if } \max_{t \le 1} S_t > K, \\ \max_{t \le 1} S_t & \text{if } \max_{t \le 1} S_t \le K. \end{cases}$$

10.15. **Another lookback option.** Consider the usual Black-Scholes model with $S_0 = 1$, $r > 0$, and

$$dS_t = \mu S_t\,dt + \sigma S_t\,dB_t,$$

where $\sigma > 0$. We want to price the following derivative with payoff at expiration T given by

$$O_T = \begin{cases} \max_{t \le T} S_t & \text{if } \max_{t \le T} S_t > K, \\ 0 & \text{otherwise.} \end{cases}$$

We denote the value of this option by $(O_t, t \in [0,T])$.

(a) Is O_t as a function of K a decreasing or an increasing function of K?

(b) Use risk-neutral pricing to show that the price at time 0 is given by

$$O_0 = e^{-rT}\widetilde{\mathbf{E}}\left[e^{\max_{t \le T}\{\sigma\widetilde{B}_t + (r - \sigma^2/2)t\}}\mathbf{1}_{\{\max_{t \le T}\{\sigma\widetilde{B}_t + (r - \sigma^2/2)t\} > \log K\}}\right],$$

where $\widetilde{\mathbf{P}}$ is the risk-free probability and $(\widetilde{B}_t, t \le T)$ is a $\widetilde{\mathbf{P}}$-Brownian motion.

(c) You are lucky and for the risky asset we can take the approximation $r = \sigma^2/2$! In this case, show that

$$O_0 = e^{-rT}\sqrt{\frac{2}{\pi}}\int_{\sigma^{-1}\log K}^{\infty} e^{\sigma y}\frac{e^{-y^2/(2T)}}{T^{1/2}}\,dy\,.$$

(d) Use the above to quickly find U_0 for the option with payoff

$$U_T = \begin{cases} \max_{t\le T} S_t & \text{if } \max_{t\le T} S_t \le K, \\ 0 & \text{otherwise.} \end{cases}$$

10.16. Geometric Asian option. Consider the Black-Scholes model with

$$S_t = S_0\exp(\sigma B_t + \mu t),\quad D_t = e^{-rt}.$$

Consider the exotic option with maturity $T = 1$ with payoff given by the geometric mean of the risky asset

$$O_1 = \exp\left(\int_0^1 \log S_t\,dt\right).$$

The goal of this exercise is to price this option.
(a) Let $(\widetilde{B}_t, t \le 1)$ be a Brownian motion under the risk-neutral probability $\widetilde{\mathbf{P}}$ of the model. Argue that we have for any $t \ge 0$

$$\int_0^t \widetilde{B}_s\,ds = tB_t - \int_0^t s\,dB_s.$$

(b) Use the above to deduce that the random variable $\int_0^1 \widetilde{B}_t\,dt$ is Gaussian with mean 0 and variance 1/3.
(c) Use risk-neutral pricing to show that the price at time 0 is

$$O_0 = S_0 e^{-r/2 - \sigma^2/12}.$$

Compare your answer with the price obtained in the Numerical Project 10.4. *Hint: Question (b) is useful as the MGF of the random variable should appear in the pricing formula.*
(d) Without any calculations, can you argue that the price of this option is *smaller* than the price of the Asian option with payoff $\int_0^1 S_t\,dt$?

10.17. Solution to the Hull-White model. Use Itô's formula to check that

$$R_t = R_0 e^{-\int_0^t b(s)\,ds} + \int_0^t \left\{e^{-\int_u^t b(v)\,dv}\right\}a(u)\,du + \int_0^t \left\{e^{-\int_u^t b(v)\,dv}\right\}\sigma(u)\,d\widetilde{B}_u$$

is the solution to the SDE

$$dR_t = (a(t) - b(t)R_t)\,dt + \sigma(t)\,d\widetilde{B}_t.$$

This verifies the claim made in Example 10.40.

10.18. Zero-coupon bond in the Hull-White model. In this exercise, we prove an explicit form for the solution of the PDE (10.49) in the Hull-White model with terminal value $f(T, x) = 1$.

(a) Show that the PDE has the form

$$\frac{\partial f}{\partial t} + \frac{\sigma^2(t)}{2}\frac{\partial^2 f}{\partial x^2} + (a(t) - b(t)x)\frac{\partial f}{\partial x} - xf = 0.$$

(b) Consider a solution of the form $f(t, x) = e^{-xh(t)-g(t)}$. Show that h and g must satisfy the ODE

$$h'(t) = b(t)h(t) - 1,$$

$$g'(t) = -a(t)h(t) + \frac{1}{2}\sigma^2(t)h(t).$$

(c) Show that the functions

$$h(t) = \int_t^T e^{-\int_t^s b(u)\,du}\,ds,$$

$$g(t) = \int_t^T \left(a(s)h(s) - \frac{1}{2}\sigma^2(s)h(s)^2\right)ds$$

are solutions of the ODEs and that $f(T, x) = 1$ as desired.
The computation $h'(t)$ is a good exercise in taking derivatives of an integral with respect to a parameter....

10.19. A model for interest rate.

Consider the following model for an interest rate $(R_t, t \geq 0)$:

$$dR_t = \frac{1}{2}R_t^{1/3}\,dt + R_t^{2/3}\,dB_t, \quad R_0 = 1.$$

(We suppose that $R_t > 0$ for all t.)

(a) Find a PDE for the function $v(t, r)$ satisfied when the process $v(t, R_t)$ is a martingale.

(b) Check if each of the two processes below are martingales or not:

$$\log R_t, \qquad t - 3R_t^{2/3}\,.$$

(c) Suppose that $v(t, R_t)$ represents the price of some derivative on the underlying interest rate R_t. Find a PDE satisfied by $v(t, r)$.
(Do not forget discounting!)

10.20. A market model with a risky asset and a random interest rate. Consider the following market model for an interest rate $(R_t, t \geq 0)$ and a discounted asset $(S_t, t \geq 0)$ under the risk-neutral probability $\tilde{\mathbf{P}}$:

$$dR_t = -R_t\,dt + d\tilde{B}_t, \quad R_0 = 0.01, \quad \tilde{S}_t = \sigma\tilde{S}_t\,d\tilde{B}_t, \quad \tilde{S}_0 = 1.$$

(a) Find a PDE for the price of a derivative of the form $O_t = f(t, R_t)$.
(Do not forget discounting.... This PDE should be an equation of a function of two variables $f(t, r)$.)

(b) Let $\mathcal{B}_t(T)$ be the price of a zero-coupon bond with maturity T and $B_T(T) = 1$ in that model. Consider a put $(P_t, t \leq T)$ and a call $(C_t, t \leq T)$ with strike price K. Argue that we have

$$C_t - P_t = S_t - K\mathcal{B}_t(T).$$

(c) Find a PDE for the price of a derivative of the form $O_t = f(t, R_t, S_t)$.
   This PDE should be an equation of a function of three variables $f(t, r, x)$.

10.11. Historical and Bibliographical Notes

The mathematical theory of derivative pricing goes back to Bachelier in *Théorie de la spéculation* [**Bac00**]. The model considered by Bachelier was a symmetric random walk. As mentioned in Chapter 3, he was able to derive the corresponding equation for Brownian motion in the continuous-time limit. A Bachelier model now refers to a market model where the process of a risky asset is given by a Brownian motion. The history of derivatives goes back way before Bachelier. Indeed, the notion of *forward contract* for example is very useful in commodity trading where a farmer might be interested in selling their harvest at a given price at a future delivery date to hedge the risk factors such as weather. We suggest [**Web09**] for an interesting account of the history of derivatives.

The revolution of Black-Scholes and Merton is the introduction of the no-arbitrage and replication assumption. This laid the foundation of a purely mathematical theory of option pricing. The original paper of Black and Scholes is [**BS73**], and similar ideas were proposed independently by Merton [**Mer73**]. Scholes and Merton received the Nobel Prize in Economics in 1997 for their work (Black passed away in 1995). For a primer in the mathematics of finance, we highly suggest [**Ste11**]. The magical identity (10.27) comes from Lemma 3.15 there. For multifactor market models with more risky assets, the reader is referred to [**Shr04**]. The converse of Theorem 10.35 is true in most models. The state-of-the-art on this is given in [**DS94**].

For more on stochastic volatility models and implied volatility, the standard reference is [**Gat12**]. The Heston model was introduced in [**Hes93**].

Bibliography

[Bac00] L. Bachelier, *Théorie de la spéculation* (French), Ann. Sci. École Norm. Sup. (3) **17** (1900), 21–86. MR1508978

[Bil95] Patrick Billingsley, *Probability and measure*, 3rd ed., Wiley Series in Probability and Mathematical Statistics, A Wiley-Interscience Publication, John Wiley & Sons, Inc., New York, 1995. MR1324786

[BS73] Fischer Black and Myron Scholes, *The pricing of options and corporate liabilities*, J. Polit. Econ. **81** (1973), no. 3, 637–654, DOI 10.1086/260062. MR3363443

[BSS09] Laurent Bienvenu, Glenn Shafer, and Alexander Shen, *On the history of martingales in the study of randomness*, J. Électron. Hist. Probab. Stat. **5** (2009), no. 1, 40. MR2520666

[CIR85] John C. Cox, Jonathan E. Ingersoll Jr., and Stephen A. Ross, *A theory of the term structure of interest rates*, Econometrica **53** (1985), no. 2, 385–407, DOI 10.2307/1911242. MR785475

[CM44] R. H. Cameron and W. T. Martin, *Transformations of Wiener integrals under translations*, Ann. of Math. (2) **45** (1944), 386–396, DOI 10.2307/1969276. MR10346

[CM15] F. Comets and T. Meyre, *Calcul stochastique et modèles de diffusions - 2e éd.*, Mathématiques appliquées pour le Master/SMAI, Dunod, 2015.

[Cra76] Harald Cramér, *Half a century with probability theory: some personal recollections*, Ann. Probability **4** (1976), no. 4, 509–546, DOI 10.1214/aop/1176996025. MR402837

[dF19] Antoine de Falguerolles, *Cholesky and the Cholesky decomposition: a commemoration by an applied statistician* (English, with English and French summaries), J. SFdS **160** (2019), no. 2, 83–96. MR3987791

[DFS03] D. Duffie, D. Filipović, and W. Schachermayer, *Affine processes and applications in finance*, Ann. Appl. Probab. **13** (2003), no. 3, 984–1053, DOI 10.1214/aoap/1060202833. MR1994043

[DKM70] Tore Dalenius, Georg Karlsson, and Sten Malmquist (eds.), *Scientists at work*, with a preface by Torgny T. Segerstedt, Festschrift in honour of Herman Wold on the occasion of his 60th birthday, Almqvist & Wiksell, Stockholm, 1970. MR0396141

[dM67] A. de Moivre, *The doctrine of chances: A method of calculating the probabilities of events in play*, Chelsea Publishing Co., New York, 1967. MR0216923

[DS94] Freddy Delbaen and Walter Schachermayer, *A general version of the fundamental theorem of asset pricing*, Math. Ann. **300** (1994), no. 3, 463–520, DOI 10.1007/BF01450498. MR1304434

[Dur96] Richard Durrett, *Probability: theory and examples*, 2nd ed., Duxbury Press, Belmont, CA, 1996. MR1609153

[Edw83] A. W. F. Edwards, *Pascal's problem: the "gambler's ruin"* (English, with French summary), Internat. Statist. Rev. **51** (1983), no. 1, 73–79, DOI 10.2307/1402732. MR703307

[Ein05] A. Einstein, *Über die von der molekularkinetischen theorie der wärme geforderte bewegung von in ruhenden flüssigkeiten suspendierten teilchen*, Annalen der Physik **322** (1905), no. 8, 549–560.

[Fis11] Hans Fischer, *A history of the central limit theorem: From classical to modern probability theory*, Sources and Studies in the History of Mathematics and Physical Sciences, Springer, New York, 2011, DOI 10.1007/978-0-387-87857-7. MR2743162

[Fou09] Jean Baptiste Joseph Fourier, *Théorie analytique de la chaleur* (French), reprint of the 1822 original; previously published by Éditions Jacques Gabay, Paris, 1988 [MR1414430], Cambridge Library Collection, Cambridge University Press, Cambridge, 2009, DOI 10.1017/CBO9780511693229. MR2856180

[Gat12] Jim Gatheral, *The volatility surface*, John Wiley & Sons, Ltd, 2012.

[Gau11]　Carl Friedrich Gauss, *Theoria motus corporum coelestium in sectionibus conicis solem ambientium* (Latin), reprint of the 1809 original, Cambridge Library Collection, Cambridge University Press, Cambridge, 2011, DOI 10.1017/CBO9780511841705.010. MR2858122

[Gir60]　I. V. Girsanov, *On transforming a class of stochastic processes by absolutely continuous substitution of measures* (Russian, with English summary), Teor. Verojatnost. i Primenen. **5** (1960), 314–330. MR0133152

[Hes93]　Steven L. Heston, *A closed-form solution for options with stochastic volatility with applications to bond and currency options*, Rev. Financ. Stud. **6** (1993), no. 2, 327–343, DOI 10.1093/rfs/6.2.327. MR3929676

[Ito44]　Kiyosi Itô, *Stochastic integral*, Proc. Imp. Acad. Tokyo **20** (1944), 519–524. MR14633

[Kak44a]　Shizuo Kakutani, *On Brownian motions in n-space*, Proc. Imp. Acad. Tokyo **20** (1944), 648–652. MR14646

[Kak44b]　Shizuo Kakutani, *Two-dimensional Brownian motion and harmonic functions*, Proc. Imp. Acad. Tokyo **20** (1944), 706–714. MR14647

[KK15]　Mikhail G. Katz and Semen S. Kutateladze, *Edward Nelson (1932–2014)*, Rev. Symb. Log. **8** (2015), no. 3, 607–610, DOI 10.1017/S1755020315000015. MR3388737

[Kol50]　A. N. Kolmogorov, *Foundations of the Theory of Probability*, Chelsea Publishing Company, New York, N. Y., 1950. MR0032961

[KP95]　P. E. Kloeden and E. Platen, *Numerical methods for stochastic differential equations*, Nonlinear dynamics and stochastic mechanics, CRC Math. Model. Ser., CRC, Boca Raton, FL, 1995, pp. 437–461. MR1337938

[KS91]　Ioannis Karatzas and Steven E. Shreve, *Brownian motion and stochastic calculus*, 2nd ed., Graduate Texts in Mathematics, vol. 113, Springer-Verlag, New York, 1991, DOI 10.1007/978-1-4612-0949-2. MR1121940

[L39]　Paul Lévy, *Sur certains processus stochastiques homogènes* (French), Compositio Math. **7** (1939), 283–339. MR919

[L40]　Paul Lévy, *Le mouvement brownien plan* (French), Amer. J. Math. **62** (1940), 487–550, DOI 10.2307/2371467. MR2734

[L65]　Paul Lévy, *Processus stochastiques et mouvement brownien* (French), suivi d'une note de M. Loève, deuxième édition revue et augmentée, Gauthier-Villars & Cie, Paris, 1965. MR0190953

[Lap95]　Pierre-Simon Laplace, *Théorie analytique des probabilités. Vol. I* (French), Introduction: Essai philosophique sur les probabilités. [Introduction: Philosophical essay on probabilities]; Livre I: Du calcul des fonctions génératrices. [Book I: On the calculus of generating functions], reprint of the 1819 fourth edition (Introduction) and the 1820 third edition (Book I), Éditions Jacques Gabay, Paris, 1995. MR1400402

[Man09]　Roger Mansuy, *The origins of the word "martingale"*, translated from the French [MR2135182] by Ronald Sverdlove, J. Électron. Hist. Probab. Stat. **5** (2009), no. 1, 10. MR2520661

[McK62]　H. P. McKean Jr., *A Hölder condition for Brownian local time*, J. Math. Kyoto Univ. **1** (1961/62), 195–201, DOI 10.1215/kjm/1250525056. MR146902

[Mer73]　Robert C. Merton, *Theory of rational option pricing*, Bell J. Econom. and Management Sci. **4** (1973), 141–183. MR496534

[MP10]　Peter Mörters and Yuval Peres, *Brownian motion*, with an appendix by Oded Schramm and Wendelin Werner, Cambridge Series in Statistical and Probabilistic Mathematics, vol. 30, Cambridge University Press, Cambridge, 2010, DOI 10.1017/CBO9780511750489. MR2604525

[Nel67]　Edward Nelson, *Dynamical theories of Brownian motion*, Princeton University Press, Princeton, N.J., 1967. MR0214150

[Nel87]　Edward Nelson, *Radically elementary probability theory*, Annals of Mathematics Studies, vol. 117, Princeton University Press, Princeton, NJ, 1987, DOI 10.1515/9781400882144. MR906454

[Nel95]　È. Nel'son, *Radikal' no èlementarnaya teoriya veroyatnosteĭ* (Russian), translated from the 1987 English original and with a preface by A. A. Ruban and S. S. Kutateladze, Izdatel'stvo Rossiĭskoĭ Akademii Nauk, Sibirskoe Otdelenie, Institut Matematiki im. S. L. Soboleva, Novosibirsk, 1995. MR1449396

[Øks03]　Bernt Øksendal, *Stochastic differential equations: An introduction with applications*, 6th ed., Universitext, Springer-Verlag, Berlin, 2003, DOI 10.1007/978-3-642-14394-6. MR2001996

[P21]　Georg Pólya, *Über eine Aufgabe der Wahrscheinlichkeitsrechnung betreffend die Irrfahrt im Straßennetz* (German), Math. Ann. **84** (1921), no. 1-2, 149–160, DOI 10.1007/BF01458701. MR1512028

[PW87]　Raymond E. A. C. Paley and Norbert Wiener, *Fourier transforms in the complex domain*, reprint of the 1934 original, American Mathematical Society Colloquium Publications, vol. 19, American Mathematical Society, Providence, RI, 1987, DOI 10.1090/coll/019. MR1451142

[Sen16]　Eugene Seneta, *Markov chains as models in statistical mechanics*, Statist. Sci. **31** (2016), no. 3, 399–414, DOI 10.1214/16-STS568. MR3552741

[Shr04]　Steven E. Shreve, *Stochastic calculus for finance. II*, Continuous-time models, Springer Finance, Springer-Verlag, New York, 2004. MR2057928

[Ste01]　J. Michael Steele, *Stochastic calculus and financial applications*, Applications of Mathematics (New York), vol. 45, Springer-Verlag, New York, 2001, DOI 10.1007/978-1-4684-9305-4. MR1783083

[Ste11]　Dan Stefanica, *A primer for the mathematics of financial engineering*, 2nd ed., FE Press, 2011.

[Str64]　R. L. Stratonovič, *A new form of representing stochastic integrals and equations* (Russian, with English summary), Vestnik Moskov. Univ. Ser. I Mat. Meh. **1964** (1964), no. 1, 3–12. MR0160262

[Tal08] Nassim Nicholas Taleb, *The black swan: The impact of the highly improbable*, 1st ed., Random House, London, 2008.

[UO30] G. E. Uhlenbeck and L. S. Ornstein, *On the theory of the brownian motion*, Phys. Rev. **36** (1930), 823–841.

[Wal12] John B. Walsh, *Knowing the odds: An introduction to probability*, Graduate Studies in Mathematics, vol. 139, American Mathematical Society, Providence, RI, 2012, DOI 10.1090/gsm/139. MR2954044

[Web09] Ernst Juerg Weber, *A short history of derivative security markets*, Springer Berlin Heidelberg, Berlin, Heidelberg, 2009, pp. 431–466.

[Wie76] Norbert Wiener, *Collected works. Vol. I*, Mathematical philosophy and foundations; potential theory; Brownian movement, Wiener integrals, ergodic and chaos theories, turbulence and statistical mechanics; with commentaries; edited by P. Masani, Mathematicians of Our Time, vol. 10, MIT Press, Cambridge, Mass.-London, 1976. MR0532698

Index

Selected Published Titles in This Series

For a complete list of titles in this series, visit the
AMS Bookstore at **www.ams.org/bookstore/amstextseries/**.